PREFACE

In the last 20 years, freeze fracture has emerged from an obscure and highly controversial beginning into a standard electron microscopy sample preparation method. Today, one can hardly open a cell biology journal without seeing freeze-fracture electron micrographs. The impact of the freeze-fracture technique on our understanding of membrane structure cannot be exaggerated. The structural basis for the "fluid mosaic" model of membranes owes much to the early findings of freeze-fracture electron microscopy. As a morphological tool, freeze fracture rivals thin sections as major electron microscopy techniques for sample preparation. Within 10 years of its birth, quantitative data have been derived from freeze-fracture electron micrographs through the use of morphometric methods specially developed to analyze freeze-fracture results.

The technology took a giant leap forward in the next 10 years when rapid freezing and labeling techniques were developed. The labeling techniques, pioneered by Pinto da Silva and Rash independently, elevated freeze fracture from a morphological tool to a cytochemical tool. The frustration over the years of not being able to identify intramembrane particles is now replaced by the joy of seeing the light at the end of the tunnel. In combination with morphometric methods, freeze-fracture electron microscopy is no longer regarded as a descriptive method, but a quantitative and analytical method. Being a microscopic method, it provides localized and ultrastructural information not obtainable by averaging measurements such as spectroscopy and chemical analysis.

The rapid slam-freezing method developed by Heuser and Reese, and the sandwich-freezing method developed by Gulik-Kryzwicki and Costello enabled samples to be frozen without chemical treatment. The short freezing time also opened up the possibility to arrest events which last for only a very short duration (milliseconds), and to differentiate chronologically stages of a rapid kinetic sequence. Through the application of rapid freezing, electron microscopy has evolved into a dynamic tool from a static one.

Along with the above development, improving stereoscopy, surface replication, and "deep-etching" techniques allowed more information to be derived from the sample. The use of colloidal gold as labels on membrane *surfaces* and on fracture *faces* greatly facilitated the analysis. The combination of these techniques enables researchers to visualize and identify structural features at the molecular level. All these developments, which seemed unreal 20 years ago when freeze fracture was in its infancy, are now within reach of researchers.

This volume is a kind of celebration of the progress of freeze-fracture electron microscopy in recent years. Many of the authors are leaders of the advancing front. Instead of offering an instruction manual of how to perform new techniques or a review to cover what has happened in the past in respective fields, both of which are abundantly available as journal articles and monographs, this volume is a collection of personal testimonies as examples of the power of the new freeze-fracture technology. Since each chapter also centers around a specific biological problem, it also serves to illustrate how much the understanding of the problem has been advanced by the new freeze-fracture methodology, which in most cases is developed by the author(s) themselves. A characteristic of frontier development is that many chapters are dealing with controversial subjects. The inclusion of these subjects in the volume represents the dynamic nature of the subject as viewed by the authors rather than the final verdict of the subject matter. It is not our intention to cover debates nor to endorse opinions within this volume.

Chapter 1 is a recollection of the early days of freeze fracture that leads to the methodology attaining its present status. An excellent survey of the history is found in the Introduction of Chapter 3. In general, Chapters 2 to 5 give different aspects of freeze fracture as a cytochemical tool, its usefulness and pitfalls, as well as the finesse and precision associated with this application. Chapters 6 to 10 are primarily examples of rigorous time- or temperature-dependent ultrastructural studies which enable reaction kinetics to be studied with high spatial and temporal resolution. Ample reference in each chapter provide readers with the background of the field and the technique.

I hope this book will offer satisfactory reading for those who wish to update their understanding of what freeze-fracture electron microscopy has, and continues to contribute, to the understanding of membrane structure.

Sek Wen Hui

THE EDITOR

Dr. Sek Wen Hui, Ph.D., is a cancer research scientist (VI) and director of the Membrane Biophysics Laboratory at the Roswell Park Memorial Institute, New York State Department of Health, Buffalo, and a Research Professor in Biophysics at the State University of New York at Buffalo.

Dr. Hui graduated in 1964 from the University of Western Australia, Perth, Australia, with a B.Sc. degree and received a Ph.D. degree in physics from the Monash University at Melbourne, Australia in 1968. He was appointed lecturer in Physical Sciences at the Flinders University of South Australia after his graduate study. In 1969, he came to the Carnegie-Mellon University in Pittsburgh, PA, and worked in the physics and biological sciences departments under an NIH Special Fellowship. He joined the Roswell Park Memorial Institute in 1972 as a cancer research scientist (I) and was promoted through the years to his present position. He was appointed assistant research professor in Biophysics, State University of New York at Buffalo in 1976; associate research professor in 1979, and research professor in 1984.

Dr. Hui is a member of the Electron Microscopy Society of America, and served as its director from 1985 to 1987. He is also a member of the Biophysical Society, American Society for Cell Biology, American Physical Society, and the New York Academy of Science.

Among other awards, Dr. Hui was a recipient of an NIH Special Fellowship and was an Awardee of a Research Career Development Award (1975 to 1980). He served as a member of the advisory committee for the NIH regional high voltage electron microscopy facility, and study sections for instrumentation and small business innovation research programs at the NIH. He is currently a member of the NIH Biophysical Chemistry Study Section.

Dr. Hui has presented over 30 invited lectures at various conferences and in universities and institutes worldwide. He has published over 100 research papers in refereed journals and 16 review chapters in various books. He has been the recipient of research and instrumentation grants from the NIH, American Cancer Society, the Charles E. Culpepper Foundation, and the Fannie E. Rippel Foundation. His current research interests are the molecular organization in membranes and the molecular mechanism of membrane fusion. He is also studying the roles of lipids on membrane functions, especially on signal transduction.

CONTRIBUTORS

Douglas E. Chandler, Ph.D.
Associate Professor
Department of Zoology
Arizona State University
Tempe, Arizona

Joseph E. Dinchuk, Ph.D.
Postdoctoral Fellow
Department of Pathology
Colorado State University
Fort Collins, Colorado

Daniel S. Duch
Postdoctoral Fellow
Department of Anesthesiology
Cornell Medical Center
New York, New York

Knute A. Fisher, Ph.D.
Associate Adjunct Professor
Department of Biochemistry and Biophysics
Associate Research Biochemist
Cardiovascular Research Institute
University of California
San Francisco, California

Sek Wen Hui, Ph.D.
Head
Membrane Biophysics Laboratory
Department of Biophysics
Roswell Park Memorial Institute
Buffalo, New York

Timothy J. A. Johnson, Ph.D.
Assistant Professor
Department of Anatomy and Neurobiology
Colorado State University
Fort Collins, Colorado

James K. Koehler, Ph.D.
Professor
Department of Biological Structure
University of Washington
Seattle, Washington

Carolyn Larabell, Ph.D.
Postdoctoral Fellow
Department of Biochemistry and Biophysics
University of California
Davis, California

Simon Rock Levinson, Ph.D.
Associate Professor
Department of Physiology
University of Colorado Health Sciences Center
Denver, Colorado

Rodman G. Miller, Ph.D.
Department of Oral Biology
University of Washington
Seattle, Washington
Assistant Professor
Department of Anatomy
University of Calgary
Calgary, Alberta, Canada

Pedro Pinto da Silva, Ph.D.
Chief
Membrane Biology Section
Laboratory of Mathematical Biology
National Cancer Institute
Frederick, Maryland

Helmut Plattner, Ph.D.
Professor
Faculty of Biology
University of Konstanz
Konstanz, Federal Republic of Germany

John E. Rash, Ph.D.
Professor
Department of Anatomy and Neurobiology
Colorado State University
Fort Collins, Colorado

Horst Robenek, Ph.D.
Priv.-Doz.
School of Medical Cytobiology
University of Münster
Münster, Federal Republic of Germany

Arindam Sen, Ph.D.
Cancer Research Scientist I
Department of Biophysics
Roswell Park Memorial Institute
Buffalo, New York

TABLE OF CONTENTS

Chapter 1
Freeze-Fracture Images: 1967-1971 1
Pedro Pinto da Silva

Chapter 2
Membrane-Splitting Analyses of Membrane-Spanning Proteins........ 11
Knute A. Fisher

Chapter 3
Labeling Intramembrane Particles in Freeze-Fracture Replicas 41
John E. Rash, Timothy J. A. Johnson, Joseph E. Dinchuk, Daniel S. Duch, and S. Rock Levinson

Chapter 4
Distribution and Mobility of Receptors in the Plasma Membrane 61
Horst Robenek

Chapter 5
Freeze-Fracture Sterol Localization with Filipin........ 87
Rodman G. Miller

Chapter 6
Polymorphic Phase Behavior of Membrane Lipids 103
Arindam Sen

Chapter 7
The Kinetics of Artificially Induced Membrane Fusion........ 129
Sek Wen Hui

Chapter 8
Freeze-Fracture (-Etching) Analysis of Exo-Endocytosis 147
Helmut Plattner

Chapter 9
The Vitelline Layer and Fertilization Envelope of Echinoderm and Amphibian Eggs: Visualization of a Cell Surface-Anchored Extracellular Matrix 175
Carolyn A. Larabell and Douglas E. Chandler

Chapter 10
The Plasma Membrane of the Spermatozoon........ 201
James K. Koehler

Index 215

Chapter 1

FREEZE-FRACTURE IMAGES: 1967—1971

Pedro Pinto da Silva

TABLE OF CONTENTS

I. Introduction 2

II. Freeze-Fracture Images: 1967—1971 2

References 8

"A sciência não é senão um jogo de creanças no crepúsculo, um querer apanhar sombras de aves e parar sombras de hervas ao vento."

Fernando Pessoa, O livro do desassossego*

I. INTRODUCTION

Morphology is a science of anticipation, an analysis of what often are the first signals of a new reality. Etymology may order the analysis of form, but the morphologist finds himself "reading" the faintest of signals. Through him reality becomes. This almost divinitory function is distinct from the systematization of form, which is done later, after the new form becomes part of the visual realm.

As the poet, the morphologist will often be the first to see (or to foresee). But, once reality becomes closer, clearer, and its more intimate corners available to more prosaic, complete, Apollonian inspection, the dreams that led to the new reality tend to be cast aside as useless, inopportune.

The importance of freeze fracture in the development of what are today the basic concepts in the topology, the structure, and the dynamics of biological membranes cannot be overestimated. Freeze fracture came into being at a time of impasse for biochemical, immunological, and spectroscopic approaches. Between 1966 and 1971, it was freeze-fracture work that allowed and proposed the synthesis of antithetic views of biological membranes (see References 2 and 3 for reviews). Any inspection of the literature of the time will show a bewildering array of membrane models. Spectroscopic and biochemical approaches provided averaging views of a supramolecular system where nonaverageable elements played crucial roles. This was clearest in the definition of the structure of myelin. X-ray diffraction studies by some of the most illustrious and competent researchers at the time failed to show the existence of transmembrane units intercalated across the bilayer, as they failed also to show the importance of these elements (today known to represent the glycoprotein P_o, the major structural protein of the peripheral myelin sheath[4]) in maintaining the structural integrity of myelin. But the observation of freeze-fracture replicas quickly revealed what are, to this day, the basic elements of its topology and structure.[5,6] Yet, it has been difficult for the Apollonian majority to accept that what they could not see with all the power at their disposal could easily be disposed as almost a *fait divers* by the morphologist.

After a period of descriptive activity that led to extensive recovering of much of the anatomy of cells and tissues, and after the rediscovery of etching (now rechristened as "deep-etching"), freeze-fracture images begin to be characterized through new methods of molecular cytochemistry.[7,9]

In the text that follows, I did not attempt to give an account of the development of the ideas that led to the interpretation of the freeze-fracture image or to the concepts in the topology, structure, and dynamics of biomembranes. I only attempted to write some personal notes relative to the 5 years that went from 1966 to 1971.

II. FREEZE-FRACTURE IMAGES: 1967—1971

How did these images of freeze-fracture membranes fascinate me! There they were in large glossy prints affixed one morning as I entered a laboratory class in cytology. It was the fall of 1966, and I had just arrived from Portugal where, after 3 years in the Navy (including 2 Dostoyevskian years as a navigation officer off the Angolan coast), I had spent a year trying to adapt autoradiography to define the metabolic and nonmetabolic spaces of ion uptake in plant cells.

* Science is nothing but a game of children in the twilight, a wish to catch the shadows of birds in flight, to stop the shadows of herbs under the wind. (In *The Book of Disquiet*, by Bernardo Soares, a heteronym of Fernando Pessoa[1]).

The images of freeze-fracture cells instantly appealed to me for two clear reasons: one, their incredible aesthetic beauty, and the other, their nonaveraged nature. Biological membranes appeared as a stunning bimodal reality — particles spread over smooth areas. Regardless of interpretation, I was immediately drawn to the thought that the particles might be related to sites of ion absorption. Early in 1967, I went to see Daniel Branton, the author of the micrographs, and inquired as to the possibility of coming to his laboratory as a research assistant.

I had in mind the comparative morphology of the membranes in the root cells of ecotypes of species growing in magnesium-rich serpentine soils (an idea taken from my father's work as a plant ecologist/taxonomist). Branton, also a plant physiologist with an interest in ion uptake, proposed a clearer project; I was to isolate plant vacuoles and use them as a membrane system. Starting in June 1967, I spent several months chopping vegetables previously immersed in hypertonic solutions. I used a variety of razor blades as well as assorted culinary contraptions. The hope was that sufficient numbers of vacuoles would be released from the cut tissues. My chopping abilities proving insufficient, we proceeded to design what could be best described as a Cuisinart for ultra-"yuppies". On inauguration day, it promptly reduced several pounds of beets to the thinnest of chips, the whole thing swirling in about a gallon of borscht-deluxe. Almost a year to design and build and there it was performing faster than rabbits making love.

After everybody had left, I took a good look at the soup. With the light microscope the beet vacuoles looked more like asteroids lost in space. I had no experience or interest in the isolation of organelles. I knew that to concentrate the errant vacuoles in that soup I would need the elutriation centrifuge that, at the time, was still being developed at Stanford University, Palo Alto, CA. So, I went to Branton and told him that I had no intention of becoming the world's expert in the isolation of plant vacuoles.

The principal problem in Branton's laboratory was to locate the fracture plane of membranes. He had proposed that membranes were split along a hydrophobic interior,[10] a hypothesis that ran against that of Mohr and Mühlethaler[11] (reviewed in Reference 2), who argued that fracture followed the surface of membranes. Branton had advanced his hypothesis based on the morphological interpretation of specimens impregnated in glycerol, freeze-fractured, and then subjected to a period of sublimation ("etching"). This approach proved less than convincing, in part, because it appeared circular (new rules of interpretation were used to formulate new rules of interpretation) and because, should membrane splitting occur, it implied a membrane model that failed to align itself with any of the then prevailing models of membrane structure: on one side Daneilli,[12] Robertson,[13] and all of those who proposed biological membranes as comprised of planar bilayers of amphipatic lipids coated with adsorbed proteins; on the other, those who held that biomembranes were planar aggregates of lipoprotein structures (among these Sjostrand,[14] Green and Fleischer,[15] Benson and Singer,[16] and others, many working with mitochondrial membranes).

While the arguments for splitting of bilayers made sense, we could advance other, equally sensible arguments, to account for a fracture plane at the surface. In particular, for us plant physiologists interested in ion uptake mechanisms, a surface cleavage plane could be envisaged as formed as the front of freezing advanced to layers of organized water thought to exist at the surfaces of membranes.

Deamer and Branton[17] used artificial planar bilayers formed by the apposition of two lipid monolayers, but these were far from natural bilayers, further still from biomembranes. Branton attempted next to perform true etching experiments where he hoped to show that the surface of the membranes could display (after sublimation) natural markers; in particular, to look at preparations of isolated rough endoplasmic reticulum to show that ribosomes could only be seen over the areas of the membrane that were exposed by sublimation. About this time, Leigh Engstrom, a graduate student from the Department of Zoology (University of California, Berkeley), started to freeze fracture erythrocyte ghost membranes and to observe the surfaces exposed by sublimation. By that time, after learning freeze fracture from Sue Whytock, I had started to freeze fracture solutions of ferritin: Deamer et al.[18] were just finishing an elegant study of lipid phases and I wanted to see what contributions proteins could bring to the freeze-fracture image. To this end, I freeze-fractured ferritin solutions and used very low angles of shadow to show that the shadowed hemispherical bumps contained the electron dense core of ferritin. Reading literature on ferritin led me to a paper by Hasselbach, where ferritin was covalently conjugated to the surface of mitochondria. It immediately became clear to me that I could try a similar approach to label the surfaces of erythrocyte membranes. Within minutes I made diagrams of the experiments and of the results that I expected to obtain. I showed these to Branton.

My first attempt to produce the conjugating agent promptly resulted in the synthesis of a thick and magnificent yellow pigment that might have led to fortune, if not fame, had I the foggiest idea of what I was doing. I tried then toluene-2,4di-isocyanate a bifunctional conjugation reagent used by Shick and Singer.[19] I remember taking my first replica of labeled ghosts to the microscope. As I arrived, the triumphal march of Aïda was playing on the radio of an adjoining darkroom. Soon after inserting the specimen, there they were: the surfaces of erythrocyte membranes exposed by etching were coated with the magnificently shadowed "pearls" of ferritin.

This was wonderful. Much less wonderful was the fact that the controls (experiments performed with unmodified ferritin) looked equally beautiful and . . . labeled. To be sure, these experiments were a partial success because they showed that a macromolecular marker could be seen on the etched surface of a membrane. They showed also that, during freezing in distilled water, macromolecules could be excluded by the growing ice crystals and be slammed against the membrane (I had first seen the exclusion phenomenon during freezing of ferritin solutions). Aïda's march had never convinced me (memories of a teenager slumbering through 19th century operatic syrup). To this day, I positively dread this march as the worst of omens.

I redesigned my experiments to include thin section observations, a modified conjugation method, extensive washing of the erythrocyte ghosts, and lower centrifugation speeds. While the results were not half as pretty, they were now correct. Thin sections showed that the ferritin had indeed bound to both surfaces of the membrane (although at the time I was puzzled by the much higher binding of ferritin to the inner surface). We prepared a manuscript where I managed to include the idea of the membrane as a bilayer continuum interrupted by intercalated proteins.

Branton gave the manuscript to Leon Wofsy who sent it to Jon Singer. Singer liked it, but objected that we should not refer to a "modified Danielli's model" but instead, to a "modified Lenard-Singer's model". By this time, Singer was on his second membrane model. The first, proposed with Benson (at the Scripps Institute of Oceanography, La Jolla, CA) viewed membranes as planar aggregates of lipoproteins.[16] The second, based on optical rotatory dispersion and circular dichroism, placed proteins across membranes comprised of a "trilayer"; the usual bilayer was now separated by a middle layer of apolar components.[20] We retained the reference to a modified Danielli's model and sent the manuscript to the Journal of Cell Biology.[21] Soon after, we received for review the manuscript of Tillack and Marchesi reporting the visualization of actin molecules on the etched outer surfaces of erythrocyte ghosts. As it was written in this version, the actin could have been slammed onto the membrane surfaces by the exclusion mechanism that I referred to above. In addition, there was no labeling of the inner surface, a point crucial to prove membrane splitting. The published version[22] assumed that brief glutaraldehyde fixation (0.1% for 3 min in the cold) had covalently conjugated the actin to the membrane surface.

In the summer of 1969, we presented our results at the International Botanical Congress[23] in Seattle, WA (see Figure 1), where I became involved with a group of botany graduate students from Berkeley in presenting a motion on overpopulation and food resources. Mühlethaler was present at the session where I delivered our paper. He graciously accepted our evidence.

We had killed two rabbits with one blow: proof of membrane splitting made it possible to interpret correctly the freeze-fracture image; therefore, the freeze-fracture image of all biological membranes could only be interpreted to mean that membranes were comprised of a bilayered continuum interrupted by nonbilayered intercalations. We hypothesized that these interruptions of the bilayer corresponded to membrane proteins (Figure 1). These etching experiments provided also the technique to solve the next questions. They concerned the chemical composition and topology of the membrane particles.

In Portugal, I had received an excellent mathematical education from Professors Sebastião e Silva (functional analysis) and Varennes e Mendonça (vectorial calculus). From them I had learned how to situate a problem in terms of domains and to think about geometric topology before I approached structure. This was handy because the next problem was topological; were the particles the representation of molecules sequestered within the apolar interior of the bilayer, or could they, as I thought, represent structures positioned across the membrane? My experiments, as well as those of Tillack and Marchesi, had shown that freeze-etching could be used as a powerful cytochemical approach to determine the distribution of receptors or antigens at the cell surface.

As a plant biologist I had only the faintest notion of blood groups, antibodies, and the like. Being at Berkeley, that wonderful and "unstuffiest" of campuses, I went to see Leon Wofsy whom I remember asking how big was an antibody. He was very kind, indeed; he sent me reading and we tried to see if individual IgM molecules from a primary response might be seen bound to the surfaces of erythrocytes. Later in 1969, Bernie Gilula (then learning to freeze fracture) introduced me to Steve Douglas, an immunologist then in Hugh Fudenberg's lab.

To Steve and I, AB(H) blood group antigens appeared as prime candidates. Their number on the surface of human erythrocytes was close to that of the membrane particles. In addition, antibody preparations were commercially available. We made some ammonium sulphate precipitations to partly purify the globulins, drew blood from each other, worked nonstop for 3 days and 2 nights, blew the filaments of two electron microscopes, crossed the Bay Bridge at dawn, and went to the University of California at San Francisco where we pirated another microscope. Finally, we could see platinum replicas of freeze-etched erythrocyte ghosts that appeared coated by a layer of "stuff" (Figure 2). This "stuff" was present only over regions of the etched outer surface that corresponded to the continuation of the aggregates of membrane particles on the fracture face (I had learned from Leigh Engstrom that short treatment of erythrocyte ghosts with distilled water led to particle aggregation). We were happy.

Steve had to leave for New York to a new position at Mount Sinai School of Medicine. When I showed our results at a laboratory seminar, people were skeptical. I also had shown them that when membranes were frozen in distilled water, the membrane particles (whether surface protrusions of intercalated structures or deformations caused by molecules sequestered within the apolar matrix) could also be observed on etched surfaces. The denser appearance of the "stuff" on the etched surfaces might, therefore, be due to the aggregates of the particles themselves. Clearly, I needed an unequivocal marker; I needed a ferritin conjugated antibody. I enrolled in immunology classes taught by Ann Good and Leon Wofsy. This ended up including night classes in the home of Leon Wofsy; remember that Berkeley was for 1 month under dawn-to-dusk surveillance by "spotter" planes; remember that CS gas [proven in Vietnam] was poured on campus by helicopters; remember . . . .

By the end of the Spring of 1970 I had performed successful labeling experiments. We sent an abstract to the cell biology meetings[24] in San Diego, CA and a manuscript was

-5-

ferritin molecules to the outside and also to the inside of the membrane.

The controls of this experiment (which contained ferritin but no conjugating agent can now be seen:

• 10 next slide, please (SLIDE 10 f-e of control)

where we cannot see ferritin either on the fracture face or on the etched face. Sections of this material,

• 11 next slide, please (SLIDE 11 section of control)

do not show any ferritin association with the membranes.

LIGHTS PLEASE

Conclusion *%*%*%*%*%*%*%*%*%*%*%*%*%*%*CONCLUSION###########

Let us see now what to conclude as far as what are the fracture faces visualised by the F-E technique and what kind of structure should a membrane have in order to account for these results.

→ As to the first point it seems improbable that the fracture face visualised by the F-E technique represents ~~any~~ either surface of the membrane.

What should then the fracture plane be?

The only plausible candidate seems to be the one earlier hypothesized by BRANTON : The fracture would proceed along the plane formed by the meeting ends of the fatty acid chains from a lipid bylayer and so, along the hydrophobic interior of the membrane

~~[illegible]~~

This seems probable since the forces of attraction would be of the London - Van der Waals Dispersion type and so extremely weak by compariso with the electrostatic forcea and hydrogen bonds that are likely to exist at the hydrophillic membrane surfaces ~~(about 2 orders of magnitude)~~

0.1 kcal/mole / vs ~5 kcal/mole

London-VdW disp ....) 0.1 kcal/mole
Electrostatic forces...... ~5 kcal/mole

-6-

→ As far as the structure of the membrane is concerned the results of the present experiment do not support the view of a membrane made up of one layer of subunits.

A model which accounts for the results obtained so far using the F-E technique might be the following:

• 12 next slide, please (SLIDE 12 membrane model)

where the membrane is seen as an extended bylayer of lipid and protein and where the membrane particles visualised by the F-E technique are seen as protein molecules (although at this moment we only have indirect evidence for this). The appearance of the particles on the fracture face would result from stretching during the fracture process.

Further protein molecules most probably exist but we have litle indication as to their exact location. THANK YOU.

~~We are currently engaged in this study.~~

FIGURE 1. The last two pages of the text in my presentation of the proof of membrane splitting to the International Botanical Congress.[23]

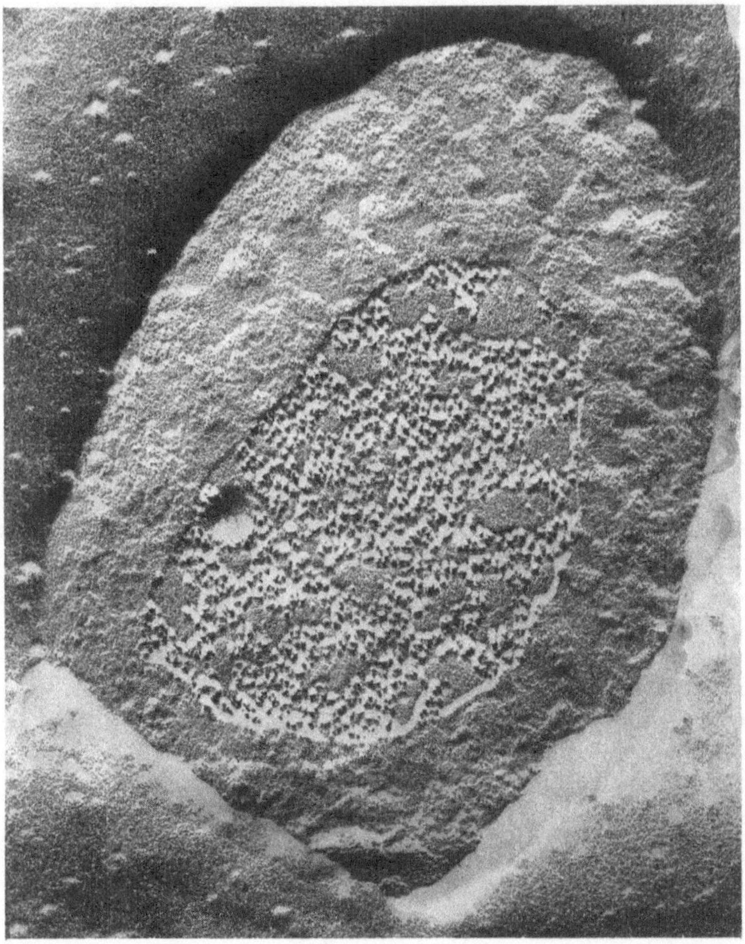

FIGURE 2. Group A human erythrocyte ghost treated with anti-A immunoglobulins, washed in distilled water, fractured, and etched. Compact plaque-like structures continue, on the etched surface, the aggregates of membrane particles on the fracture face. (Magnification × 100,000.) (Douglas, S. D. and Pinto da Silva, P., unpublished.)

written with Steve Douglas (while attending a Gordon Conference) and sent to *Nature (London).*[25] It was swiftly rejected because "the membrane particles and the antigen sites might have coincidently coaggregated". By this time, Branton had left Berkeley and, after a summer stay in Marchesi's laboratory (then at the National Institutes of Health), had left for a sabbatical year at Aaron Klug's and Mark Bretschter's laboratories at the Medical Research Council (MRC), Cambridge, England.

I was bitterly disappointed with the rejection of the manuscript, but Daniel Mazia consoled me and encouraged me to insist, which I promptly did. The manuscript stayed in *Nature* for many months and was accepted following a letter from Aaron Klug. In the Fall of 1970, I presented the work at the cell biology meeting (San Diego, November 1970) only to be stunned by a similar presentation by Tillack and Marchesi using influenza virus as a marker (a virus that I had unsuccessfully tried to obtain from the NIH), and trypsin as an agent to induce the particle aggregation essential to the interpretation of the results. The work reported was unrelated to that described on their abstract.[26] It had apparently been done after the abstract was sent.

By the fall of 1970, I was exposing erythrocyte ghost membranes to buffers with different pH values to show that the membrane particles, which I now knew corresponded to transmembrane proteins, could move translationally along the plane of the membrane and that this movement was rapid, reversible, and could be prevented by glutaraldehyde. After the cell biology meetings, I visited Jon Singer with whom I discussed my results on translational mobility.

Finally, sometime in March of 1971 I received the galley proofs of the *Nature* paper. On June 23, the paper by Bretschter proposing glycophorin as a transmembrane protein was published in *Nature*.[27] It had been received in *Nature's* office on April 19, i.e., after the galley proofs of our paper were mailed to me. Somehow, it had managed to be published first. That year, I presented the work showing pH-dependent aggregation of membrane particles as an abstract to the cell biology meeting of 1971.[28] The abstract was rejected for platform presentation. It did report the first instance of translational movement of a transmembrane protein.[29]

The major topological features of the membrane were now established. Just as Robertson had used electron microscopy to show that thin sections of biological membranes displayed the trilaminar profile typical of erythrocyte membranes, so could the results of freeze fracture be extrapolated to other biological membranes. The work of Hubbel and McConnell[30] and others had established the main features of the molecular motion of membrane lipids; Cone had demonstrated the rotational mobility of membrane proteins;[31] my work had shown reversible, translational displacement of transmembrane proteins.[29] The biochemical work of Steck et al.,[32,33] and the proposal of the terms intrinsic and extrinsic or integral and peripheral to classify membrane proteins closed this chapter in the definition of what are to this day the main features of the topology, structure, and dynamics of biological membranes. For me, it was both a time of incredible intellectual stimulation and also of exposure to the politics of science. Membranes had become big business.

REFERENCES

1. **Pessoa, F.,** O livro do desassossego, por Bernardo Soares, Atica Editora, Lisboa, 1982.
2. **Pinto da Silva, P.,** Topology, dynamics and molecular cytochemistry of integral membrane proteins: a freeze-fracture view, in *Electron Microscopy of Proteins,* Vol. 6, Harris, J. R. and Horne, R. W., Eds., Academic Press, New York, 1987, 2.
3. **Pinto da Silva, P.,** Seeing and thinking biological membranes. In *Biomembrane and Receptor Mechanisms,* Fidia Research Series, Vol. 7, Bertoli, E., Chapman, D., Canibria, A., and Scapagrini, U., Eds., Liviana Press, Padova, Italy, 1987, 11.
4. **Lemke, G. and Axel, R.,** Isolation and sequence of a cDNA encoding the major structural protein of peripheral myelin, *Cell,* 40, 501, 1985.
5. **Pinto da Silva, P. and Miller, R. G.,** Membrane on the fracture faces of frozen myelin, *Proc. Natl. Acad. Sci. U.S.A.,* 72, 4046, 1975.
6. **Franks, N. P.,** Myelin structure, *Nature (London),* 259, 447, 1976.
7. **Pinto da Silva, P., Barbosa, M. L. F., and Aguas, A. P.,** A guide to fracture-label: cytochemical labeling of freeze-fractured cells, in *Advanced Techniques in Biological Microscopy,* Vol. 3, Koehler, J. K., Ed., Springer-Verlag, Heidelberg, 1986, 201.
8. **Pinto da Silva, P.,** Molecular cytochemistry of freeze-fracture cells: freeze-etching, fracture-label, fracture-permeation, and label-fracture, in *Advances in Cell Biology,* Vol. 1, Miller, K. R., Ed., JAI Press, Greenwich, CT, 1987, 157.
9. **Pinto da Silva, P.,** Visual thinking of biological membranes: from freeze-etching to label-fracture, in *Immunogold Labeling Handbook,* Verkleij, A. J., Ed., CRC Press, Boca Raton, FL, in press, 1988.
10. **Branton, D.,** Fracture faces of frozen membranes, *Proc. Natl. Acad. Sci. U.S.A.,* 55, 1048, 1966.
11. **Mohr, H. and Mühlethaler, K.,** Fine structure of frozen-etched yeast cells, *J. Cell Biol.,* 17, 609, 1963.
12. **Danielli, J. F.,** The present position in the field of facilitated diffusion and selective ion transcript, *Proc. Symp. Colston Res. Soc.,* 7, 1, 1954.
13. **Robertson, J. D.,** The unit membrane, in *Electron Microscopy in Anatomy,* Boyd, J. D., Johnson, F. R., and Lever, J. D., Eds., Williams & Wilkins, Baltimore, 1961, 74.

14. **Sjostrand, F.**, A comparison of plasma membranes, cytomembranes and mitochondrial membrane elements with respect to ultrastructural features, *J. Ultrastruct. Res.*, 9, 561, 1963.
15. **Green, D. and Fleischer, S.**, The role of lipids in mitochondrial electron transfer and oxidative phosphorylation, *Biochim. Biophys. Acta*, 70, 554, 1963.
16. **Benson, A. A. and Singer, S. J.**, The lipoprotein model for membrane structure, in *Abstracts, 150th National Meeting*, American Chemical Society, Atlantic City, NJ, 1965, 8.
17. **Deamer, D. W. and Branton, D. W.**, Fracture phases in an ice-bilayer model membrane system, *Science*, 158, 655, 1967.
18. **Deamer, D. W., Leonard, R., Tardieu, A., and Branton, D.**, Lamellar and hexagonal lipid phases visualized by freeze-etching, *Biochim. Biophys. Acta*, 219, 47, 1970.
19. **Shick, A. and Singer, S. J.**, On the formation of covalent linkages between protein molecules, *J. Biol. Chem.*, 236, 2477, 1961.
20. **Lenard, J. and Singer, S. J.**, Protein conformation in cell membrane preparations as studied by optical rotatory dispersion and circular dichroism, *Proc. Natl. Acad. Sci. U.S.A.*, 56, 1828, 1966.
21. **Pinto da Silva, P. and Branton, D.**, Membrane splitting in freeze-etching. Covalently bound ferritin as a membrane marker, *J. Cell Biol.*, 45, 598, 1970.
22. **Tillack, T. W. and Marchesi, V. T.**, Demonstration of the outer surface of freeze-etched red blood cell membranes, *J. Cell Biol.*, 45, 649, 1970.
23. **Pinto da Silva, P. and Branton, D.**, Freeze-etch studies of natural and artificial membranes using specific markers, Proc. 11th Int. Botanical Congr., Seattle, 1969, 171.
24. **Pinto da Silva, P., Douglas, S. D., and Branton, D.**, Localization of A_1 antigen sites on the human erythrocyte membrane, *J. Cell Biol.*, 47, 159A, 1970.
25. **Pinto da Silva, P., Douglas, S. D., and Branton, D.**, Localization of A antigen sites on human erythrocyte ghosts, *Nature (London)*, 232, 194, 1971.
26. **Tillack, T. W., Scott, R. E., and Marchesi, V. T.**, Studies on the chemistry and function of the intramembranous particles observed by freeze-etching of red cell membranes, *J. Cell Biol.*, 47, 213a, 1970.
27. **Bretchter, M. S.**, Major human erythrocyte glycoprotein spans the membrane, *Nature (London) New Biol.*, 231, 229, 1971.
28. **Pinto da Silva, P.**, Translational mobility of membrane intercalated particles of human erythrocyte ghosts, Proc. 11th Annual Meeting of Am. Soc. Cell Biol., 1971.
29. **Pinto da Silva, P.**, Translational mobility of the membrane intercalated particles of human erythrocyte ghosts, pH-dependent, reversible aggregation, *J. Cell Biol.*, 53, 777, 1972.
30. **Hubbel, W. L. and McConnell, H. M.**, Spin-label studies of the excitable membranes of nerve and muscle, *Proc. Natl. Acad. Sci. U.S.A.*, 61, 12, 1968.
31. **Cone, R. A.**, Rotational diffusion of rhodopsin in the visual receptor membrane, *Nature (London) New Biol.*, 236, 39, 1972.
32. **Fairbanks, G., Steck, T. L., and Wallach, D. F. H.**, Electrophoretic analysis of the major polypeptides of the human erythrocyte membrane, *Biochemistry*, 10, 2606, 1971.
33. **Steck, T. L., Fairbanks, G., and Wallach, D. F. H.**, Disposition of the major proteins in the isolated erythrocyte membrane. Proteolytic dissection, *Biochemistry*, 10, 2617, 1971.

Chapter 2

MEMBRANE-SPLITTING ANALYSES OF MEMBRANE-SPANNING PROTEINS

Knute A. Fisher

TABLE OF CONTENTS

I. Membrane Splitting as a Fractionation Technique 12
A. Splitting the Lipid Bilayer 12
B. Bilayer Splitting and Membrane Proteins 14
1. Peripheral Proteins 15
2. Integral Proteins 15

II. Freeze Fracture of Planar Monolayers 16
A. Principle 16
B. Practice 17
1. Sample Preparation 17
2. Planar Surface Preparation 17
3. Monolayer Preparation 17
4. Sandwich Preparation, Freezing, and Splitting 17

III. Effects of Splitting on Transmembrane Proteins 18
A. *Halobacterium halobium* Purple Membrane Bacteriorhodopsin 20
B. Red Cell Plasma Membrane Band 3 and Glycophorin A 21
1. Whole Cell Studies 24
a. Qualitative Aspects of Cell Splitting 24
b. Quantitative Aspects of Cell Splitting 28
2. Isolated Membrane Studies 29
a. Qualitative Aspects of Membrane Splitting 29
b. Quantitative Aspects of Membrane Splitting 31

IV. Synopsis of Protein Partitioning 33
A. Regulation of Partitioning Patterns 33
B. Relevance of Asymmetric Partitioning 34

Acknowledgments 34

References 35

I. MEMBRANE SPLITTING AS A FRACTIONATION TECHNIQUE

The freeze-fracture technique is routinely used as a morphological and cytochemical method.[1,2] However, it can also be viewed as a physical fractionation or separation technique, analogous to differential centrifugation, chromatography, or gel electrophoresis. Because freeze fracture physically separates the lipid bilayer of the membrane, it can be used to produce two fractions: one enriched in cytoplasmic leaflets and the other enriched in extracellular leaflets. Techniques of planar cell and membrane freeze fracture have been developed to produce split membrane fractions in quantities suitable for direct chemical and physical analysis.[3-11] In addition, a variety of quantitative methods have been developed to measure the transbilayer distributions of nanogram amounts of native and radiolabeled lipids[4] and proteins in split membranes (for a recent review see Reference 12). Although planar membrane preparations can be and have been used for many different types of studies of membrane structure, this chapter is limited to biochemical studies of split cell and membrane monolayers, especially the effect of splitting on proteins that span the bilayer.

A. Splitting the Lipid Bilayer

Three hypotheses underlie the membrane-splitting approach. The first is the lipid bilayer hypothesis of Gorter and Grendel,[13] now in its 64th year of general acceptance. Few structural hypotheses in biology have been as durable. The second is that different classes of lipids are asymmetrically arranged across the bilayer. Certainly, such asymmetry has been repeatedly shown for the human erythrocyte plasma membrane.[14,15] The third generally accepted view is that the process of freeze fracture physically separates the bilayer producing split "half" membrane leaflets[16,17] (Figure 1). The membrane leaflet closest to the cytoplasm has been called the inner leaflet, the intracellular leaflet, the cytoplasmic leaflet, or the protoplasmic leaflet. The membrane leaflet closest to the exterior of the cell has been called the outer leaflet, extracellular leaflet, exoplasmic leaflet, or exofacial leaflet.

Microscopic, spectroscopic, and fluorometric methods have been developed to quantify the amount of each split membrane leaflet produced.[4,5] Some of these methods are briefly described below. Although this chapter focuses on membrane proteins rather than membrane lipids, a few examples of the application of membrane-splitting analyses to amphipathic and hydrophobic molecules are given as an introduction (Figure 2).

The transbilayer distributions of several lipid-soluble molecules, including cholesterol, dipalmitoylphosphatidylcholine (DPPC), and the tumor promoter, 12-*O*-tetradecanoyl-phorbol-13-acetate (TPA), have been measured by membrane splitting.[18] In our early studies more than 10 years ago, split membrane fractions were quantified by light and electron microscopic methods.[3,4] Cholesterol distribution studies represented the first biochemical application of planar membrane-splitting methods to a study of the transmembrane distribution of native lipids in the human red blood cell (RBC).[4] Native cholesterol was found to distribute asymmetrically across the plasma membrane; twice as much cholesterol was found in the extracellular leaflet as in the cytoplasmic leaflet. Subsequently, we developed a more rapid, precise, and general method, named double-labeled membrane splitting (DBLAMS),[5,19] to examine the distribution of radiolabeled molecules, including tritiated cholesterol. The DBLAMS examination of radiolabeled cholesterol confirmed our previous findings with native cholesterol. At steady state, ^{3}H-cholesterol distributed asymmetrically across the bilayer of the intact human RBC plasma membrane; approximately twice as much ^{3}H-cholesterol was found in the outer leaflet of the bilayer as in the inner leaflet (Figure 2A, A′).

The transbilayer distribution of tritiated dipalmitoylphosphatidylcholine (^{3}H-DPPC) was also recently evaluated by membrane splitting.[116] As a control, red cells were incubated with aqueous dispersions of ^{3}H-DPPC for 4 h at 37°C, cooled to 0 to 4°C, and thoroughly washed

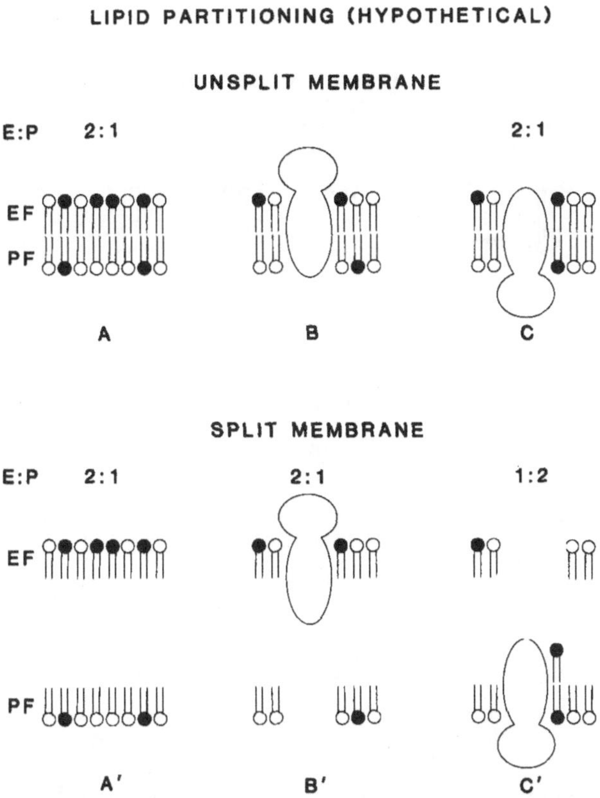

FIGURE 1. Hypothetical lipid partitioning during membrane splitting. Potential distribution patterns of free (A) and protein-associated (B, C) lipids. Three examples of asymmetrically distributed lipid with outside to inside distribution (E:P) are all 2:1. (A, A′) Lipid bilayer separated by freeze fracture. (B, B′) Distribution of lipid unaffected by association with integral membrane proteins. (C, C′) Distribution of lipid altered by protein associations. This latter possibility is probably limited to trace lipids, since bulk lipids in the bilayer are expected to partition independently of integral membrane protein as in A, A′, and the proportion of protein-associated lipid to free lipid is small. Other patterns of partitioning could be derived from combinations of these three modes.

prior to monolayer formation. DBLAMS analysis revealed a significantly different asymmetry for DPPC relative to cholesterol: about 80% partitioned with the extracellular leaflet vs. 20% with the inner leaflet. Similar values were determined for incubations ranging up to 24 h. Thus the asymmetry probably represents the equilibrium distribution of ^{3}H-DPPC (Figure 2B, B′). Quantification of ^{3}H-DPPC distribution demonstrated that DBLAMS techniques were suitable for examining radiolabeled phospholipids as well as the neutral lipid cholesterol. In addition, the DPPC studies validated qualitative and quantitative applications of the DBLAMS methods, since other biochemical and biophysical assays had demonstrated a similar transbilayer asymmetry of the choline-containing phospholipids of the red cell membrane, including sphingomyelin and phosphatidylcholine.[14,15]

Even more recently, we examined the transbilayer distribution of the lipophilic tumor promoter, TPA, by membrane splitting.[18] In the phosphatidylinositol pathway of transmembrane signaling, the second messenger diacylglycerol (DAG) is thought to be associated with the cytoplasmic leaflet of the membrane bilayer because DAG activates cytosolic protein kinase C (PKC).[20] Since TPA also activates PKC, presumably on the cytoplasmic side of

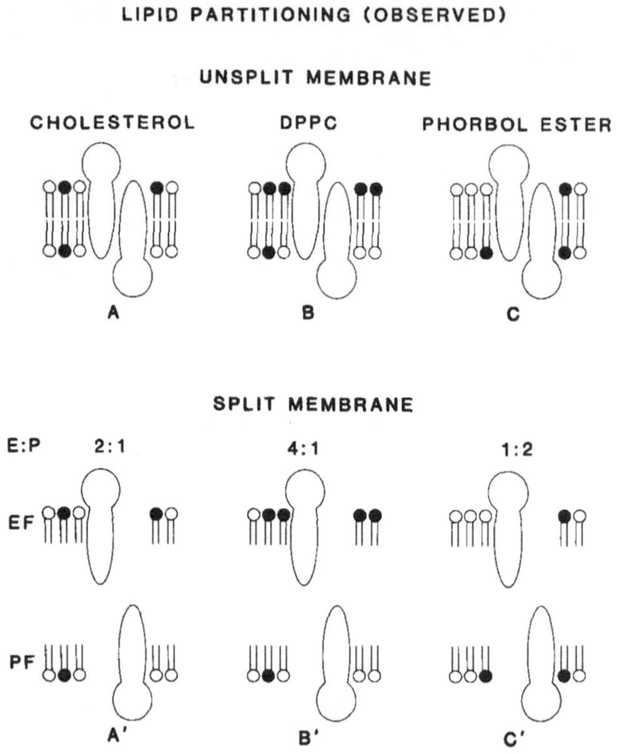

FIGURE 2. Experimental determination of distribution of hydrophobic molecules in split plasma membranes of the human red cell. Both ^{3}H-cholesterol (A, A′) and ^{3}H-dipalmatoylphosphatidylcholine (DPPC) (B, B′) can distribute asymmetrically across the bilayer — cholesterol 2:1, (extracellular leaflet: cytoplasmic leaflet) (A′); DPPC 4.5:1 (B′). In contrast, the tumor promoter TPA (phorbol ester) preferentially partitions with the cytoplasmic leaflet at a ratio of 1:2, (extracellular leaflet:cytoplasmic leaflet) (C′).

the membrane, TPA must be at least transiently associated with the cytoplasmic leaflet of the bilayer. However, until the membrane-splitting studies,[18] there was no direct evidence for the association of TPA with the inner leaflet of the membrane, nor any quantitative data about its distribution. The distribution of TPA in the plasma membrane of the human RBC was quantified under conditions where TPA was associated with the membrane and activated PKC as measured by induced phosphorylation of membrane proteins.[18] TPA was found to distribute asymmetrically across the membrane bilayer; 33 ± 11% (mean ± SD, n = 39 split pairs) of the label was associated with the extracellular leaflet and 66% with the cytoplasmic leaflet (Figure 2C, C′). It is likely that TPA partitions preferentially to the cytoplasmic leaflet of the bilayer consistent with models that depict TPA resident in that leaflet. Thus, membrane-splitting methods can uniquely be applied to evaluating and quantifying transbilayer distribution not only of neutral lipids and phospholipids, but also of amphiphilic or lipophilic second messenger analogs such as TPA.

B. Bilayer Splitting and Membrane Proteins

For a discussion of the effects of membrane splitting on membrane proteins, it is useful to think in terms of peripheral and integral proteins.[21,22] Peripheral proteins are those that can be relatively easily dissociated from membranes by mild treatments, such as high salt or chelating agents, and after extraction contain little bound lipid and are soluble in aqueous buffers.[21] Integral proteins require disruption of the lipid bilayer by treatment with detergents, bile acids or protein denaturants, often retain bound lipid, and are highly insoluble in aqueous buffers around neutral pH.[21]

The effect of membrane splitting on peripheral membrane proteins has not been disputed. It is assumed that the proteins lie outside the fracture plane and are thus unaffected by the separation of the lipid bilayer. The effect of membrane splitting on integral membrane proteins, however, has been disputed. Because integral proteins usually span the bilayer, some researchers have suggested that membrane splitting cleaves covalent bonds.[9-11,23] However, our experiments evaluating the effect of membrane splitting on membrane-spanning integral membrane proteins[24-26] have consistently been at odds with those findings. Although it seems theoretically possible, we have no experimental evidence for covalent bond cleavage for transmembrane proteins of either *Halobacterium halobium* purple membrane[27,28] or human red cell plasma membranes.[26]

1. Peripheral Proteins

The location of peripheral proteins, relative to the plane of fracture, can *a priori* predict their partitioning behavior (Figure 3). If they are located on the surfaces of membranes, not intercalated into the lipid bilayer and not tightly interacting with transmembrane proteins, membrane splitting would not be expected to have an effect on the polypeptide backbone. As a well-studied example, the membrane skeleton of the human red cell contains a variety of peripheral membrane proteins[29-31] whose covalent bonds would not be expected to rupture. These cytosolic peripheral proteins interact with each other and with the cytosolic domains of integral membrane proteins and include all the cytoskeletal components: spectrin, bands 1 and 2; ankyrin, band 2.1; bands 4.1a, 4.1b, 4.2, 4.9; actin, band 5; and glyceraldehyde-3-phosphate-dehydrogenase, band 6. Because bands 2.1 and 4.1 interact with the cytoplasmic domain of band 3 as well as with spectrin, they would be expected to partition intact with the cytoplasmic leaflet of the membrane. And, in fact, that partitioning behavior is found in split membrane fractions from both native and radiolabeled red cells. Band 4.1, however, also interacts with the cytoplasmic domain of glycophorin A,[32] and although 4.1 partitions with the cytoplasmic leaflet, glycophorin A preferentially partitions with the extracellular leaflet.

2. Integral Proteins

As with peripheral membrane proteins, the location of integral proteins relative to the plane of fracture is important in determining the effect of membrane splitting. However, the effects of membrane splitting on integral membrane proteins are less predictable than the effects on peripheral proteins (Figure 3). Integral polypeptides span the bilayer and are simultaneously exposed to two hydrophilic environments, the extracellular and cytoplasmic surfaces of the membrane. Moreover, data from different laboratories has generated some controversy about the effect of freeze fracture on the polypeptide backbone of integral membrane proteins. Both in early studies of the RBC membrane[10,23,33] and in more recent studies,[11] researchers have proposed that covalent bonds of the membrane-spanning sialoglycoproteins were cleaved (Figure 3B) and that covalent bonds may be broken.[9,34] In contrast, we have never found any evidence for covalent bond cleavage either in qualitative or in semiquantitative studies. Rather, we consistently find that integral membrane proteins of the RBC membrane partition intact and preferentially with either the inner or the outer leaflet of the split membrane (Figure 3A). Moreover, the pattern of distribution of integral membrane proteins appears to be more a function of the tertiary structure of the polypeptide chains than a function of interactions and associations between integral and peripheral membrane proteins.

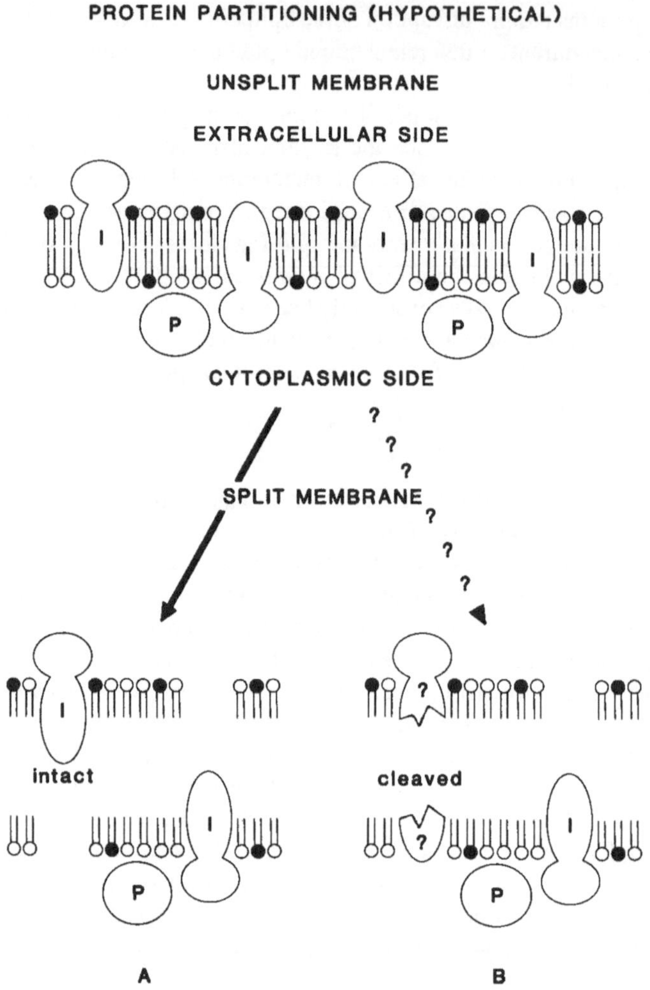

FIGURE 3. Hypothetical effects of membrane splitting on membrane proteins. Two effects have been proposed: bilayer splitting without cleavage of peptide bonds (A) or with cleavage (B). We have only observed case A. Peripheral (P) and integral membrane proteins (I) are indicated. The regulation of uncleaved protein monomers and multimers partitioning is thought to be a function of surface hydrophilicity, mass, and/or intermolecular associations.

II. FREEZE FRACTURE OF PLANAR MONOLAYERS

Freeze-fracture researchers recognized the potential of membrane splitting as a fractionation technique, and there were several early attempts at bulk fracture.[35-37] The planar freeze-fracture approach, however, provided the first and only method for isolating split membrane fractions in quantities suitable for direct biochemical evaluation.[3,4] The planar cell and membrane monolayer preparations and splitting approaches are simple both in concept and in practice.

A. Principle

The basic principle underlying membrane leaflet isolation by membrane splitting is based on the observation that membrane bilayers oriented parallel to the plane of fracture will fracture in preference to those of other orientations.[3] To convert this principle into practice required methods to orient cells and membranes, to freeze fracture those oriented preparations, and to recover the split leaflets for chemical and physical evaluation.

B. Practice

The practice of planar monolayer freeze fractures involves four steps: (1) sample preparation; (2) planar surface preparation; (3) monolayer preparation; and (4) sandwich preparation, freezing, and splitting.

1. Sample Preparation

The single most important variable in a biological sample preparation is the removal of soluble protein. Both isolated cells and isolated membranes must be well washed and concentrated prior to application to the planar surface. Soluble protein effectively competes with cells or membranes for binding sites on cationic planar surfaces, diminishing or preventing complete coverage of the surface and flattening of the membranes. We consistently find that washing and sample concentration are critical for successful bulk freeze-fracture studies. Details of sample preparation for red blood cells[3,4,9,11] and purple membranes isolated from *Halobacterium halobium*[6,27,28,38,39] have been published.

2. Planar Surface Preparation

Both physical and chemical approaches have been used to tether and flatten cells to form planar cell and membrane monolayers.[7,8,12,40-42] Most of these approaches have been used for morphological studies,[43-49,57-59] fewer for cytochemical, notable electron microscope autoradiographic studies,[50-55] and fewer still for biochemical analyses.[4,9-11,26,34,35] For morphological studies, surfaces have been chemically modified with aldehydes[56-59] or lectins.[56,117] For biochemical studies, cell and membrane attachment is routinely mediated by polycationic surface modification. Both glass[6,10,11,26] and mica[8,39] have been used as planar surfaces for biochemical studies. The negative charge on clean glass or mica surfaces is converted to a net positive charge by titration either with low molecular weight poly-1-lysine[3,10,11,35,43,44,60] or with Alcian blue.[7,46,49,61]

3. Monolayer Preparation

Isolated cells and membranes bear a net negative charge at physiological pH and will readily and tenaciously attach to positive surfaces. To form closely packed cell and membrane monolayers, however, samples must be concentrated. For example, applying a slurry of pelleted red cells resuspended in one half volume of buffer insures formation of a closely packed monolayer of cells suitable for planar monolayer freeze fracture.[3,9-11] Subsequent thorough washing away of unattached cells or membranes is required. Alternatively, cells or ghost membrane monolayers can be "sheared" with a stream of buffer to form a membrane monolayer whose cytoplasmic surface is exposed.[25,45,47,49,57,58] For accurate biochemical analysis, it is necessary to quantify microscopically the amount of overlapped or folded membrane. Single membrane monolayer preparations must be completely devoid of overlapped or folded membrane to ensure that membrane splitting will produce a pure fraction of cytoplasmic leaflets.

4. Sandwich Preparation, Freezing, and Splitting

For biochemical preparations (bulk splitting), cationic cover glasses or mica containing cell or membrane monolayers are placed against a second glass plate to form a sandwich. Originally a copper plate was used to form a glass/copper sandwich for cholesterol[4,35] and protein evaluations.[10] However, to minimize interactions of proteins with copper ions the copper was replaced with brass,[9] heavy glass,[19] or, more recently, with starch-coated glass microscope slides.[11] In our experience, untreated heavy glass/cover glass sandwiches have fracturing properties similar to the starch-coated glasses.[11]

There has been some confusion about the distance between the cover plate and the cell monolayer. For example, it has been suggested that the copper plate has an "uneven surface for the attachment of cells", and that starch polysaccharides have a "slight affinity" for erythrocytes, implying close proximity of cover plate to cells.[11] In our experience, the covering plate, whether copper or glass, is usually tens of microns from the cell surface, as can be verified by macroscopic examination of split sandwiches. The height of plateaus that occasionally occur in unfractured areas on the cationic glass surface is clearly more than a few microns. The exact height of the plateau can be determined by shadowing, given the angle and length of shadow.

The sandwiches are frozen in Freon-22 cooled with liquid nitrogen, and transferred to liquid nitrogen for splitting. Sandwiches are pried apart and split under liquid nitrogen using a razor blade.[4,9,11,26] The plane of fracture preferentially runs through the attached and flattened membranes.

Planar membrane-splitting studies can be grouped into two categories: cell monolayer splitting and membrane monolayer splitting. In general, cell monolayers have been used to isolate extracellular leaflets, and membrane monolayers have been used to isolate cytoplasmic leaflets. For qualitative analyses, planar monolayers of cells have also been used for DBLAMS experiments to quantify transmembrane distributions of lipids and proteins (Figure 4). Planar single membrane monolayers (SMM) have been used for both protein topography and single membrane monolayer splitting (SMMS) studies. In SMM, the cytoplasmic surface of the membrane is totally exposed, and thus peripheral membrane proteins and/or the cytoplasmic domain of integral membrane proteins can be readily modified or removed by chemical extraction or enzymatic labeling. SMM have also been used for monolayer freeze-fracture autoradiography (MONOFARG). MONOFARG is an electron microscopic cytochemical approach for determining the distribution of radioisotopes in the plane of intact and split membranes at high resolution.[50,51,55] MONOFARG can also be used to evaluate the trans-bilayer distribution of radiolabeled molecules in isolated membranes and has been used to quantify the transmembrane partitioning patterns of radiolabeled transmembrane polypeptides.

III. EFFECTS OF SPLITTING ON TRANSMEMBRANE PROTEINS

Both theoretical considerations and experimental observations support the hypothesis that membrane-spanning polypeptides remain intact during membrane splitting.

A theoretical analysis of the energies required to cleave the peptide backbone of transmembrane helices suggests that the probability of bond cleavage is slight. Kell et al. calculated the most likely planes of membrane separation by considering which intra- and intermolecular bonds might be disrupted.[62] They used equilibrium energies as an approximation to the real energy of attraction between various molecular species in the frozen bilayer and modeled the membrane in rapidly frozen tissue preparations as a gel state, two dimensional, isotropic lipid bilayer containing integral membrane proteins that spanned the thickness of the bilayer. They calculated the work required to rupture and separate to infinity one of the weakest protein structures, the 3.6-residue-per-turn α-helix, such as that found in glycophorin A,[63] and found the total energy to be approximately 92 kcal/mol. Of that amount, about 80 kcal/mol represented the dissociation energy for the C–C bond.[64] Since the dispersion energy for separation of the lipid bilayer into two lipid leaflets is about 0.5 kcal/mol, rupture of even a single weak α-helix requires about 200 times more energy (see Table 2 in Reference 62). Moreover, integral membrane proteins, such as bacteriorhodopsin and RBC band 3, are composed of multiple helices that span the bilayer many times, and thus the likelihood of such structures being ruptured during bilayer splitting seems vanishingly small.

The dissociation energies most important in considerations of splitting effects on mem-

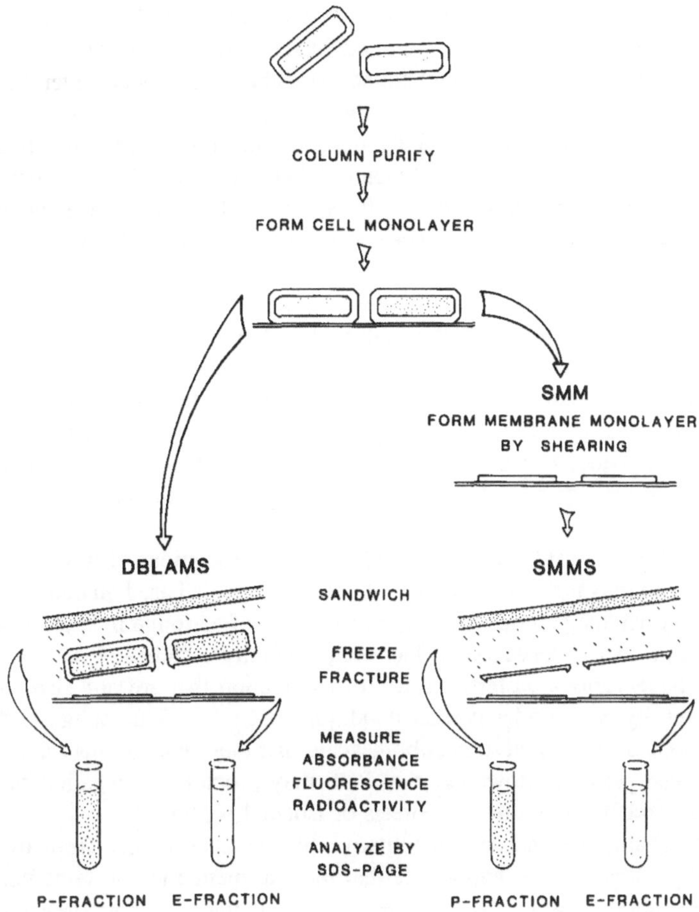

FIGURE 4. Monolayer splitting methods for analyzing "half" membrane leaflets. In double-labeled membrane splitting (DBLAMS), intact cells are labeled with FITC-Con-A, cell monolayers formed by attachment to a polylysine-treated cover glass, second glass plate applied, the sandwich frozen, and glasses pried apart. The fracture preferentially passes through the flattened membrane producing a fraction enriched in extracellular leaflet. In single membrane monolayer splitting (SMMS), cell or ghost monolayers are prepared, sheared with buffer to produce single planar membranes (SMM) with cytoplasmic surfaces exposed, and sandwiches freeze-fractured. After splitting, cytoplasmic leaflets partition to the unbound side of the membrane. Both intact and split cell and membrane fractions have been analyzed by a variety of techniques, including fluorometry, spectroscopy, electrophoresis, and light and electron microscopy (e.g., monolayer freeze-fracture autoradiography).

brane-spanning proteins include intermolecular forces between integral and peripheral proteins, electrostatic forces between lipid head groups or ice-entrapped ions, and hydrogen bonding to ice. The dispersion energy for protein-protein separation perpendicular to the bilayer was estimated at 9.4 kcal/mol and bilayer-ice separation at 10 kcal/mol. The total separation of plasmalemma-ice interface was calculated to be approximately 18 kcal/mol (see Reference 62, page 119) where the sum included oligosaccharide H bonds. In other words, the sum of the energies of multiple interactions may, in fact, be equal to or greater than the energy of a single α-helix. However, most integral membrane proteins are composed of many membrane-spanning helices and moreover are usually associated as oligomers, for example, in the RBC membrane as dimers and tetramers. It is thus likely that the interfacial energy represents only a fraction of the energy present in multiple covalent structures.

Cytochemical studies provide a second line of experimental evidence which suggests that membrane-spanning polypeptides can partition intact to one or the other split membrane leaflet. For example, Pinto da Silva and collaborators have developed several cytochemical methods that combine freeze fracture with colloidal gold labeling.[65] Using colloidal gold modified with lectins they have repeatedly observed labeling of intramembrane particles (IMPs) on the cytoplasmic fracture face. Fracture-label studies of red cell membranes showed that IMPs on cytoplasmic fracture faces (P-faces) were labeled by concanavalin A, whose receptor is located on band 3, and by wheat germ agglutinin (WGA), whose receptors are located on the sialoglycoproteins. Because lectins specifically recognize oligosaccharides present exclusively at the extracellular surface of the intact plasma membrane, detection of label on the P-faces suggests the IMPs must contain those oligosaccharides. The simplest interpretation is that membrane-spanning integral membrane proteins can partition intact with the cytoplasmic leaflet carrying their carbohydrate residues with them. It should also be noted, however, that in the fracture-label studies, samples had been glutaraldehyde-fixed and glycerol-impregnated, fractured, and thawed prior to labeling. Such chemical modification and physical processing are significantly different from the biochemical studies discussed below.

Further indirect experimental support for oligomeric associations of RBC membrane sialoglycoproteins, and thus for lack of covalent bond cleavage, is based on detergent extraction data. It is well known that glycophorins A, B, and C are often found in at least two positions in SDS gels. From that observation, it has been generally assumed " . . . that these sialoglycoproteins have various preferred states of association that are not disrupted by weak detergent or even by SDS under certain conditions . . . ".[63] This suggests that protein-protein interactions occur in native membranes and that they remain intact during detergent isolation. It is also possible, however, that SDS may promote aggregation of monomeric species as demonstrated by re-electrophoresis of isolated peptides.

Finally, both qualitative and semiquantitative biochemical studies from my laboratory have consistently supported the hypothesis that integral membrane proteins partition intact with the split membrane leaflet. Nevertheless, early qualitative membrane-splitting studies of radiolabeled membrane-spanning glycopeptides suggested that the covalent backbone was cleaved.[10,11] In the first published study of the effect of freeze fracture on RBC membrane proteins, it was reported that "Well defined fragments of the sialoglycoprotein were produced by the freeze-fracture procedure indicating that selected covalent bonds of these transmembrane proteins were broken",[10] and more recently that " . . . only sialoglycoprotein and their fragments are retained in the external face", " . . . consistent with results . . . which showed that covalent bonds may be broken during the procedure".[11] A third laboratory also examined the effect of membrane splitting red cell membrane proteins and concluded that " . . . new bands (indicative of fragmentation of IMP) appear inconsistently . . . ".[42]

Although some qualitative membrane-splitting studies of radiolabeled membrane-spanning glycopeptides suggest that the covalent backbone of membrane-spanning proteins is cleaved,[10,11] neither theoretical considerations nor our quantitative experimental studies have confirmed those findings, either for intact cells[26] or isolated membranes.[66,67] We have examined whole red cell monolayers as well as isolated purple membrane and red cell ghost membrane monolayers.

A. *Halobacterium halobium* Purple Membrane Bacteriorhodopsin

Although the biochemistry of membrane splitting has focused almost exclusively on the human erythrocyte plasma membrane, the effects of membrane splitting on purple membranes of *Halobacterium halobium* were also examined in early biochemical studies.[27,28,116] Purified purple membrane sheets contain the single integral membrane protein, bacteriorhodopsin (bR), 75% by weight, and several classes of glyco-, sulfo-, and phospholipids, 25% by weight. The bR molecule is composed of seven transmembrane helices,[68] is not glycosolated, has a small cytoplasmic doman, and is closely packed in a paracrystalline lattice. Highly oriented single membrane monolayers were prepared, freeze fractured, and the amount of membrane splitting was quantified by electron microscopy.[28] Gel electrophoresis of freeze-fractured planar purple membrane monolayers revealed that bR partitioned intact to the cytoplasmic side of the membrane without detectable peptide bond cleavage.[27] Because of the unique structural features of bR, the lack of covalent bond cleavage was not surprising. However, our initial unpublished findings on RBC membrane proteins contrasted sharply with a report,[10] and thus we focused our attention on the integral proteins of the RBC plasma membrane.

B. Red Cell Plasma Membrane Band 3 and Glycophorin A

Biochemical investigations of split planar monolayer preparations of red cell plasma membrane proteins have included both whole cell and isolated membrane studies, and both qualitative and semiquantitative studies. For each of these types of investigations, both native and radiolabeled membrane proteins have been examined. In the discussion to follow, examples are given of native proteins as well as chemically modified, radioisotopically labeled, membrane proteins (Figure 5) from unsplit and split membranes. I have included representative data, both published (Figures 6 and 9) and unpublished (Figures 7 and 8), of SDS polyacrylamide gel analyses of whole red cells (Figures 6 and 7) and isolated RBC membranes (Figures 8 and 9).

It is clear from numerous symposia and reviews that RBC membrane proteins have received and continue to receive much attention.[31,69-73] Despite its limited functions and structural peculiarities, such as the presence of an elaborate membrane cytoskeleton,[31] the red cell serves as a model of membrane organization for other plasma membranes. Although the biological relevance of this practice may be questioned, for studies at the pauci-molecular level the red cell membrane is without peer. From a biochemical point of view, this is the result of the application of SDS-polyacrylamide gel electrophoresis (SDS PAGE) to the analysis of chemically and enzymatically modified red cell polypeptides. The general organization of membrane polypeptides proposed in the early 1970s has been repeatedly confirmed.[74,75] Gel electrophoresis of washed ghosts reveals seven to nine major and a large number of minor Coomassie-staining bands, plus several PAS-positive bands (designated PAS 1, 2, 2′, 3, 4). These membrane polypeptides fall primarily into two categories, following the nomenclature of Steck:[75] those that are predominately cytoskeletal and located on the cytoplasmic side of the bilayer (bands 1, 2, 2.1, 4.1 a, 4.1b, 4.2, 4.9, 5, 6) and those that are transmembrane (notably band 3 and glycophorin A). During the past 10 years both the cytoskeletal and transmembrane polypeptides have received detailed structural and functional analysis.[29,31,73,76-85]

The current picture for the membrane cytoskeleton is one where the spectrin (bands 1 and 2) polypeptides associate as heterodimers (one band 1 plus one band 2) that link end-to-end forming tetramers whose free-ends interact with oligomers of band 5 (f-actin) and band 4.1. Band 2 of spectrin is associated with the integral membrane protein band 3 via ankyrin (band 2.1 and derivatives), and spectrin may also interact with membrane aminophospholipids.[86] Ankyrin interacts with the cytoplasmic domain of band 3 at a ratio of one ankyrin per four band-3 polypeptides. Besides mediating spectrin-actin association, band 4.1 interacts with the cytoplasmic domain of glycophorin A.[32]

Band 3 is the major integral membrane protein of the human erythrocyte and comprises about one quarter of the total membrane protein.[75,87] Band 3 is thought to be composed of three domains: an outer surface hydrophilic domain of about 38,000 Da, a 17,000-Da hydrophobic fragment that spans the bilayer, and a 40,000-Da hydrophilic cytoplasmic segment.[87] The cytoplasmic domain (polyanionic amino terminus) of band 3 contains numerous binding sites[79,88-92] for band 4.2, hemoglobin, phosphofructokinase, and a variety

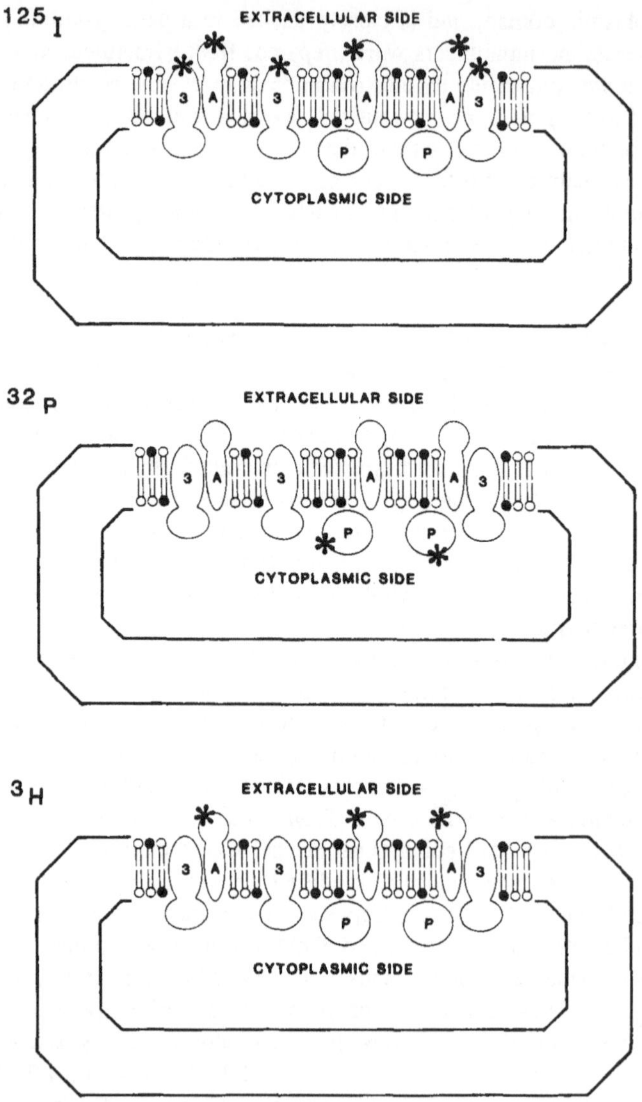

FIGURE 5. Three commonly used radioisotopic-labeling techniques for analyses of membrane splitting effects on red cell membrane proteins. Integral membrane proteins band 3 (3) and glycophorin A (A) and the peripheral membrane skeletal proteins (P) can be labeled; asterisks depict preferential sites of covalent attachment of radioisotope. Lactoperoxidase-catalyzed radioiodination (^{125}I) labels band 3 and glycophorin A. Phorbol esters stimulate phosphorylation (^{32}P) of peripheral proteins. Tritiated borohydride (^{3}H) labels sialoglycoproteins. Labeled polypeptides are then analyzed by SDS-polyacrylamide gel electrophoresis and autoradiography or fluorography.

of glycolytic enzymes including aldolase and glyceraldehyde-3-phosphodehydrogenase (G3P). It is thought that band 3 exists as dimers in the membrane.[87,93]

Glycophorin A is composed of 131 amino acids representing approximately 40% of the dry weight of the molecule; the remaining 60% of the molecule is composed of oligosaccharide.[63,77,94] Glycophorin A, like band 3, is composed of three domains. The extracellular domain, the amino terminus, contains the carbohydrate and is also rich in threonine and serine residues. Carbohydrates are linked through O-glycosidic bonds. An asparagine residue at position 26 is glycosolated through an N-glycosidic bond. In the membrane-spanning hydrophobic domain, 23 amino acid residues, 73 to 95, contain a high proportion of nonpolar amino acids. The cytoplasmic domain contains a cluster of basic amino acids around residue 100, as well as an acidic region. The total calculated molecular weight of a glycophorin A subunit is 31,000 Da.

Although glycophorin A is often depicted as dimers in the membrane,[95] the evidence for oligomeric association is indirect. Glycophorin A migrates in two postions, PAS 1 and PAS 2, on SDS PAGE gels, and can be reversibly converted from PAS 1 to PAS 2 as a function of protein and detergent concentration, temperature, time of incubation, and ionic strength.[94,96-98] Also, carboxylmethylation of a single methionine residue located in the hydrophobic region of the polypeptide chain will promote dissociation of dimers into monomers. In addition, a hydrophobic, tryptic peptide, T6, isolated from the membrane-spanning region of the polypeptide binds to glycophorin A monomers and also prevents reassociation of the subunits to the dimeric form.[94] Agglutination of RBCs by myxoviruses and hemoglutinin A provides further indirect evidence for the aggregation of glycophorin or of sialic acid residues in the intact RBC membrane.[99] It is known that sialic acid residues are required for RBC interaction with viruses, and for lectin labeling, and isolated RBC sialoglycoproteins can prevent hemoglutinin-induced RBC agglutination. However, to be effective isolated sialoglycoproteins must be highly aggregated, with molecular weights in the range of 1.2×10^7 Da.[100] This suggests that the sialoglycoproteins are also aggregated *in vivo*. Of course this may also mean that isolated sialoglycoproteins have low affinity for viral or hemoglutinin receptors.

Despite the foregoing observations, the conformation of glycophorin A and its quaternary associations in the native membrane remain unknown. Nevertheless, there is evidence that glycophorin A can associate with band 3.[101-103] Clearly it is of special interest to membrane-splitting studies to have independent observations on pauci-molecular associations before speculating on how protein-protein and protein-lipid interactions are affected by the splitting process.

The membrane-splitting biochemical studies have focused on the two RBC integral membrane proteins: band 3 and glycophorin A. The hydrophilic and hydrophobic domains of these two RBC molecules are quite different in their extracellular regions, in their membrane-spanning regions, and in their cytoplasmic regions. Band 3 exposes on the extracellular surface of the membrane several polypeptide loops that are minimally glycosylated (about 2000 Da per monomer). Glycophorin A, in contrast, exposes a single polypeptide chain with a large amount of covalently bound carbohydrate. The two molecules also differ in their membrane-spanning regions. Band 3 is thought to be composed of multiple (7 to 12) membrane-spanning helices perhaps similar to bacteriorhodopsin, whereas glycophorin A is composed of a single, 23 amino acid transmembrane helix. Finally, the cytoplasmic domains of band 3 and glycophorin A are also quite dissimilar. Band 3 possesses a 40-kDa cytoplasmic domain, whereas glycophorin A exposes 38 amino acids on its carboxy terminus.[63] As previously mentioned, the cytoplasmic domain of band 3 interacts with numerous glycolytic enzymes, peripheral membrane skeletal proteins, and hemoglobin. In contrast, the cytoplasmic domain of glycophorin A makes far fewer associations, binding predominantly to band 4.1.[32] Despite these differences, both band 3 and glycophorin A may have some similarity in their quaternary associations, in that both band 3 and glycophorin A are thought to exist as dimers as well as heteroligomers.[101] Freeze-fracture studies have also suggested that band 3 associates with the sialoglycoproteins in native membranes.[104] Similarly, in a model system glycophorin has been shown to form oligomers of 10 to 12 monomers when isolated and reconstituted into liposomes.[105] Clearly, band 3 and glycophorin A, along with bacteriorhodopsin, provide a range of integral membrane protein structures and interactions that are suitable for evaluating the effects of freeze-fracture on membrane-spanning proteins.

In membrane-splitting studies, cell monolayers are formed from intact, washed native or radiolabeled erythrocytes. Freeze fracture of cell monolayers produces a fraction enriched in the extracellular leaflet of the membrane; the complementary fraction is composed predominantly of intact membrane plus cytoplasm (mainly hemoglobin). Two experimental approaches have been taken by researchers studying whole cell monolayers: qualitative and semiquantitative. The qualitative approach has been to remove unfractured "contaminant" membranes as completely as possible from the polylysine glass side of the split-membrane preparation to produce a "pure" extracellular leaflet fraction.[10,11] The alternative approach has been to recover polypeptides from both fractions as completely as possible, and to quantify them (Figure 6).[12]

1. Whole Cell Studies

To date, three laboratories have published articles describing biochemical analyses of split red cell membrane polypeptides.[9-11,23,26,66] All of these studies were based on planar monolayers of whole red blood cells. Biochemical studies of isolated and split single membrane monolayers have also been published as abstracts,[6,25,67] as a recent techniques chapter,[12] and as the unpublished data and figures in the current review.

a. Qualitative Aspects of Cell Splitting

There is continuing controversy about the qualitative effect of planar freeze fracture on covalent bonds of transmembrane sialoglycoproteins. This section examines methodological details that may be relevant to understanding the contradictory results and conclusions. Differences include RBC preparation, planar monolayer size, sandwich materials, sample harvest, sample solubilization and concentration, and autoradiography.

RBC preparation — Whereas some workers have focused on examining chemically modified (e.g., radioistopically labeled) RBCs, we have examined native polypeptides of washed RBCs. We use SDS PAGE and silver-staining methods to evaluate nanogram levels of chemically unmodified split membrane proteins and sialoglycoproteins.[26] In studies of unmodified membranes we consistently find that all polypeptide bands in the control unsplit samples can also be detected in one or both of the split leaflet fractions. Moreover, new bands which might represent cleavage of covalent bonds of native membrane proteins have never been detected.[26] Others have argued that their new bands are not generated by proteolytic artifact and that the silver-staining techniques we have used have not been sensitive enough to detect small amounts of split polypeptides.[11] However, we have also used lactoperoxidase-mediated radiolabeling techniques[12] and still have not detected covalent bond cleavage. Although we have examined lactoperoxidase-radioiodinated samples, we have not examined oxidized and borohydride-reduced samples. This latter chemical modification procedure can reduce[106] and cleave peptide bonds.[107] In our examinations of radioiodinated (Figure 7) or phosphorylated membrane proteins we do not detect new bands in split cell monolayer fractions.

Sample size — A second difference in experimental detail pertains to the size of individual samples, and to the number of samples pooled for SDS PAGE. Some workers have routinely used large (24 × 50 mm) sandwiches[10,11] and others have used 20 × 20 mm Alcian blue-treated mica pieces.[9] In contrast we use 11 × 22 mm polylysine-treated glass. Some pool 20 to 30 samples for each of their gels,[11] and others prepare ". . . up to 20 pieces . . ." for SDS PAGE of outer membrane leaflets.[9] In contrast, we usually pool a maximum of six samples (11 × 22 mm) in a small volume that does not require concentration prior to sampling for SDS PAGE. Sample size is relevant to sandwich-freezing rate and sample number is relevant to the subsequent need for concentration prior to SDS PAGE. Both of these factors may be important in the generation of new bands.

Sandwich materials — Early researchers[10] followed the copper/glass sandwich procedure I developed for cholesterol analyses[4] for initial polypeptide analyses. However, applying

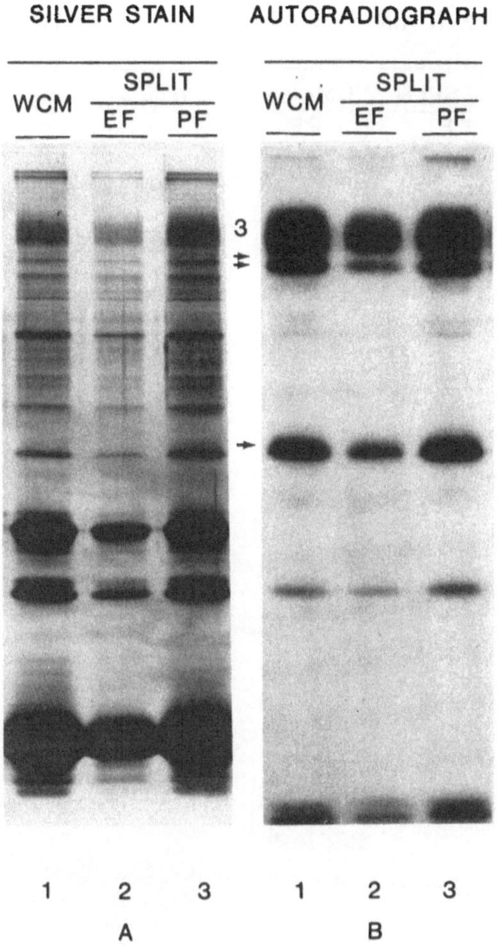

FIGURE 6. Gel and companion autoradiograph of ^{125}I-labeled intact and freeze fractured human red cells. (A) Silver-stained gel of intact whole cell monolayer (WCM), and split extracellular leaflet (EF) fraction, and cytoplasmic (protoplasmic) leaflet (PF) fraction. Note that all bands in the split fraction have matching counterparts in the unsplit control. (B) Autoradiograph of lactoperoxidase radioiodinated intact cell monolayer and complementary split leaflets, EF and PF. LPO radioiodination of intact cells labels band 3 and sialoglycoprotein bands, especially glycophorin A (PAS 1, double arrow, and PAS 2, single arrow). Note again that all radiolabeled bands in the split fractions have matching bands in the unsplit control; i.e., no new bands are present in the split fractions relative to the unsplit control, indicating the absence of cleavage of radioiodinated membrane-spanning protein band 3 and glycophorin during freeze fracture. It should also be noted, however, that the EF fraction is not pure but also contains unsplit membrane (Modified from Fisher, K. A. and Yanagimoto, K. C., *J. Cell Biol.*, 102, 551, 1986.)

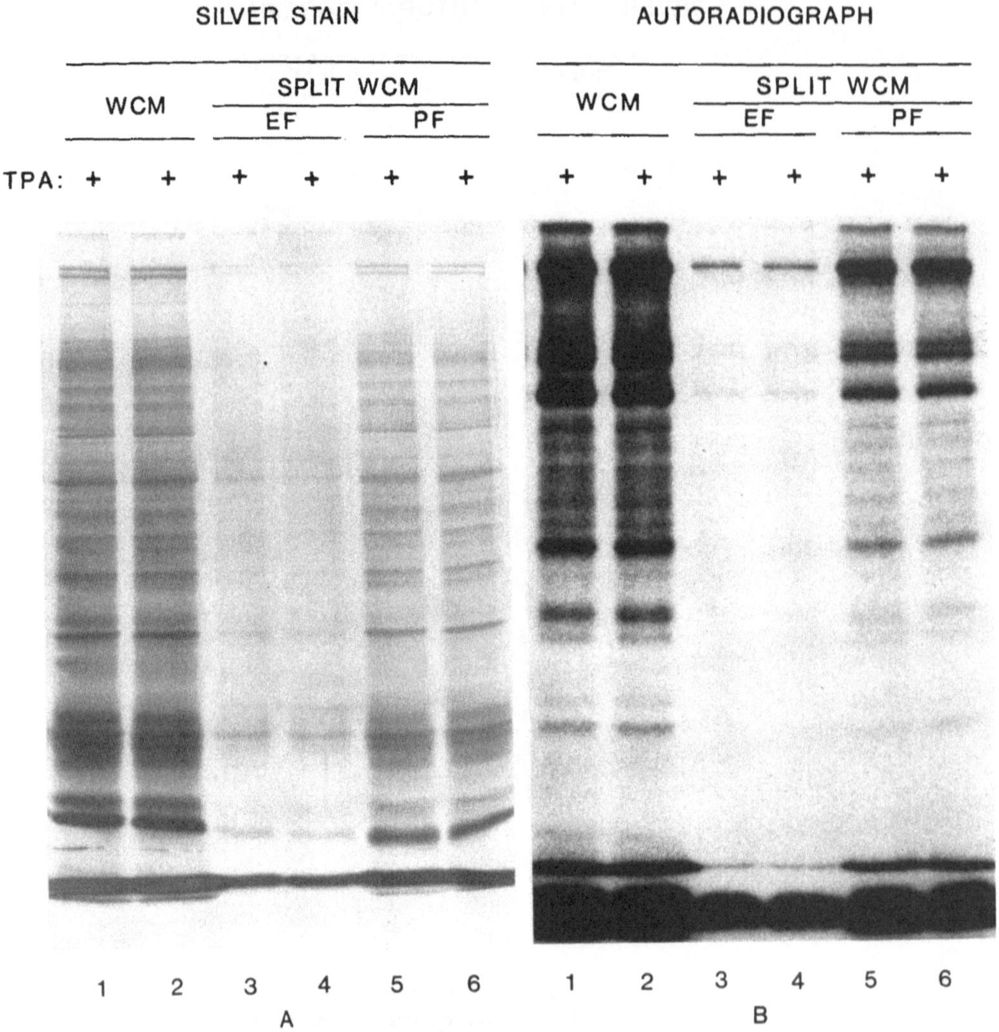

FIGURE 7. Gel and companion autoradiograph of ^{32}P-labeled intact and freeze fractured human red cells. (A) Silver-stained gel of intact and split cell monolayers metabolically labeled with ^{32}P after tumor promoter stimulation of protein kinase C. WCM and split EF and PF fractions shown in duplicate. All bands present in the split sample have counterparts in the unsplit sample. (B) Autoradiograph of (A). TPA stimulated ^{32}P-labeling of several high molecular weight bands. These proteins are cytosolic and all preferentially partition with the PF upon freeze fracture. Note again the absence of new bands in the split membrane fractions.

those procedures to polypeptide assays we found that pure copper sheet interacted with specific polypeptides and/or residual proteases and generated irreproducible data. Thus for polypeptide assays we developed a glass/glass sandwich[19] and using it for the past seven years have never detected polypeptide bond cleavage. Even so, researchers recently described a modified monolayer freeze-fracture technique in which a glass/starch-treated glass sandwich was used, producing a new band, albeit apparently at a different position than in their original report.

Sample harvest — A major difference between laboratories is in sample recovery after freeze fracture. Some attempt to produce pure EF fractions by selectively removing pink areas from the split surface; the pink areas on the cationic glass mainly contain intact cells.[4] Although the pink areas often represent less than 5 to 10% of the total surface area, because the split membrane area contains only about 33% of the total surface of the cell, the polypeptides in the unfractured membrane can contribute significantly to the total polypeptide load. For EF samples, some break off " . . . the red-tinted glass . . . " areas and place individual pieces of glass into SDS solutions.[11] PF samples are treated similarly. For EF samples, others select for clear areas: "Pink edges were usually cut off, but mica pieces with central pink spots were discarded".[9] In contrast, we attempt to recover and quantify all polypeptides from both fractions of split membranes as well as from unsplit control membranes. As others have pointed out,[11] because we retrieve all of the sample, our extracellular fraction is contaminated with intact unsplit membrane. For these reasons our single membrane monolayer splitting (SMMS) studies are especially important since SMMS produces an uncontaminated fraction of cytoplasmic leaflet.

Sample solubilization and concentration — A recent report[11] challenged our suggestion that new bands may have resulted from solubilization conditions favoring proteolysis. It was argued that " . . . samples were removed from liquid nitrogen and immediately dropped directly into boiling SDS buffer that contained phenylmethylsulfonal fluoride (PMSF) and EDTA in order to prevent proteolysis" (Reference 11, page 449). Clearly these precautions should prevent the problems inherent in an earlier report, in which sialoglycoprotein cleavage was described, and in which " . . . pink spots were absorbed on filter paper immediately on thawing. The coverslip was then refrozen . . . "[10] However, there are proteases that remain active in the presence of SDS, PMSF, and EDTA. Furthermore, additions of numerous split samples at liquid nitrogen temperature to small volumes of hot buffer could lower the temperature to a range favoring proteolysis.

Further differences between laboratories are in the number of samples pooled for SDS PAGE, and the composition and volumes of detergent solutions to which the pooled samples are added. For example, some add 20 to 30 pieces of glass to 5 ml of sample buffer containing 0.01% SDS for EF fraction analysis, but for PF fraction analysis add the same number of pieces to 20 ml sample buffer. In contrast, we add one to six glasses to 0.6 or 1.0 ml of sample buffer containing 0.1% SDS for both split PF and EF fractions and the same for unsplit fractions. We have found that for reproducible comparison of EF and PF fractions by gel electrophoresis, it is important to use identical volumes of sample buffer for each split fraction. Moreover, we have found that the concentration of SDS must be at least 0.1% to remove all polypeptides from the PL glass surface. In addition, in an earlier study complementary copper plates were immersed " . . . in a hypotonic sodium phosphate solution (7mM, pH 7.4), and the resulting membranes were isolated by centrifugation at 27,000 × G for 10 min" (Reference 10, page 1344). Dilute buffer, and the time in it during centrifugation could contribute to proteolysis. Moreover, the low-speed centrifugation would probably pellet intact membrane vesicles but not microvesicles derived from split membrane P-faces and thus be a highly selective procedure.

Assuming that proteolysis due to solubilization in dilute detergent and/or subsequent concentration by dialysis is not a problem, what other explanations are there for the additional bands seen in the autoradiographs of tritiated and radioiodinated EF fractions (for example, Figure 3, lane C, Reference 10)? One possibility is that concentration by lyophilization generates new bands. Although some researchers lyophilize their samples to reduce volumes for gel electrophoresis, we and others have found that lyophilization often generates high molecular weight bands that fail to enter the gel and/or low molecular weight bands (see Figure 8 in Reference 11, page 446). Another possibility lies with the different degrees of dilution of each of the two split samples and of the intact membrane sample. Such dilution effects are known to alter sialoglycoprotein associations.[63,108,109]

Autoradiography — A further explanation for the new bands may be found in different approaches to autoradiographic detection of labeled bands, notably the time of autoradiographic exposure. For example, in Figure 3, lane C of Edwards et al.,[10] tritiated and radioiodinated membrane proteins of the EF fraction were exposed four (tritiated) and nine (radioiodinated) times longer than their split (PF) and unsplit counterparts. In contrast we apply split and unsplit fractions to a single gel and expose each gel for a fixed time. Clearly, the gels of Edwards et al. show radiolabeled bands in the split EF fractions that are not seen in their complementary PF fractions or presumably in their intact membrane. But qualitative inspection of autoradiographs can lead to erroneous conclusions. For example, although we have also found minor bands in overexposed autoradiographs of split fractions, comparison with identical overexposed, unsplit fractions has consistently revealed the same minor bands. In response to the claim that "Radiolabeling . . . is more sensitive than the silver-staining method used by Fisher . . . ",[11] I should reiterate that we have also examined lactoperoxidase radioiodinated membranes and have never observed new bands in split-membrane fractions relative to unsplit controls.

In summary, some researchers have observed new bands derived from radioactive polypeptides and have stated that the covalent backbone of glycophorin A is cleaved. In contrast, we have been unable to detect new bands in either native or radiolabeled split-cell membranes (Figures 6 and 7). Despite these problems there has been some general agreement among laboratories. There is a consensus among the published studies that band 3 tends to partition preferentially with the cytoplasmic leaflet. There is also general agreement that glycophorin A partitions preferentially with the extracellular leaflet.

b. Quantitative Aspects of Cell Splitting

In contrast to qualitative biochemical approaches, we have focused on a semiquantitative approach to cell monolayer splitting studies. For example, the DBLAMS technique was designed to quantify the amount of each leaflet present in the split fractions.[5] This avoids the problem of contamination by direct measurement of the amount of each half membrane in the split fractions. Similarly, the electron microscope technique of monolayer freeze-fracture autoradiography (MONOFARG) was designed as a quantitative cytochemical method to examine the location of radiolabeled molecules in the plane of the split membrane, and can also be used to quantify transbilayer distributions.

Usually our first question is whether or not the sum of the splits equals the unsplit control. For example, in our cholesterol studies,[4] did the amount of cholesterol measured in the split cytoplasmic fraction plus the split extracellular fraction equal the amount measured in the intact control monolayer? Clearly if the sum is significantly greater or less than 100% of the control, correct quantification of the transbilayer partitioning patterns would be jeopardized. We have continued to apply the summation test in our polypeptide analyses. For example, we have measured the radioactivity of gel bands derived from radioiodinated membrane and have found no new bands under conditions where the sum of the split activity equals the unsplit activity.

A second line of quantitiation has been to measure what fraction of membrane-spanning proteins partitions to each of the two split leaflets. We have measured the relative partitioning in split whole cell monolayers of integral polypeptide bands, focusing on band 3 and glycophorin A. Initially, we asked a simple question, "Can we detect a difference in the ratios of polypeptides in split versus unsplit fractions?". Intact RBCs were exposed to lactoperoxidase-catalyzed radioiodination, labeling both band 3 and sialoglycoproteins. Cell monolayers were prepared and split, and proteins solubilized in SDS and separated by gel electrophoresis. Bands were autoradiographed and/or excised and solubilized for measurement of radioactivity by liquid scintillation counting (LSC). We calculated the ratio of band 3 radioactivity to the sum of radioactivity of the two glycophorin bands (PAS 1 plus PAS 2) and found differences in ratios among extracellular leaflet, cytoplasmic leaflet, and whole cell fractions.[67] The ratio of glycophorin A to band 3 was consistently lower in the protoplasmic fraction (e.g., 0.49) and higher in the extracellular fraction (0.60) relative to unsplit cell monolayers (0.58). These data suggested an asymmetric partitioning of band 3 relative to glycophorin A: more band 3 partitions with the cytoplasmic leaflet and/or more glycophorin A with the extracellular leaflet. These tentative conclusions from whole cell splitting studies were extended and confirmed by the membrane-splitting experiments described below. It should be noted, however, that a significant amount of band 3 also partitioned to the extracellular leaflet and a small amount of glycophorin A (about 16%) to the cytoplasmic leaflet.

Clearly these semiquantitative findings, based on recovery of total membrane polypeptides, support qualitative findings based on selected EF-enriched preparations. In addition, our findings are also in agreement with the cytochemical labeling experiments of Pinto da Silva and collaborators.[110,111] They found more WGA binding sites, specific for sialoglycoproteins, associated with the extracellular leaflets[110] and more concanavalin A binding sites, specific for band 3, associated with the cytoplasmic leaflet.[111] This agreement is surprising since the biochemical and cytochemical approaches are extraordinarily different. In contrast to simple adsorption of washed cells to a planar cationic surface followed by freeze fracture, the cytochemical method involves fixing cells in glutaraldehyde, embedding them in 30% bovine serum albumen, again fixing with glutaraldehyde, impregnating with 30% glycerol, freezing, fracturing, thawing in glycerol/glutaraldehyde, washing, labeling at 37°C, fixing, dehydrating with ethanol, critical point drying, and replicating.[110] Furthermore, they reported that 70 to 90% of the WGA label was associated with the extracellular leaflet of fracture-labeled cells, a number that brackets our 85% average of split single membranes (see below). Lectin binding sites in the native cell are on the extracellular surface. WGA-labeled sites, however, were associated with cytosolic leaflets, suggesting that intact oligosaccharides could be pulled through the bilayer during fracturing.

2. *Isolated Membrane Studies*

Two types of preparations have been used for biochemical studies of isolated and split single-membrane monolayers. In the first, nonvesicular membrane fragments, purple membrane sheets from *Halobacterium halobium,* were examined.[27,118] In the second, red cell monolayers were used to isolate and produce single membrane monolayers.

In the purple membrane studies, overlapped or folded membranes were removed by sonication.[6] Oriented single membrane samples have also been prepared from isolated purple membrane fragments by selective pH titration of surface charge prior to and during adsorption to the cationic surfaces.[6,12]

In the red cell membrane studies, planar intact cell monolayers were rapidly sheared with a forceful stream of buffer.[25] Unattached membranes and cytoplasm were hydraulically sheared away, exposing the cytoplasmic surfaces of the residual attached membranes. This technique was originally used by Clarke et al.[45] for structural analysis, and the approach has been dubbed the "lysis-squirting technique" by Nermut.[7,47] Given careful shearing procedures, it is possible to remove all overlapped, unattached membrane, producing a highly oriented preparation (Reference 12, Figure 2).

a. Qualitative Aspects of Membrane Splitting

Single membrane monolayer splitting (SMMS) differs from whole cell splitting in two ways: in SMMS, cytosol is removed before splitting, and splitting produces a highly purified cytoplasmic leaflet fraction. Thus, whereas whole cell preparations generate a fraction enriched in extracellular leaflets bound to the cationic surface and variably contaminated with intact membrane, single membrane preparations generate a fraction enriched in cytoplasmic leaflets derived from the unattached side of the planar membrane. Removal of cytosolic proteins by shearing simplifies the interpretation of the effect of splitting on membrane-spanning polypeptides, since shearing reduces the total number and overlap of polypeptide bands on SDS PAGE gels. However, hydraulic shearing does not remove peripheral membrane polypeptides. SDS PAGE reveals that all membrane polypeptides are present, both

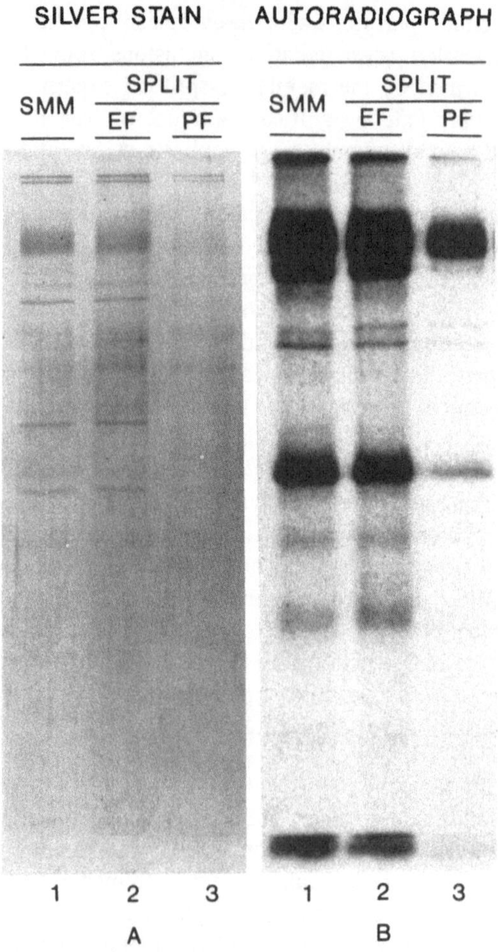

FIGURE 8. Gel and companion autoradiograph of ^{125}I-labeled single membrane monolayers (SMM) and split membrane fractions. (A) Silver-stained gel of SMM and EF and PF fractions. The EF fraction contains intact membrane bilayers also enriched in EF leaflets. In contrast, the PF fraction is composed of the split inner leaflets exclusively. Like radiolabeled whole cell monolayers, splitting SMM produces no new bands. The PF (lane 3) contains peripheral membrane proteins, as well as band 3. The vertical streaking is an artifact of silver staining. (B) Autoradiograph of (A). SMM and a complementary EF and PF pair. Note the conservation of all radiolabeled bands in the split fractions and the qualitative enrichment of band 3 relative to the glycophorin bands (PAS 1 and PAS 2). Enrichment has been verified and quantified by excising gel bands, solubilizing, and measuring radioactivity by liquid scintillation counting. Such analyses consistently reveal an enrichment of glycophorin A peptides in the EF, confirming asymmetric partitioning of glycophorin A with the extracellular leaflet.

peripheral and integral, and present in the same proportion as found in fresh RBC ghost preparations.[112] The qualitative effect of SMMS on native membrane-spanning polypeptides is the same as for whole cell preparations (Reference 12, Figure 6). There is no covalent cleavage of the polypeptide backbone; no new bands are detected (Figures 8 and 9). Given that there is no folded or overlapped membrane and that the planar membrane splits, then

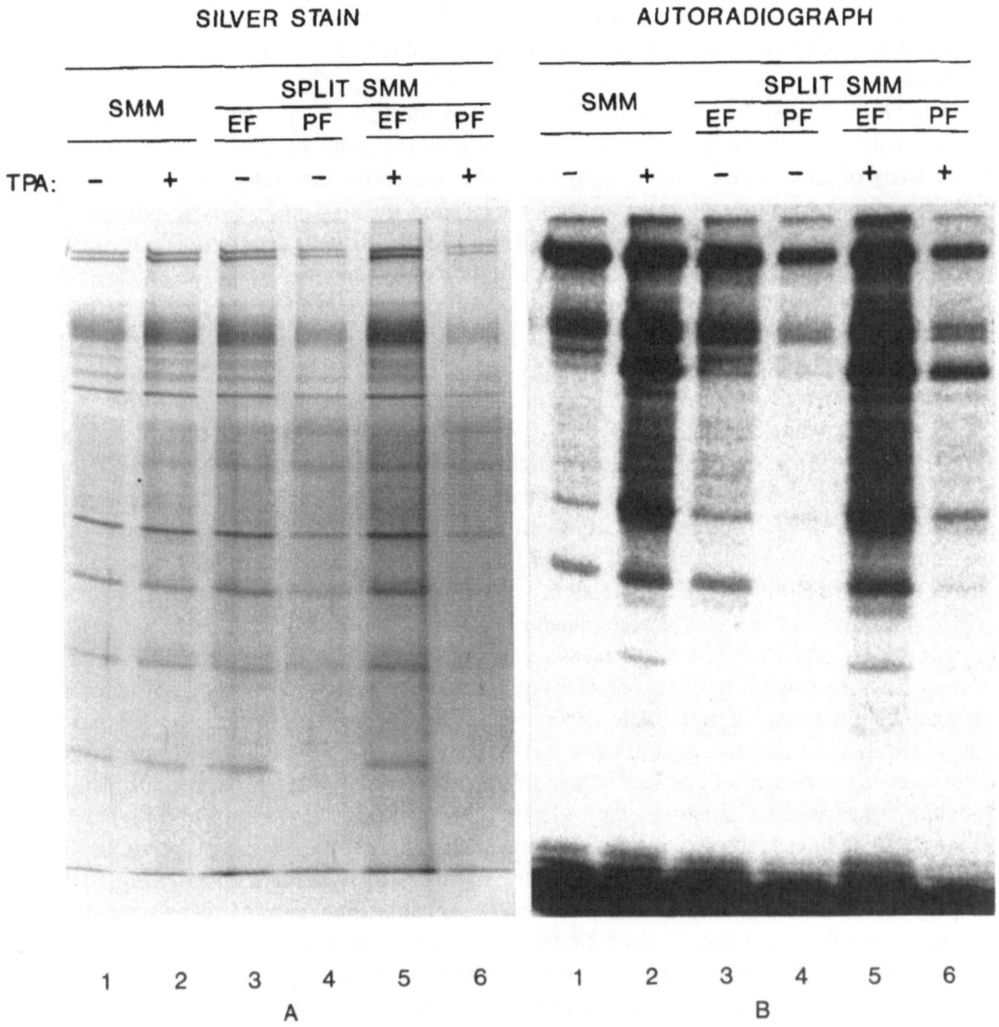

FIGURE 9. Gel and companion autoradiograph of ^{32}P-labeled SMM and split leaflet fractions. Intact cells were incubated with $^{32}P_i$ for 4 h at 37°C, treated with TPA for 10 min, washed, stored at 4°C for 15 h, washed, and lysed. Ghosts were washed 5 times, and planar ghost monolayers and SMM prepared. TPA-treated (+) and untreated (−) SMM were frozen; some SMM were split to produce two fractions: pure cytoplasmic leaflets (*PF*) and a fraction enriched in extracellular leaflets (*EF*). Six 11 × 22 mm SMM were added to 600 μl of 70°C sample buffer for 3 min, cooled, sonicated, and 100-μl portions applied to each lane. Two gels were processed simultaneously, one silver stained (A) and the other, dried and autoradiographed (B). As with previous split cell and membrane samples, there is no evidence for covalent cleavage of either peripheral or membrane-spanning polypeptides in the ^{32}P-labeled preparations whether untreated or treated with TPA to stimulate PKC-mediated phosphorylation of membrane proteins.

the cytoplasmic leaflet fraction contains essentially 100% split membrane. The total absence of new bands in this fraction lends strong experimental support to the hypothesis that the covalent backbones of membrane-spanning polypeptides remain intact during freeze fracturing.

b. Quantitative Aspects of Membrane Splitting

In contrast to the DBLAMS approach for quantifying split membrane fractions from cell monolayers, no similar spectroscopic approach has been developed for quantifying split membrane fractions from single membrane monolayers. Nevertheless, using a combination of PAGE and microscopic techniques it has been possible to quantify the transbilayer partitioning patterns of both lipids and proteins. ^{3}H-Cholesterol and ^{125}I-labeled band 3 and glycophorin A (Figure 8) and ^{32}P-phosphorylated membrane proteins (Figure 9) have been examined.

We have used SDS PAGE plus LSC to measure the activity in each polypeptide band and calculate the percentage of radioactivity in a single band as a function of the total radioactivity of all summed bands. Using monolayer freeze-fracture autoradiography (MONOFARG) of single membrane monolayers we measured the amount of radioactivity in each leaflet. Given the combination of SDS PAGE and MONOFARG data, we determined the transbilayer partitioning distribution of band 3 and of glycophorin A. For example, using SDS PAGE we found that the percentage distribution of ^{125}I in the pure split cytoplasmic leaflet fraction (total polypeptide activity of 550 cpm) was 76% in band 3 (418/550 cpm) and 24% in glycophorin A (PAS 1, 47 cpm, plus PAS 2, 85 cpm; 132/550 cpm). Then using MONOFARG to measure an identical sample, we found that the unsplit membrane contained 925 silver grains per 2145 μm^2 (0.43 grains/μm^2) and the split areas 900 grains per 6076 μm^2 (0.15 grains/μm^2). The extracellular leaflet contained 35% of the radioactivity (15/43 grains/μm^2) and the complementary cytoplasmic leaflet 65%. Thus in these preliminary experiments the cytoplasmic leaflet contained 49% of the band 3 polypeptides (65% total cpm times 76% band 3 cpm). The fact that this represents about a 1:1 distribution of band 3 between the two membrane leaflets came as a surprise. IMPs are known to contain band 3 protein[104] and P-faces of cytoplasmic leaflets of unfixed, uncryoprotected red cell ghost contain 3250 IMPs, whereas E-faces contain 604 IMPs.[113] Clearly, further experiments are necessary to establish if the degree of preferential partitioning in isolated membrane preparations is the same for intact cell preparations as it is for isolated membrane preparations.

In a similar quantitative evaluation of glycophorin A by SMMS we found that 84% partitioned with the extracellular leaflet, a highly asymmetric, five to one partitioning pattern. Although these data are tentative, they illustrate that the transbilayer partitioning patterns of integral membrane proteins can also be evaluated in single membrane preparations.

We have also taken a second, but similar, approach to quantification of SMMS. We compare the radioactivity measured in cut bands to densitometer scans of silver-stained gels of spectrin (band 1 plus band 2) and actin (band 5), two markers for the cytoplasmic surface of the membrane. We assume that as peripheral proteins on the cytoplasmic surface of the membrane, these polypeptides partition with that surface during freeze fracture. We tested the assumption that spectrin and actin could serve as quantitative markers for split cytoplasmic leaflets by plotting areas from densitometer-scanned gel bands vs. hemoglobin absorbance (a cytosolic marker) and found a linear relationship. We also verified that the areas of bands 1 plus 2 and band 5 in silver-stained gels were linearly proportional to concentration in a range suitable for membrane-splitting analysis. We concluded that freeze fracture had a similar effect on the membrane skeletal proteins, spectrin and actin, as it did on cytosolic hemoglobin, and that the peripheral membrane molecules could serve as quantitative markers for the cytoplasmic leaflet.

Based in part on previous observations, we made two additional testable assumptions. First, we assumed that in SMM after splitting the heavy glass side of the sandwich contained 100% inner leaflet. And second, we assumed that the sum of the splits equaled the intact sample with respect both to radioactivity and to areas measured by densitometry. Using this quantitative approach we found results similar to the SDS PAGE plus MONOFARG study. Glycophorin A partitioned preferentially with the extracellular leaflet: 86% to the E-leaflet. We also found that a significant fraction of band 3 partitioned with the extracellular leaflet. Furthermore, the amount of membrane splitting could be measured and appeared to be adequate. The amount of spectrin and actin found in the P-fraction showed that membrane splitting was about 40%; i.e., 40% of the total SMM bands 1, 2, and 5 partitioned with the cytoplasmic leaflet fraction.

Clearly there are significant numerical differences between IMP partitioning patterns of unfixed whole cells or isolated ghosts[113] and the quantification of discrete polypeptide band partitioning reported here for SMMS. Nevertheless, the overall conclusion from our biochemical analyses that there are significant numbers of both band 3 and glycophorin A proteins that can partition with the extracellular leaflet is consistent with the observation of IMPs on that leaflet by conventional freeze fracture, with freeze-fracture cytochemistry,[111] and with indirect indication of the association of glycophorin A and band 3 by rotational diffusion studies of antibody-immobilized glycophorin A.[101]

Although determining the chemical nature of the IMP is in fact of some interest, the planar freeze-fracture biochemical techniques only provide relative information about IMP composition. Bulk fracture is an averaging technique. It can give qualitative information about what classes of molecules may be present in or on each split membrane leaflet, and semiquantitative information about their transbilayer amounts. At best it can indirectly and only partially support or negate IMP models that designate specific molecules as their constituents. In contrast to bulk fracture, the electron scattering metal replicas of IMPs represent molecular associations orders of magnitude removed from the number of molecules needed to produce a single gel band. Nevertheless, there has been a consistent attempt to relate biochemical findings derived from membrane-splitting experiments to morphological findings derived from freeze-fracture experiments. For example, the occurrence of glycophorin A in the external leaflet fraction has been correlated with a particular morphological characteristic of E-face particles.[9] Similarly, it has been suggested that " . . . protein 3 may account for most of the intermembrane particles found on the P-face . . . ," and that " . . . protein 4.5 may account for the particles on the E-face of the freeze-fracture human erythrocyte membrane."[23]

Perhaps this attempt to describe the chemical nature of IMPs had its origins in the earliest attempts in freeze fracture to determine whether IMPs were even composed of protein![114] During the intervening 20 years, there have been a variety of approaches to examine the chemical composition of an IMP. For example, optical diffraction patterns of freeze-fractured particle patches of purple membrane have been correlated to high resolution diffraction patterns of bR,[27] and a variety of integral membrane proteins such as rhodopsin,[115] calcium dependent ATPase, or glycophorin A[105] have been reconstituted into liposomes, and freeze fractured, with IMP number correlated to protein concentration.

IV. SYNOPSIS OF PROTEIN PARTITIONING

Given an asymmetric partitioning of membrane-spanning polypeptides, what influences the partitioning behavior? For the red cell, why does a significant fraction of glycophorin partition with the extracellular leaflet? Is it a function of the molecule itself, for example, a function of the presence of a highly hydrophilic glycosylated extracellular domain? And what about band 3? Does the large cytoplasmic domain affect its partitioning to the cytoplasmic leaflet? What about peripheral membrane protein interactions with membrane-spanning proteins?

A. Regulation of Partitioning Patterns

The tendency to interpret the electron microscopic structures that are seen in shadowed replicas of freeze-fractured surfaces in molecular terms has been extended to interpretations of the regulation of partitioning. There has been a continuing tendency to describe IMP partitioning in terms of molecular interactions, even in the absence of specific chemical information about what molecules might be involved. Those involved in the biochemical studies of split membranes have also participated in this practice. For example, it has been suggested that because the interaction of red cell glycophorin with the cationic surface is

". . . stronger than that of band 3 and also stronger than their own interactions on the inner surface of the plasma membrane . . . most glycophorins might be retained in the OML; whereas most band 3 molecules might separate with the inner leaflet."[9] Similarly other have suggested that "The transmembrane proteins distribute asymmetrically around their anchorage on the surface of the membrane" (Reference 23, page 31). In fact, although logical, these speculations have essentially no experimental support.

Although data are preliminary, there is some evidence that the hydrophilic and hydrophobic domains of membrane-spanning molecules control their partitioning patterns. For example, we found that extraction of peripheral membrane proteins prior to SMMS experiments failed to alter significantly the partitioning pattern of band 3 or glycophorin A. It might also be mentioned that bacteriorhodopsin in isolated purple membrane sheets partitions preferentially with the cytoplasmic leaflet,[27,28] even in the absence of other peripheral or integral membrane proteins.

B. Relevance of Asymmetric Partitioning

Plasma membrane asymmetries are an important feature of the interface between the living cell and its environment. The cytosolic side of the membrane is certainly quite dissimilar in composition, structure, and function from the extracellular side. Whereas, the extracellular leaflet serves as a barrier to restrict the ready exchange of extracellular molecules with the cell, the cytosolic leaflet expedites permeability and exchange processes. Membrane splitting and semiquantitative analyses also demonstrate polypeptide partitioning asymmetries. For example, there is a significant enrichment of glycophorin A polypeptides in the extracellular leaflet fraction of the split bilayer. What is the relevance, if any, of this correlation between composition and function? We simply do not know.

It should be emphasized that the quantitative measurements of isolated membrane proteins described in this article were derived from cells and membranes artificially flattened by absorption to polylysine polymers. Thus it is certainly possible that polylysine attachment and/or lysis may alter the partitioning pattern of glycophorin A and of band 3. However, the surface concentration of glycosylated polypeptides and polylysine polymers predicts that only a small fraction of red cell surface residues interact directly with the ϵ amino groups of the polylysine. Moreover, in studies where we examined the effect of fixation, cryoprotection, and polylysine on IMP partitioning, we found no effect of poly-1-lysine on particle distribution patterns in the absence of cryoprotection.[113]

In synopsis, our membrane-splitting analyses of membrane-spanning proteins of the human red cell have shown that integral membrane polypeptides partition intact to one or both sides of the membrane: band 3 to both sides and glycophorin A predominantly to the extracellular side (Figure 10). Furthermore, preliminary quantitative studies suggest that the regulation of this partitioning behavior is dominated by the hydrophilicity or hydrophobicity of the surface domain of transmembrane molecules. However, it is possible that partitioning is also controlled by associations with other membrane-spanning proteins. The planar splitting approaches just described should help in establishing the features important in the asymmetric partitioning of membrane-spanning proteins.

ACKNOWLEDGMENTS

I thank Kathleen C. Yanagimoto for technical help, Eleanor G. Crump for both technical and editorial help, and Karen Schwelm for secretarial assistance. Research cited in this review was supported by National Institutes of Health Grant GM 31517.

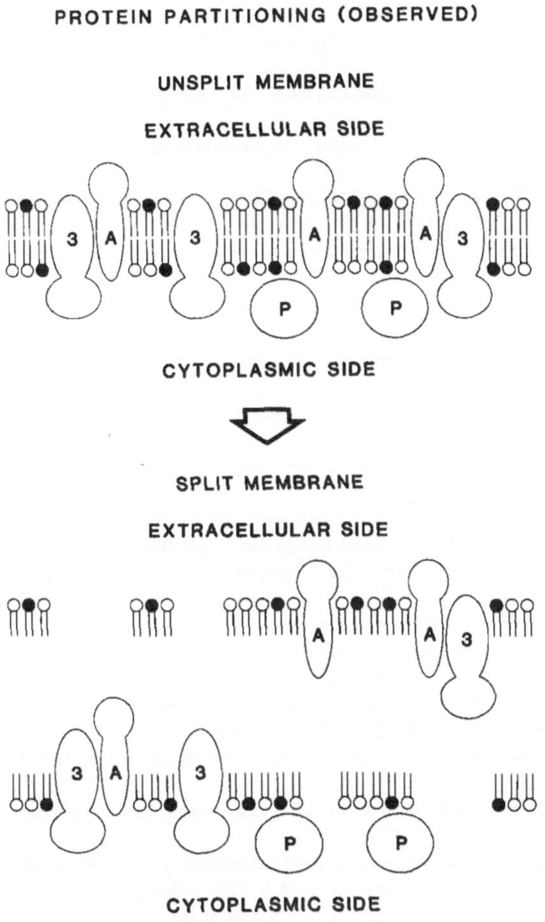

FIGURE 10. Synopsis of the effect of planar membrane splitting on RBC membrane-spanning polypeptides. Glycophorin A (A) partitions preferentially with the extracellular leaflet and band 3 (3) with the cytoplasmic leaflet. Preliminary experiments show that the presence or absence of peripheral membrane proteins (P) may have a minor effect on the partitioning behavior of integral membrane proteins, suggesting that the hydrophilic domains of membrane-spanning polypeptides and/or oligomeric associations may explain their asymmetric partitioning patterns. Although this cartoon depicts heterologous associations of band 3 and glycophorin A as a means of regulating partitioning to the extracellular or cytoplasmic leaflet, the degree to which this occurs is unknown.

REFERENCES

1. **Fisher, K. A. and Branton, D.,** Application of the freeze-fracture technique to natural membranes, *Methods Enzymol.*, 32, 35, 1974.
2. **Pinto da Silva, P.,** Topology, dynamics, and molecular cytochemistry of integral membrane proteins: a freeze-fracture view, in *Electron Microscopy of Proteins, 6 Membranous Structures,* Acadmemic Press, London, 1987, 1.
3. **Fisher, K. A.,** "Half" membrane enrichment: verification by electron microscopy, *Science,* 190, 983, 1975.

4. **Fisher, K. A.,** Analysis of membrane halves: cholesterol, *Proc. Natl. Acad. Sci. U.S.A.*, 73, 173, 1976.
5. **Fisher, K. A.,** Spectroscopic assays for measuring quantities of erythrocyte membrane "halves", *J. Cell Biol.*, 92, 44, 1982.
6. **Fisher, K. A.,** Preparation of planar membrane monolayers for spectroscopy and electron microscopy, *Methods Enzymol.*, 88, 230, 1982.
7. **Nermut, M. V.,** The "cell monolayer technique" in membrane research, *Eur. J. Cell Biol.*, 28, 160, 1982.
8. **Nermut, M. V.,** The cell monolayer technique—an application of solid-phase biochemistry in ultrastructural research, *Trends Biochem. Sci.*, 8, 303, 1983.
9. **Nermut, M. V.,** Ultrastructural and chemical analysis of the outer membrane leaflet of the human red blood cell, *Eur. J. Cell Biol.*, 31, 296, 1983.
10. **Edwards, H. H., Mueller, T. J., and Morrison, M.,** Distribution of transmembrane polypeptides in freeze fracture, *Science*, 203, 1343, 1979.
11. **Edwards, H. H., Mueller, T. J., and Morrison, M.,** Monolayer freeze fracture—a modified procedure, *J. Electron Microsc. Tech.*, 3, 439, 1986.
12. **Fisher, K. A.,** The analysis of fractured membrane "halves", in *Advances in Cell Biology*, Miller, K. R., Ed., JAI Press, Greenwich, Cn, 1987, 1.
13. **Gorter, E. and Grendel, F.,** On bimolecular layers of lipids on the chromocytes of the blood, *J. Exp. Med.*, 41, 439, 1925.
14. **Op den Kamp, J. A. F.,** Lipid asymmetry in membranes, *Annu. Rev. Biochem.*, 48, 47, 1979.
15. **van Deenen, L. L. M.,** Structural organization and dynamics of phospholipids in red cell membranes, in *Normal and Abnormal Red Cell Membranes*, Lux, S. E, Marchesi, V. T, and Fox, C. F., Eds., Alan R. Liss, New York, 1979, 451.
16. **Branton, D.,** Fracture faces of frozen membranes, *Proc. Natl. Acad. Sci. U.S.A.*, 55, 1048, 1966.
17. **Pinto da Silva, P. and Branton, D.,** Membrane splitting in freeze-etching. Covalently bound ferritin as a membrane marker, *J. Cell Biol.*, 45, 598, 1970.
18. **Fisher, K. A. and Yanagimoto, K. C.,** Transmembrane distribution of a tumor promoter: phorbol diester analysis by membrane splitting, *J. Cell Biol.*, 103, 344a, 1986.
19. **Fisher, K. A.,** "Half" membranes: rapid quantitative concentration assays, *Fed. Proc. Fed. Am. Soc. Exp. Biol.*, 39, 1985, 1980.
20. **Nishizuka, Y.,** The role of protein kinase C in cell surface signal transduction and tumour promotion, *Nature, (London)*, 308, 693, 1984.
21. **Singer, S. J.,** The molecular organization of biological membranes, in *Structure and Function of Biological Membranes*, Rothfield, L. I., Ed., Academic Press, New York, 1971, 145.
22. **Singer, S. J. and Nicolson, G. L.,** The fluid mosaic model of the structure of cell membranes, *Science*, 175, 720, 1972.
23. **Morrison, M., Mueller, T. J., and Edwards, H. H.,** Protein architecture of the erythrocyte membrane, in *The Function of Red Blood Cells: Erythrocyte Pathobiology*, Wallach, D. F. H., Ed., Alan R. Liss, New York, 1981, 17.
24. **Fisher, K. A.,** Freeze-fractured RBC membrane peptides and glycopeptides: analysis by SDS gel electrophoresis and silver-staining, *J. Cell Biol.*, 95, 250a, 1982.
25. **Fisher, K. A.,** Sheared planar membrane monolayers: analysis of bilayer splitting, *Biophys. J.*, 41, 118a, 1983.
26. **Fisher, K. A. and Yanagimoto, K. C.,** Effect of membrane splitting on transmembrane polypeptides, *J. Cell Biol.*, 102, 551, 1986.
27. **Fisher, K. A. and Stoeckenius, W.,** Freeze-fractured purple membrane particles: protein content, *Science*, 197, 72, 1977.
28. **Fisher, K. A., Yanagimoto, K. C., and Stoeckenius, W.,** Oriented adsorption of purple membrane to cationic surfaces, *J. Cell Biol.*, 77, 611, 1978.
29. **Bennett, V.,** The molecular basis for membrane-cytoskeleton association in human erythrocytes, *J. Cell Biol.*, 18, 49, 1982.
30. **Bennett, V.,** The membrane skeleton of human erythrocytes and its implications for more complex cells, *Annu. Rev. Biochem.*, 54, 273, 1985.
31. **Bennett, V., Cohen, C. M., Lux, S. E., and Palek, J., Eds.,** *Membrane Skeletons and Cytoskeletal-Membrane Associations*, Alan R. Liss, New York, 1986, 1.
32. **Anderson, R. A.,** Glycophorin-protein 4.1: a transmembrane protein complex that may regulate the erythrocyte membrane skeleton, in *Membrane Skeletons and Cytoskeletal-Membrane Associations*, Vol. 38, Bennett, V., Cohen, C. M., Lux, S. E., and Palek, J., Eds., Alan R. Liss, New York, 1986, 223.
33. **Edwards, H. H., Mueller, T. J., and Morrison, M.,** Fractionation of membrane polypeptides by freeze fracture, *J. Cell Biol.*, 79, 226a, 1978.
34. **Nermut, M. V.,** The cell monolayer technique in chemical and ultrastructural analysis of red blood cell ghosts, *Proc. R. Microsc. Soc.*, 16, 178, 1981.

35. **Fisher, K. A.,** Analysis of cholesterol distribution in bound erythrocyte membranes, *J. Cell Biol.*, 63, 100a, 1974.
36. **Jacobson, B. S., Welt, M., and Branton, D.,** Purification of half-membranes by bulk freeze-fracturing, *J. Cell Biol.*, 63(2,pt2), 153a, 1974.
37. **Park, R. B. and Pfeifhofer, A.,** Chemical composition of fractured membrane halves, in *Membrane Proteins in Transport and Phosphorylation*, Azzone, G. F., Klingenberg, M. E., Quagliariello, E., and Siliprandi, N., Eds., North-Holland, Amsterdam, 1974, 97.
38. **Nermut, M. V. and Williams, L. D.,** Two freeze-fracture patterns in the purple membrane of *Halobacterium halobium*, *Cell Biol. Int. Rep.*, 5, 667, 1981.
39. **Nermut, M. V. and Williams, L. D.,** Freeze-fracturing of monolayers (capillary layers) of cells, membranes and viruses: some technical considerations, *J. Microsc.*, 110, 121, 1977.
40. **Fisher, K. A.,** Split membrane lipids and polypeptides, in *Electron Microscopy 1978*, Vol. 3, Proc. 9th Int. Congr. Electron Microscopy, Sturgess, J. M., Ed., Toronto, 1978, 521.
41. **Fisher, K. A.,** Split membrane analysis, *Annu. Rev. Physiol.*, 42, 261, 1980.
42. **Nermut, M. V.,** The cell monolayer technique in ultrastructural studies of cell membranes, in *The Science of Biological Specimen Preparation*, Revel, J. P., Barnard, T., and Haggis, G. H., Eds., SEM, Inc., AMF, O'Hare, IL, 1984, 181.
43. **Mazia, D., Sale, W. S., and Schatten, G.,** Polylysine as an adhesive for electron microscopy, *J. Cell Biol.*, 63(2,pt2), 212a, 1974.
44. **Mazia, D., Schatten, G., and Sale, W.,** Adhesion of cells to surfaces coated with polylysine. Applications to electron microscopy, *J. Cell Biol.*, 66, 198, 1975.
45. **Clarke, M., Schatten, G., Mazia, D., and Spudich, J. A.,** Visualization of actin fibers associated with the cell membrane in amoebae of *Dictyostelium diocoideum*, *Proc. Natl. Acad. Sci. U.S.A.*, 72, 1758, 1975.
46. **Nermut, M. V. and Lang, R. D. A.,** In situ studies of the red blood cell membrane skeleton, *Cell. Biol. Int. Rep.*, 4(8), 789, 1980.
47. **Lang, R. D. A., Nermut, M. V., and Williams, L. D.,** Ultrastructure of sheep erythrocyte plasma membranes and cytoskeletons bound to solid supports, *J. Cell Sci.*, 49, 383, 1981.
48. **Lang, R. D. A. and Nermut, M. V.,** Visualization of band 4 protein in sheep red cell membranes, *Cell Biol. Int. Rep.*, 4, 751, 1980.
49. **Nermut, M. V.,** Visualization of membrane skeleton in human erythrocytes by freeze etching, *Eur. J. Cell Biol.*, 25, 265, 1981.
50. **Nermut, M. V. and Williams, L. D.,** Freeze-fracture autoradiography of the red blood cell plasma membrane, *J. Microsc.*, 118, 453, 1980.
51. **Fisher, K. A.,** Autoradiography of membrane "halves": tritiated-cholesterol labeled erythrocytes, *J. Cell Biol.*, 70, 218a, 1976.
52. **Fisher, K. A.,** Monolayer freeze-fracture autoradiography: preliminary estimates of 125-I resolution, *J. Cell Biol.*, 91, 249a, 1981.
53. **Fisher, K. A.,** Monolayer freeze-fracture autoradiography, *Proc. R. Microsc. Soc.*, 16, 172, 1981.
54. **Fisher, K. A.,** Monolayer freeze-fracture autoradiography: quantitative analysis of the transmembrane distribution of radioiodinated concanavalin A, *J. Cell Biol.*, 93, 155, 1982.
55. **Fisher, K. A.,** Monolayer freeze-fracture autoradiography: origins and directions, *J. Microsc.*, 126, 1, 1982.
56. **Alpin, J. D. and Hughes, R. C.,** Protein-derivatised glass coverslips for the study of cell-to-substratum adhesion, *Anal. Biochem.*, 113, 144, 1981.
57. **Buechi, M. and Bachi, T.,** Immunofluorescence and electron microscopy of the cytoplasmic surface of the human erythrocyte membrane and its interaction with Sendai virus, *J. Cell Biol.*, 83, 338, 1979.
58. **Buechi, M. and Bachi, T. H.,** Microscopy of internal structures of Sendai virus associated with the cytoplasmic surface of host membranes, *Virology*, 120, 349, 1982.
59. **Bachi, T. and Buechi, M.,** Internal structural differentiation of the plasma membrane in sendai virus maturation, in *The Replication of Negative Strand Viruses*, Bishop, H. L. and Compans, R. W., Eds., Elsevier/North-Holland, Amsterdam, 1981, 553.
60. **Nevo, A., DeVries, A., and Katchalsky, A.,** Interaction of basic polyamino acids with the red blood cell. I. Combination of polylysine with single cells, *Biochim. Biophys. Acta*, 17, 536, 1955.
61. **Sommer, J. R.,** To cationize glass, *J. Cell Biol.*, 75, 245a, 1977.
62. **Kell, M. J., Defelice, L. J., and Goolsby, W.,** Equilibrium energy analysis of freeze fracture planes in membranes, *J. Theor. Biol.*, 115, 103, 1985.
63. **Furthmayr. H.,** Glycophorin A: a model membrane glycoprotein, in *Biology of Carbohydrates*, Vol. 1, Ginsburg, V. and Robbins, P., Eds., John Wiley & Sons, New York, 1981, chap. 4.
64. **Sanderson, R. T.,** *Chemical Bonds and Bond Energy*, Academic Press, New York, 1976.
65. **Pinto da Silva, P.,** Molecular cytochemistry of freeze-fractured cells: freeze-etching, fracture-label, fracture-permeation, and label fracture, in *Advances in Cell Biology*, Miller, K. R., Ed., JAI Press, Greenwich, Conn., 1987, 157.

66. **Fisher, K. A. and Yanagimoto, K. C.,** Planar freeze-fracture of radioiodinated erythrocyte membranes: effects of splitting on transmembrane polypeptides, *J. Cell Biol.*, 99, 286a, 1984.
67. **Fisher, K. A. and Yanagimoto, K. C.,** Cytoplasmic "half" membranes isolated from freeze-fractured human erythrocytes: glycophorin A/band 3 ratios, *J. Cell Biol.*, 101, 307a, 1985.
68. **Henderson, R. and Unwin, P. N. T.,** Three-dimensional model of purple membrane obtained by electron microscopy, *Nature (London)*, 257, 28, 1975.
69. **Brewer, G. J.,** *The Red Cell*, Alan R. Liss, New York, 1981, 1.
70. **Kruckeberg, W. C., Eaton, J. W., and Brewer, G. J.,** *Erythrocyte Membranes 2.*, Alan R. Liss, New York, 1981, 1.
71. **Lassen, U. V., Ussing, H. H., and Wieth, J. O.,** *Membrane Transport in Erythrocytes*, Munksgaard, Copenhagen, 1980, 1.
72. **Wallach, D. F. H.,** *The Function of Red Blood Cells: Erythrocyte Pathobiology*, Alan R. Liss, New York, 1981, 1.
73. **Gratzer, W. B.,** The red cell membrane and its cytoskeleton, *Biochem. J.*, 198, 1, 1981.
74. **Fairbanks, G., Steck, T. L., and Wallach, D. F. H.,** Electrophoretic analysis of the major polypeptides of the human erythrocyte membrane, *Biochemistry*, 10, 2606, 1971.
75. **Steck, T. L.,** The organization of proteins in the human red cell blood cell membrane, *J. Cell Biol.*, 62, 1, 1974.
76. **Marchesi, V. T.,** Stabilizing infrastructure of cell membranes, *Annu. Rev. Cell Biol.*, 1, 531, 1985.
77. **Marchesi, V. T.,** Functional proteins of the human red cell membrane, *Semin. Hematol.*, 16, 3, 1979.
78. **Marchesi, V. T., Furthmayr. H., and Tomita, M.,** The red cell membrane, *Annu. Rev. Biochem.*, 45, 667, 1976.
79. **Pasternack, G. R., Anderson, R. A., Leto, T. L., and Marchesi, V. T.,** Interactions between protein 4.1 and band 3. An alternative binding site for an element of the membrane skeleton, *J. Biol. Chem.*, 260, 3676, 1985.
80. **Lux, S. E.,** Spectrin-actin membrane skeleton of normal and abnormal red blood cells, *Semin. Hematol.*, 16, 21, 1979.
81. **Leto, T. L., Correas, I., Tobe, T., Anderson, R. A., and Horne, W. C.,** Structure and function of human erythrocyte cytoskeletal protein, in *Membrane Skeletons and Cytoskeletal-Membrane Associations*, Vol. 38, Bennett, V., Cohen, C. M., Lux, S. E., and Palek, J., Eds., Alan R. Liss, New York, 1986, 201.
82. **Weinstein, R. S., Khodadad, J. K., and Steck, T. L.,** The band 3 protein intramembrane particle of the human red blood cell, in *Membrane Transport in Erythrocytes*, Lassen, U. V., Ussing, H. H., and Wieth, J. O., Eds., Munksgaard, Copenhagen, 1979, 35.
83. **Branton, D., Cohen, C. M., and Tyler, J.,** Interaction of cytoskeletal proteins on the human erythrocyte membrane, *Cell*, 24, 24, 1981.
84. **Cohen, C. M.,** The molecular organization of the red cell membrane skeleton, *Semin. Hematol.*, 20, 141, 1983.
85. **Cohen, C. M. and Foley, S. F.,** Organization of the spectrin actin band 4.1 ternary complex and its regulation by band 4.1 phosphorylation, in *Membrane Skeletons and Cytoskeletal-Membrane Associations*, Vol. 38, Bennett, V., Cohen, C. M., Lux, S. E., and Palek, J., Eds., Alan R. Liss, New York, 1986, 211.
86. **Haest, C. W. M., Plasa, G., Kamp, D., and Deuticke, B.,** Protein-lipid interactions in the erythrocyte membrane: relevance for structural properties, in *Membrane Transport in Erythrocytes*, Lassen, U. V., Ussing, H. H., and Wieth, J. O., Eds., Munksgaard, Copenhagen, 1980, 108.
87. **Steck, T. L.,** The band 3 protein of the human red cell membrane: a review, *J. Supramol. Struct.*, 8, 311, 1978.
88. **Gillies, R. J.,** The binding site for aldolase and G3PHD in erythrocyte membranes, *Trends Biochem. Sci.*, 7, 41, 1982.
89. **Murthy, S. N. P., Liu, T., Kaul, R. K., Kohler, H., and Steck, T. L.,** The aldolase-binding site of the human erythrocyte membrane is at the NH_2 terminus of band 3, *J. Biol. Chem.*, 256, 11203, 1981.
90. **Richards, S., Higashi, T., and Uyeda, K.,** The interaction of phosphofructokinase with erythrocyte membranes, *Fed. Proc. Fed. Am. Soc. Exp. Biol.*, 38, 798, 1979.
91. **Salhany, J. M., Cordes, K. A., and Gaines, E. D.,** Light scattering measurements of hemoglobin binding to the erythrocyte membrane. Evidence for transmembrane effects related to a disulfonic stilbene binding to band 3, *Biochemistry*, 19, 1447, 1980.
92. **Yu, J. and Steck, T. L.,** Associations of band 3, the predominant polypeptide of the human erythrocyte membrane, *J. Biol. Chem.*, 250, 9176, 1975.
93. **Nigg, E. and Cherry, R. J.,** Dimeric association of band 3 in the erythrocyte membrane demonstrated by protein diffusion measurements, *Nature (London)*, 277, 493, 1979.
94. **Furthmayr, H. and Marchesi, V. T.,** Subunit structure of human erythrocyte glycophorin A, *Biochemistry*, 15, 1137, 1976.

95. **Fowler, V. M.,** An actomyosin contractile mechanism for erythrocyte shape transformations, in *Membrane Skeletons and Cytoskeletal-Membrane Associations,* Vol. 38, Bennett, V., Cohen, C. M., Lux, S. E., and Palek, J., Eds., Alan R. Liss, New York, 1986, 167.
96. **Silverberg, M. and Marchesi, V. T.,** The anomalous electrophoretic behavior of the major sialoglycoprotein from the human erythrocyte, *J. Biol. Chem.,* 253, 95, 1978.
97. **Marton, L. S. G. and Garvin, J. S.,** Subunit structure of the major human erythrocyte glycoprotein: depolymerization by heating ghosts with sodium dodecyl sulfate, *Biochem. Biophys. Res. Commun.,* 52, 1457, 1973.
98. **Silverberg, M., Furthmayr, H., and Marchesi, V. T.,** The effect of carboxymethylating a single methionine residue on the subunit interactions of glycophorin A, *Biochemistry,* 15, 1448, 1976.
99. **Laver, W. G. and Valentine, R. C.,** Morphology of the isolated hemagglutinin and neuraminidase subunits of influenza virus, *Virology,* 38, 105, 1969.
100. **Springer, G. F., Schwick, H. G., and Fletcher, M. A.,** The relationship of the influenza virus inhibitory activity of glycoproteins to their molecular size and sialic acid content, *Proc. Natl. Acad. Sci. U.S.A.,* 64, 634, 1969.
101. **Nigg, E. A., Bon, C., Girardet, M., and Cherry, R. J.,** Band 3-glycophorin A association in erythrocyte membranes demonstrated by combining protein diffusion measurements with antibody-induced cross-linking, *Biochemistry,* 19, 1887, 1980.
102. **Cherry, R. J. and Nigg, E. A.,** Rotational diffusion of erythrocyte membrane proteins, in *the Function of Red Blood Cells: Erythrocyte Pathobiology,* Wallach, D. F. H., Ed., Alan R. Liss, New York, 1981, 59.
103. **Cherry, R. J. and Nigg, E. A.,** Molecular interactions involving band 3: information from rotational diffusion measurements, in *Membrane Transport in Erythrocytes,* Lassen, U. V., Ussing, H. H., and Wieth, J. O., Eds., Munksgaard, Copenhagen, 1980, 130.
104. **Pinto da Silva, P. and Nicolson, G. L.,** Freeze-etch localization of Concanavalin A receptors to the membrane intercalated particles of human erythrocyte ghost membranes, *Biochim. Biophys. Acta,* 363, 311, 1974.
105. **Segrest, J. P., Gulik-Krzwicki, T., and Sardet, C.,** Association of the membrane-penetrating polypeptide segment of the human erythrocyte MN-glycoprotein with phospholipid bilayers. I. Formation of freeze-etch intramembranous particles, *Proc. Natl. Acad. Sci. U.S.A.,* 71, 3294, 1974.
106. **Paz, M. A., Henson, E., Rombauer, R., Abrash, L., Blumenfeld, O. O., and Gallop, P. M.,** Amino alcohols as products of a reductive side reaction of denatured collagen with sodium borohydride, *Biochemistry,* 9, 2123, 1970.
107. **Crestfield, A. M., Moore, S., and Stein, W. H.,** The preparation and enzymatic hydrolysis of reduced and S-carboxymethylated proteins, *J. Biol. Chem.,* 238, 622, 1963.
108. **Mueller, T. J., Dow, A. W., and Morrison, M.,** Heterogeneity of the sialoglycoproteins of the normal human erythrocyte membrane, *Biochem. Biophys. Res. Commun.,* 72, 94, 1976.
109. **Tuech, J. K. and Morrison, M.,** Human erythrocyte membrane sialoglycoproteins: a study of interconversion, *Biochem. Biophys. Res. Commun.,* 59, 352, 1974.
110. **Pinto da Silva, P. and Torrisi, M. R.,** Freeze-fracture cytochemistry-partition of glycophorin in freeze fractured human erythrocyte membranes, *J. Cell Biol.,* 93, 463, 1982.
111. **Pinto da Silva, P., Parkison, C., and Dwyer, N.,** Fracture-label: cytochemistry of freeze-fracture faces in the erythrocyte membrane, *Proc. Natl. Acad. Sci. U.S.A.,* 78, 343, 1981.
112. **Dodge, J. T., Mitchell, C., and Hanahan, D. J.,** The preparation and chemical characteristics of hemoglobin-free ghosts of human erythrocytes, *Arch. Biochem. Biophys.,* 100, 119, 1963.
113. **Pricam, C., Fisher, K. A., and Friend, D. S.,** Intramembranous particle distribution in human erythrocytes: effects of lysis, glutaraldehyde, and poly-l-lysine, *Anat. Rec.,* 189, 595, 1977.
114. **Branton, D. and Deamer, D. W.,** *Membrane Structure,* Springer-Verlag, New York, 1972, 1.
115. **Chen, Y. S. and Hubbell, W. L.,** Temperature and light dependent changes in rhodopsin lipid membranes, *Exp. Eye Res.,* 17, 517, 1973.
116. **Fisher, K. A. and Yanagimoto, K. C.,** unpublished observations.
117. **Fisher, K. A.,** unpublished observations.
118. **Fisher, K. A., Yanagimoto, K. C., and Stoeckenius, W.,** unpublished observations.

Chapter 3

LABELING INTRAMEMBRANE PARTICLES IN FREEZE-FRACTURE REPLICAS

John E. Rash, Timothy J. A. Johnson, Joseph E. Dinchuk, Daniel S. Duch, and S. Rock Levinson

TABLE OF CONTENTS

I. Introduction 42

II. Conceptual Obstacles to the Development of Freeze-Fracture "Cytochemistry" 44

III. Colloidal Gold: An Electron-Dense Label Uniquely Suited for Freeze-Fracture Replicas 44

IV. Platinum/Carbon Films: Major Sources of Nonspecific Adsorption (i.e., "False Positives") 45

V. Method for Retrieving Labeled Replicas 45
A. "Sectioned-Labeled Replicas" (SLR) 45
B. "Intact Labeled Replicas" (ILR) 47
C. "Whole Mount" Viewing Methods 49

VI. Labeling Methods 49
A. "Label-Fracture" 50
B. "Fracture-Label" 50
C. "Replica-Label" 51
D. "Shadow-Label" 55

VII. Labeling Specificity 55

VIII. Summary 57

Acknowledgments 57

References 57

I. INTRODUCTION

The freeze-fracture/freeze-etch technique, orginally introduced to examine the substructure of crystalline viruses,[1] has been applied advantageously to the analysis of biological membranes. Two concepts from freeze fracture are central to current understanding of membrane structure and function: (1) the location of the fracture plane and (2) the composition and function of the small "particles" on the fracture faces of biological membranes. Initially, it was assumed that the fracture plane represented the natural interface between the hydrophobic membrane lipids and the aqueous cytoplasmic or extracytoplasmic compartments.[2,3] This proposal was challenged based on morphological evidence for *membrane splitting*, presumably between apposed lipid leaflets.[4] Splitting of membranes was first demonstrated in artificial bilayer membranes composed of asymmetrically radiolabeled Ca^{2+} stearate.[5] After fracturing, the lipids in the monolayers were recovered without detectable intermixing of components. Compelling evidence for the splitting of natural biological membranes was obtained in combined freeze-etch/*surface-labeling* experiments.[6] Blood type antigens A and B on erythrocyte plasma membranes were labeled with ferritin-conjugated antibodies and the membrane *surfaces* exposed by deep-etching. Replicated labels were present on the etched surfaces but not on adjacent freeze-fractured membrane "faces."[6] Based on these experiments, it was widely accepted that biological membranes are split during the fracturing process, thereby exposing their internal components. Since the particles on freeze-fractured membrane faces were revealed to be *intramembrane* components, they were designated "intramembrane particles" or "IMPs".

During this period, it was widely recognized that the number and distribution of IMPs in diverse biological membranes was correlated with the biological and/or enzymatic activity of the membrane.[7,8] The correlation of IMPs with membrane proteins was strengthened by experiments demonstrating that incremental protease digestion of red blood cell membranes caused incremental removal of IMPs.[8] Conversely, IMPs were not present in replicas of membrane bilayers containing only purified lipids, but were found in replicas of artificially reconstituted membrane bilayers when purified membrane protein was included in the reconstitution mixture.[9] The size of the resulting IMPs was similar to those in the membranes from which the protein was obtained.

These data provided much of the structural basis for the "fluid mosaic" model[10] for biological membranes. In this model, the IMPs were directly correlated with intramembrane proteins. This conclusion was based in part on the incorrect assertion[10] that the surface labeling experiments of Pinto da Silva and co-workers[6] had demonstrated direct labeling and positive identification of protein in IMPs. Instead, Pinto da Silva and co-workers had carefully noted that the blood-typing antibodies used to label membrane surfaces were bound to both glycoproteins and glyco*lipids* and, thus, were able to conclude only that IMPs could be composed of either glycoproteins, glycolipids, or both.

In the succeeding years, and based on the widespread assumption that IMPs represent membrane integral proteins, a major goal of freeze-fracture electron microscopy has been to develop methods to identify individual IMPs. Three conceptual and technological developments are beginning to make that goal a reality: (1) the development of reliable methods for retrieval of labeled replicas;[11-17] (2) the introduction of colloidal gold[18,19] as an electron dense and nearly "indigestible" labeling reagent, uniquely suited to the constraints of freeze fracture;[15,20-23] and (3) the recognition that the adsorptive properties of platinum/carbon films provide the major source of nonspecific binding ("false positives"),[16,24,25] combined with the development of methods to minimize this source of nonspecific adsorption.[16,24,25]

Here, we describe methods for labeling membrane integral proteins in freeze-fracture replicas and show that individual IMPs are labeled by protein-specific ligands. Based on the growing number and complexity of freeze-fracture labeling schemes, we propose that

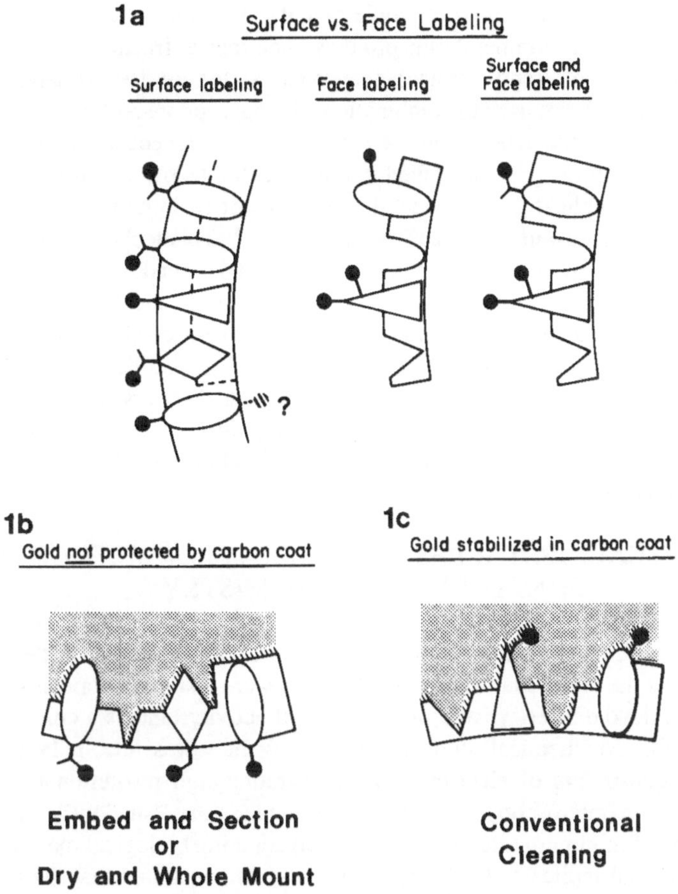

FIGURE 1. Diagram illustrating the four major regions of an intramembrane protein that have the potential to be labeled. There are two regions exposed on the two surfaces of each biological membrane — the outer or exoplasmic surface and the inner or protoplasmic surface. In addition, either of two intramembrane domains may be exposed by membrane splitting — the portion originally intercalated within the exoplasmic leaflet (plus its external determinants) is extracted from the external leaflet and exposed as a P-face IMP. Alternatively, the portion originally intercalated within the protoplasmic leaflet (plus internal determinants) may be exposed as a protuberance from the E-face. Under appropriate circumstances, one or more of these four sites may be labeled by site-specific immunological or pharmacological probes. However, access to labeling sites on the cytoplasmic surface (internal determinants) may be limited by ligand penetration into membrane-enclosed compartments.

labeling terminology coined by others to emphasize the sequence of fracturing vs. labeling steps[20,23] be modified by adding a third term designating the method and sequence for stabilizing and retrieving the labeled freeze-fracture preparations. Before describing the methods available for IMP labeling, however, we call attention to the topographical concept that each intramembrane protein consists of *four* regions — two ''surface'' domains and two intramembrane (or ''face'') domains (see Figure 1a). ''Surface'' determinants of intramembrane *proteins* extend above the lipid bilayer, and in principle, are available for labeling, *whether or not the membrane is split*. In contrast, intramembrane ''face'' determinants are exposed *only after the membrane has been fractured*. (Of course, surface determinants on the same protein continue to be exposed.)

Using conventional thin sectioning or whole mount viewing techniques, one can label the "surface" determinants of intramembrane proteins, then freeze, fracture, (etch) and replicate the intramembrane "face" (and surface) determinants (Figure 1b). If labels specific to individual classes of IMPs are used in combination with appropriate replica retrieval methods, the IMPs and their attached labels may be visualized simultaneously. (This technique, a variation of "label-fracture",[23] is described in greater detail below.) Alternatively, one may fracture, expose, and replicate the "face" determinants of an intramembrane protein, then thaw and label its unshadowed "surface" determinants.[15] (This technique, called "replica-label", is described below.) In a third technique, called "shadow-label", the platinum shadow without a stabilizing carbon coat is used to delineate the IMP face determinants (Figure 1c). The membranes are then thawed and the protein determinants are labeled on the unshadowed "lee" sides of the IMPs. (Shadow-label is also described below.) These three IMP labeling methods (plus "fracture-label" which does not allow visualization or labeling of IMPs[20,22]) provide useful insights for correlating biochemical function with individual IMPs, allow for subcellular mapping of individual classes of IMPs, and may allow mapping of submolecular domains within individual IMPs.

II. CONCEPTUAL OBSTACLES TO THE DEVELOPMENT OF FREEZE-FRACTURE "CYTOCHEMISTRY"

In the formative years of freeze fracture, little effort was expended on developing IMP labeling because of the belief that freeze-etch *replicas* were just that — metallic *impressions* of the living cell, devoid of enzymatic or biochemical activity, and as a consequence, were not appropriate for cytochemical studies. This strong bias was due in large part to the recognition that constraints of electron beam penetration/high resolution in conventional transmission electron microscopy required that the platinum "shadow" and the carbon stabilizing film be of minimum thickness, and that all adhering biological material be cleaned ("digested") from the replica.[4] However, once it was demonstrated that the biology could be exploited for labeling,[11-17,21-23] rather than simply representing electron-opaque material to be removed, the development of IMP labeling techniques was feasible.

III. COLLOIDAL GOLD: AN ELECTRON-DENSE LABEL UNIQUELY SUITED FOR FREEZE-FRACTURE REPLICAS

Following the introduction of colloidal gold as a highly visible electron-dense label for EM cytochemistry,[18,19] it was realized that the high visibility of colloidal gold made it especially useful for labeling freeze-fracture replicas.[15,16,20-23] As a "noble metal", colloidal gold was resistant to replica cleaning steps commonly employed.[15,16,20,22] This property extended the use of colloidal gold to labeling procedures in which the gold colloid was exposed by deep-etching or by freeze-drying, the labeled surfaces and faces were stabilized by rotary carbon coating, and the biological remains digested away. These labeled and replicated surfaces were similar to conventional freeze-fracture or freeze-etch replicas, but with the added feature that colloidal gold labels were closely associated with IMPs[15,16] or with lectin binding sites (glycolipids and glycoproteins).[20,22] Other electron-dense markers, such as ferritin[12-15] and other methods of tracing, such as "freeze-fracture autoradiography",[14,26,27] have been used in attempts to label IMPs in freeze-fracture replicas. However, colloidal gold provides much better electron contrast than ferritin and it can be made in a variety of sizes for multiple labeling. Colloidal gold also provides much higher resolution than autoradiographic techniques, making it an extremely useful marker for IMP labeling.

IV. PLATINUM/CARBON FILMS: MAJOR SOURCES OF NONSPECIFIC ADSORPTION (i.e., "FALSE POSITIVES")

Before showing images of specific IMP labeling, we first demonstrate that the platinum/carbon replica itself provides adsorptive surfaces that are the *the major sources of nonspecific labeling* in freeze-fracture labeling schemes.[16,17] We then show methods to minimize adsorption of labels to the platinum and carbon surfaces.[16,17]

The replica materials consist of ultraclean films, typically platinum and carbon, formed by vaporization under high vacuum conditions (typically 2900 to 3500°C and 10^{-9} to 10^{-12} atmospheres) and their deposition onto freshly cleaved surfaces.[28] The replica is composed of components similar to "finely divided platinum" and "activated charcoal". Those materials are widely used as catalysts and purifying agents because they efficiently adsorb a molecular layer of many organic and inorganic compounds.[29] It is essential to recognize that as the freshly replicated samples are returned to atmospheric pressure and then thawed, the highly adsorptive platinum and carbon surfaces are sequentially exposed to air, water vapor, and then, most importantly, to the dissolved materials in the freezing and labeling medium (e.g., serum proteins, etc.). Unless these adsorptive sites are blocked *before* exposing the replicas to labeling reagents, both platinum and carbon surfaces provide numerous sites whereby dissolved primary and secondary labels (i.e., antibodies, toxins, lectins, and immunogold) will be preferentially adsorbed during thawing and labeling steps.

To demonstrate the potential for nonspecific adsorption.[16,24,25] a preparation of washed mouse red blood cells (i.e, no free serum proteins) was freeze-fractured, replicated, and thawed in a solution containing affinity purified polyclonal antibodies directed at mouse membrane immunoglobins (mIg). The preparation was then exposed to colloidal gold that was coated with affinity-purified goat anti-rabbit antibody. The specimens were rinsed, fixed, and processed for EM cytochemistry, stained, embedded in plastic, and sectioned parallel to the replica-tissue interface. Since mIg are not known to be present in erythrocyte membranes, specific labeling was not expected. In all replicated areas, however, labels are concentrated on cell membranes, on cytoplasmic components, and on the replicated solvent (Figure 2a). Since the platinum/carbon film is heavily "labeled" regardless of material replicated, this image demonstrates an extreme example of high-density nonspecific adsorption to the replica materials. Stereoscopic viewing (Figure 2b) is useful because it reveals that gold granules are attached to both sides of the replica (i.e., to both carbon and platinum layers). A "labeling and blocking buffer" (LBB) was formulated which, by being present in the various media *prior to* and *during* exposure of samples to anti-antibody-gold conjugates,[16,24,25] significantly reduces nonspecific adsorption. (An effective LBB may be made using 50 mm glycine, 0.1% gelatin, 0.05% Tween 20 in 0.1 *M* MOPS buffer, pH 7.4.) When thawed and labeled in the presence of LBB, the Pt/C replica remains devoid of colloidal gold (Figure 2c, from Reference 24), demonstrating the effectiveness of LBB in eliminating nonspecific adsorption (see References 16, 24, and 25 for more details).

Viewing and comparing the replicas by stereoscopy in both normal and reversed perspective is useful because it reveals the sidedness of labels, as well as the "correct" relative separation of labels from the appropriate IMP. Sidedness of labels also provides a useful criterion for identifying examples of "nonspecific" adsorption to the carbon support film. In artificially reconstituted vesicles with few IMPs, close proximity of labels to IMPs and proper sidedness are two criteria for obtaining confidence that IMP labeling is specific.

V. METHOD FOR RETRIEVING LABELED REPLICAS

A. "Sectioned-Labeled Replicas" (SLR)

In 1977/1978, it was recognized that if conventional electron-dense labels (e.g., ferritin) were to be employed for labeling freeze-fracture replicas,[11,12,14] several different viewing

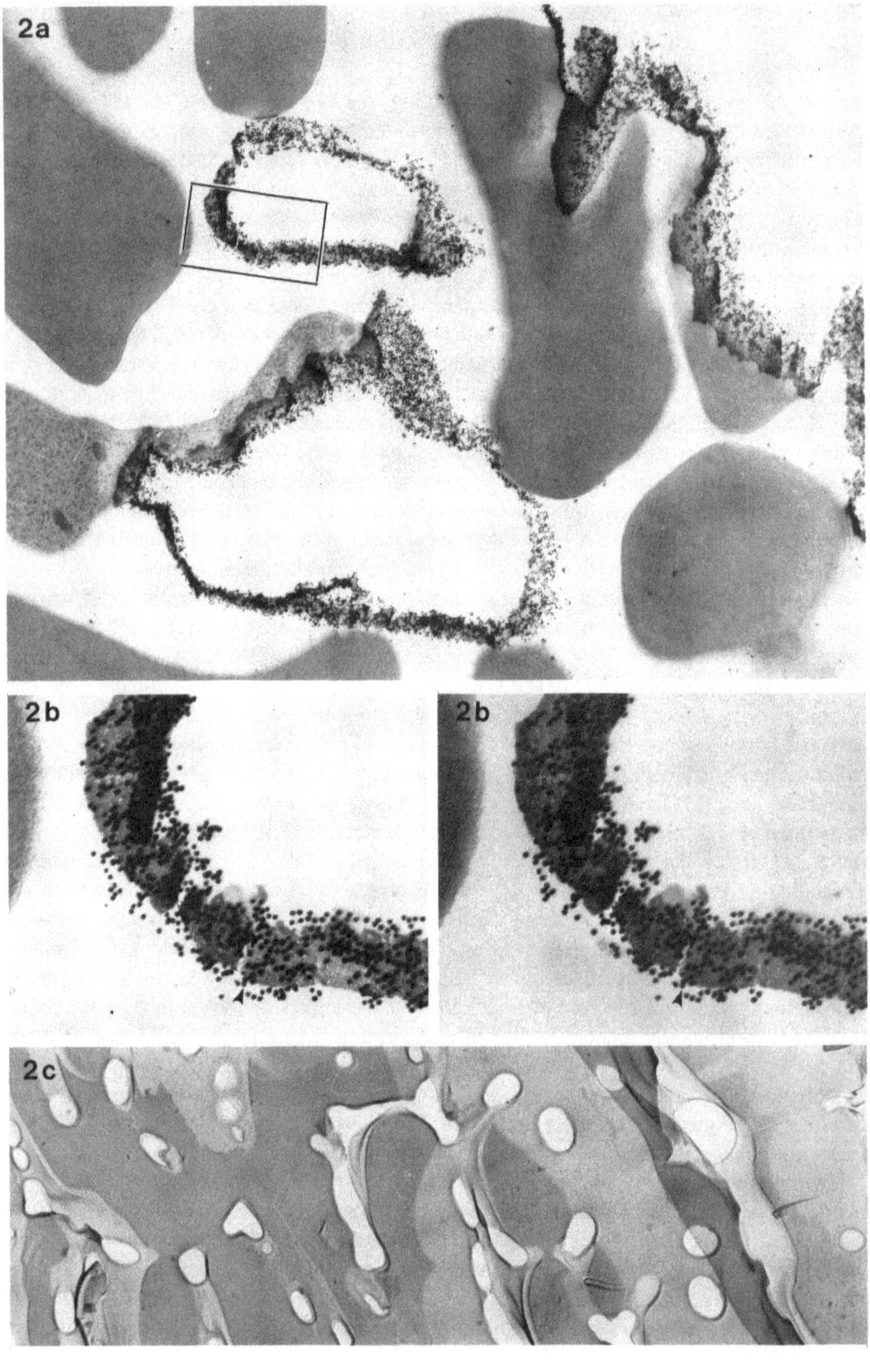

FIGURE 2.

methods would be required, either because digestion of the adhering biological remnants would dissolve the labels or their biological links (antibodies, toxins, and lectins) and release the labels into the digestion medium (see Figure 1b,), or because in conventional replicas, ligand binding sites would be obscured by Pt/C films. Thus, labeling individual IMPs in freeze-fracture replicas was first attempted based on the simple expedient of combining freeze fracture and conventional plastic embedding and sectioning techniques.[11,12,14] By avoiding tissue digestion, much of membrane biochemistry and surface morphology was retained. In "sectioned-labeled replica" (SLR) retrieval methods, tissues are labeled either before or after freeze-fracture replication, then they are fixed, stained, embedded in plastic, and conventional thin or thick sections are cut, usually parallel to the original plane of the replica.[11-13] Thin or thick sections of the labeled replica-tissue interface are viewed by conventional transmission electron microscopy (TEM)[11,12,14-17] or by high voltage EM (HVEM).[16] The colloidal gold labels are visualized at high resolution attached to the replicated structures, which remain attached to their conventionally stained thin section counterparts. Thus, SLR permits the use of conventional cytochemical labeling methods for identifying structures in freeze-fracture replicas.

B. "Intact Labeled Replicas" (ILR)

In 1980 to 1982, additional methods for retrieving and viewing colloidal gold-labeled replicas were introduced.[15,20-22] It was shown that tissues and tissue analogs (isolated cells or reconstituted membrane vesicles suspended in gelatin or agarose) had sufficient structural integrity that platinum-shadowed specimens (with or without a stabilizing carbon coat) could be thawed, labeled, and retrieved by one of several methods. In addition to embedding in plastic for SLR,[11,12,14,15] samples could also be freeze-dried (or critical-point dried) to expose the labeled surfaces (or faces), and the platinum-shadowed surfaces plus attached labels could subsequently be stabilized by carbon coating.[15,22] After conventional cleaning (see Figure 1c), these "intact labeled replicas"[15,16] (or in the sequentially mistitled "critical-point drying fracture-labeled"[22,30] cells) were then viewed as conventional replicas. Regardless of drying technique employed, samples and labels are stabilized by and the colloidal gold labels are protected within the carbon coating matrix. This property is utilized below in a new descriptive terminology that emphasizes *in sequence* the three primary steps (e.g., "label-replica-section", "replica-label-whole mount", or "label-etch-replica").

Recent images from mouse spleen B lymphocytes (Figure 3) that were labeled with anti-immunoglobin and counter labeled with immunogold *before* freeze-etching reveal that colloidal gold labels are suspended in the carbon matrix immediately above external membrane particles (EMPs) on the deep-etched E-surface (Figure 3b). [The fracture "step" from the P-face to the deep-etched and exposed E-surface is visible at the small arrow (Figure 3b).] No labels remain on freshly exposed P-faces, which have numerous IMPs and shallow holes representing etching damage at membrane "pits". (The freeze-fracture nomenclature of Branton et al.[31] is used in this research review.) In stereoscopic views, the great three-

FIGURE 2. Nonspecific adsorption of primary and secondary labels to the platinum and carbon films of a conventional freeze-fracture replica. Gold granules are most densely packed where the solvent phase was replicated, presumably because the gold granules have direct access to both sides of the replica. Nonspecific adsorption is reduced by about half where access is limited by replicated cytoplasm. In stereo pairs (Figure 2b), gold granules are seen attached to both platinum and carbon surfaces. (The gold granules that appear suspended approximately 200 Å above the platinum layer are adsorbed to the surface of the impervious but electron transparent carbon coat. In contrast, a replica of 30% ethanol in distilled water treated with "labeling-blocking buffer" (LBB) before and during exposure to labels, as well as during initial cleaning steps, is devoid of attached labels (Figure 2c). (This image is analagous to the replicated intercellular solvent phase in Figure 2a. (For additional details, see References 24 and 25.) Low molecular weight amines are included in LBB to block residual aldehyde groups, while gelatin is included because it adsorbs to exposed platinum and carbon surfaces. Thus, these reagents effectively prevent subsequent nonspecific adsorption of immunogold labels. (a; Magnification × 10,800; b; magnification × 30,800; and c; magnification × 96,800.) (Modified from Reference 16.)

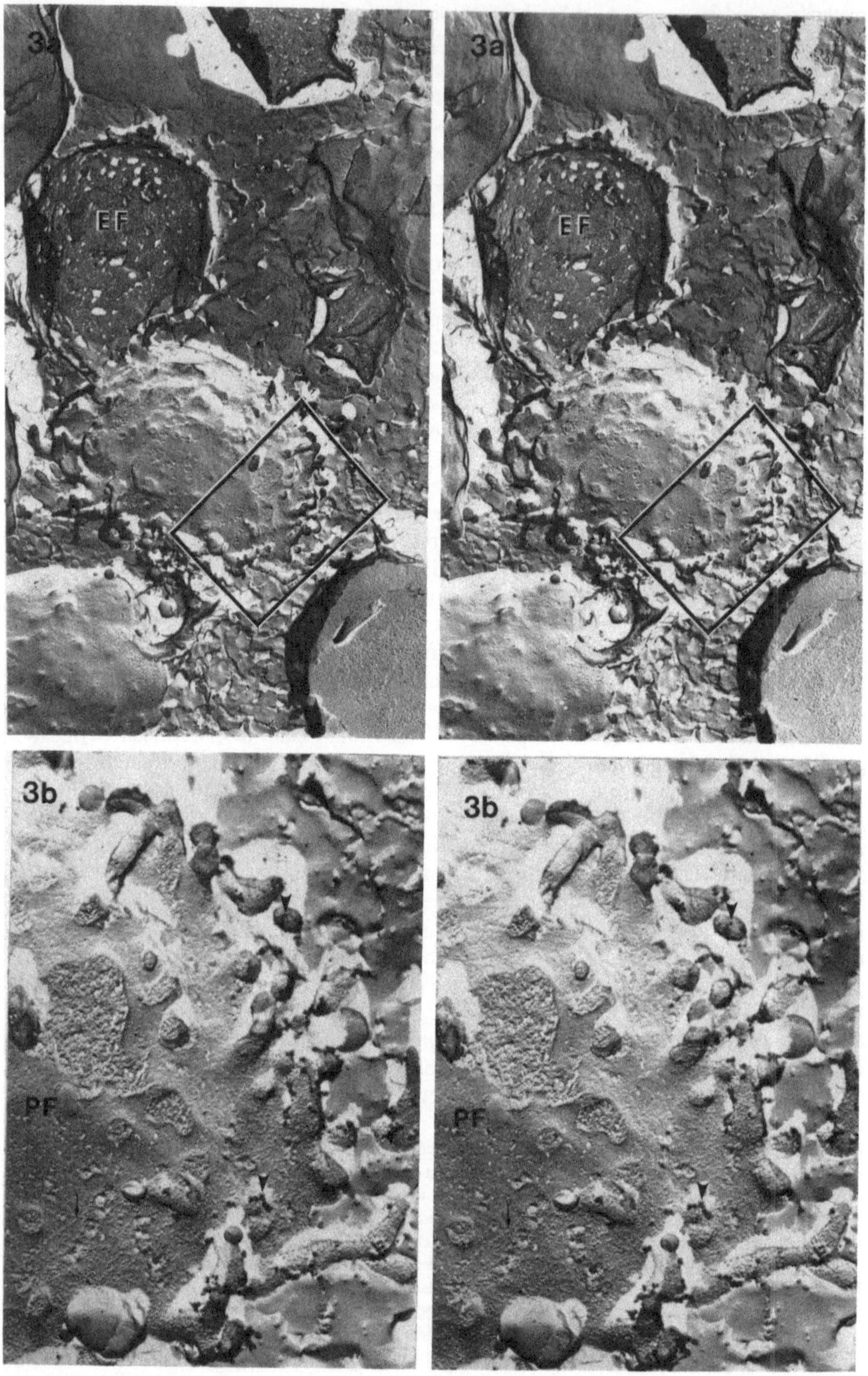

FIGURE 3.

dimensional relief (Figure 3a) and the "sidedness" of labels is easily discerned (Figure 3b). No labels are detected beneath the platinum replica, thereby demonstrating the effectiveness of blocking, cleaning, and rinsing procedures. The restriction of labels to cells having the morphology of B lymphocytes and not to erythrocytes, granulocytes, or other identifiable cell types provides confidence in labeling specificity.

C. "Whole Mount" Viewing Methods

In 1984, Pinto da Silva and Kan[23] introduced an unstained "whole mount" method of viewing the external leaflets of erythrocyte plasma membranes that had been labeled with lectin-gold before freezing. The unfixed cell suspensions were then fractured and replicated, rinsed, but not subsequently digested before viewing in the TEM. Brief rinsing removed most cells that were not replicated. Split and replicated membrane leaflets remained adherent to the replica, often with lectin-gold labels attached to surface glycoproteins and glycolipids. When air dried and viewed by TEM, the low electron density of the adherent unstained membrane leaflets was insufficient to block beam penetration, allowing direct correlation of labels with replica detail. (Since lectin-gold labels were employed, specific labeling of IMPs was not possible. However, that report established the utility of whole mount viewing for examining labeled vesicles and labeled E-leaflets in isolated cell suspensions.) The label-replica-whole mount viewing method (described below and illustrated in Figure 6) is simple, permits rapid sample preparation, and in appropriate circumstances, allows high resolution labeling by high specificity, high affinity ligands.

VI. LABELING METHODS

In theory, there are four stages in the freeze-fracture process wherein samples may be labeled (Figure 4): (1) before tissues are frozen, i.e., label, then fracture or label-fracture; (2) after fracturing but before (or without) shadowing and carbon coating, i.e., fracture-label; (3) after shadowing with platinum but before (or without) carbon coating, i.e., shadow-label, and (4) after Pt/C replication, i.e., replica-label. (For detailed descriptions of specific techniques and appropriate references, see the next four sections.)

For those investigators who are not involved in replica labeling techniques, however, this simple sequence may be misleading. The four techniques do NOT provide similar information. For example, fracture-label (Figure 4B) does not reveal IMPs and cannot be used in IMP labeling techniques. Fracture-label can be accomplished without a freeze-fracture device and without resort to replication of membrane "faces" or "surfaces". Label-fracture (Figure 4A), whether viewed as unstained whole mounts,[23] by the sectioned-labeled replica (SLR) technique,[11-16] or by "thin-section label-fracture",[30] has the potential disadvantage that, during the cleaving process, the attached labels may change the "particle partition coefficient"[32] of the labeled intramembrane proteins [i.e., the attached label may "pull" to the E-leaflet the intramembrane proteins that otherwise would have formed P-face IMPs. If so, the newly created and replicated E-face IMPs may have a different size and/or shape

FIGURE 3. A population of dissociated and washed spleen cells (several classes of leukocytes, plus erythrocytes) that was surface-labeled with immunoglobin recognizing mIg on B lymphocytes, fixed, infiltrated with 30% ethanol (an etchable cryoprotectant[44]), freeze-fractured, deep-etched, replicated with platinum/carbon, thawed, digested, mounted, and photographed for stereoscopic viewing. Smooth, flattened erythrocytes and irregular leucocytes of unknown type are intermixed with microvilli-rich B lymphocytes, identified by their high density of anti-mIg immunogold binding (Figure 3b). (See References 16 and 24 for appropriate "controls" and additional images of labeled and unlabeled cell types.) Immunogold labeling is greatest on microvilli (arrowheads), somewhat less on smooth cell surfaces, and not present on split membrane "faces". The "step" from the P-face to the E-surface is indicated by a small arrow. After deep etching, E-faces of cells ("EF", Figure 3a) appear as delicate leaflets suspended above the lowered ice table. Their delicacy is revealed by the appearance of numerous holes and by the lace-like consistency following sublimation of ice through the numerous membrane "pits". [These images represent a "label-etch-replica" preparation.] a; Magnification × 4,000; b; magnification × 12,000.)

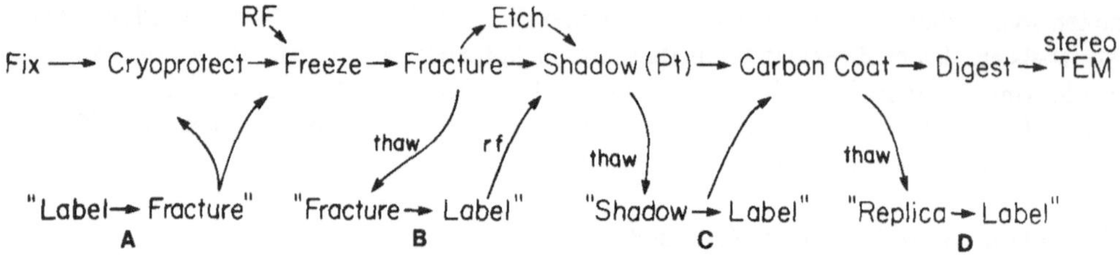

FIGURE 4. Intramembrane proteins may be labeled at one or more of four primary stages in the freeze-fracture process — (A) before samples are frozen, (B) after samples are freeze-fractured but before the metal "shadow" is deposited, (C) after shadowing, but before binding sites are obliterated by carbon coating, and (D) after "replication" (metal shadow plus carbon coat). In these four procedures, different determinants on membrane proteins may be exposed for labeling. In addition, IMPs may be preserved as in conventional replicas, their cleaving properties may be altered substantially by prior labeling, or the unshadowed membrane faces may reorganize during thawing and labeling. Each procedure has distinctive advantages and disadvantages, especially when IMPs are to be visualized and labeled.

than its antecedent P-face IMP.] Replica-label (Figure 4D) has the advantage that IMPs are replicated as in conventional replicas, but has the disadvantage that ligands have access only to *surface* determinants. Intramembrane determinants are obscured by the replica (Pt/C) films. Finally, in shadow-label (Figure 4C), the internal (or face) determinants are shadowed on one side, potentially leaving the unshadowed side exposed for labeling. However, the effects of platinum bombardment, especially on small molecules, is not yet established. Each of these techniques provides unique information concerning the identities, compositions, orientations, subunit structures, and cleaving patterns of membrane integral proteins/ IMPs. Each is described in greater detail below.

A. "Label-Fracture"

Pinto da Silva and co-workers[22] originally introduced a two-part labeling terminology (i.e., label-fracture) to describe a specific version of "freeze-fracture cytochemistry"[21] in which they designate the sequence of label, then fracture. By this term they differentiated this particular version of freeze-fracture labeling and whole mount viewing from other slightly different methods called "prefracture labeling".[11,12,14] Subsequently, however, Pinto da Silva has described the work of other groups using still other variations as label-fracture,[30] suggesting that strict adherence to terminological purity is not possible. Unfortunately, in recent reports,[30] they abandoned this designation according to temporal sequence describing their methods as "thin-section fracture-label" and "critical-point drying fracture-label." Likewise, we have described sectioned-labeled replicas,[15,16,33] contributing to the terminological confusion. Moreover, in the 15 years since surface labeling procedures were first attempted[6,34] and in the 10 years since prefracture labeling and "post-shadow labeling" of IMPs was introduced,[11,12,14] several labeling strategies and many viewing modes have been introduced.[15,16,20-27,30,35-37] With more than a dozen different versions currently in existence, a systematic nomenclature is needed. The simple descriptive terminology of Pinto da Silva[20-23] (which calls attention to the sequence of labeling vs. preservation methods) established a vital precedent because (1) it implicitly denotes whether intramembrane proteins are exposed and shadowed and (2) it establishes whether *surface* and/or *face* determinants of intramembrane proteins are available for binding during the labeling steps (see Figure 1). Thus, we use that convention designating the temporal sequence of fracturing and labeling, but add a third term designating the sequence and method of replica retrieval (e.g., "replica-label-section").

B. "Fracture-Label"

Methods have been described for fracture-label[20,22] using lectin-gold. Initially, fracture-label did *not* involve platinum shadowing or replication of IMPs,[20] and thus, was not strictly a freeze-fracture labeling technique. Subsequent versions[22] have incorporated platinum shadowing following fracturing, thawing, labeling, refreezing, and freeze-drying. In fracture-label, freshly fractured but *unshadowed* membrane faces and adjacent surfaces were thawed and labeled by lectin-gold. Unexpectedly, lectin binding sites originally present on the membrane *surface* (and thus not exposed by the fracturing process) were labeled. Thus, Pinto da Silva and co-workers reasoned that this version of "freeze-fracture cytochemistry could not provide information regarding IMP identification because the fractured but unshadowed surfaces were unstable when thawed and labeled." Split and unstabilized membranes were "replaced by rugous faces" that resulted "from reorganization of membrane components (mainly membrane lipids) upon thawing in aqueous media". Regardless, fracture-labeled specimens have provided useful information concerning the nature of the cleaving process and the reorganization of *unstabilized* half-membrane leaflets when thawed in aqueous media.

C. "Replica-Label"

In replica-label, unlabeled samples are frozen, fractured, and replicated as in conventional freeze fracture, then thawed and labeled. An advantage of replica-label over label-fracture occurs because labels are applied *after* fracturing and replication. Thus, there is no possibility of ligand-induced alteration of particle partition coefficients.[32] Two examples of replica-label are provided. In Figure 5, membrane immunoglobin molecules (mIg) in mouse B lymphocyte plasma membranes were naturally dispersed in the plasma membrane (Figure 5a) or were "patched" (Figure 5b) by reaction for 15 min with goat anti-mouse Ig (in the presence of 30 m*M* sodium azide to prevent capping and subsequent internalization of label[35]). Samples were fixed, fractured, replicated (Pt and C), thawed and labeled with rabbit anti-goat immunogold, embedded, and sectioned (i.e., replica-label-section). In images of labeled *unpatched* membranes (Figure 5a), colloidal gold labels are irregularly ("randomly") distributed in the plasma membrane, with higher densities of gold associated with microvilli (cf. Figures 5a and 3b). In the reverse stereo perspective (seen by using the two images on the right to form the stereo pair), the sidedness of gold granules is easily discerned. (No granules are attached to the carbon side of the replica.) Because of the close proximity of colloidal gold granules to IMPs of several different diameters, it is not possible from these images of unpatched mIg to ascribe labeling to any one class of IMPs or to identify mIg.

In B lymphocytes having patched mIg (Figure 5b), large areas of the plasma membrane are devoid of gold labels.[16] However, distinctive clusters of colloidal gold are found closely associated with discrete patches of 30 to 60 Å diameter IMPs. Without the easily identified colloidal gold markers and stereoscopic viewing, these clusters of minute IMPs might otherwise have been overlooked. The clusters of immunogold permit correlation of mIg with very small diameter IMPs. This correlation of the smallest discernible IMPs with mIg is consistent with the transmembrane structure of mIgs as a single α-helical polypeptide strand[39] that had been shadowed by 20 Å of platinum.

In a second version of replica-label (Figure 6), we froze artificial membrane vesicles which contained as their only protein, purified sodium channels from electrophorus electricus.[40,41] These samples were freeze-fractured, replicated with platinum and carbon, thawed in LBB, "sandwich" -labeled with rabbit anti-Na^+ channel polyclonal antibody in LBB, and counter-labeled with goat anti-rabbit antibody adsorbed to colloidal gold. The sample was rinsed carefully, air dried, and examined as a "whole mount", The platinum and carbon coats stabilized the split membranes and delineated IMPs on membrane *faces*. In conventional stereoscopic images (Figures 6a, 6c, and 6d) and in "intaglio" images (Figure 6b), gold granules are closely associated with the only class of IMPs (90 to 110 Å diameter). Very few gold granules were associated with other areas of the replica.

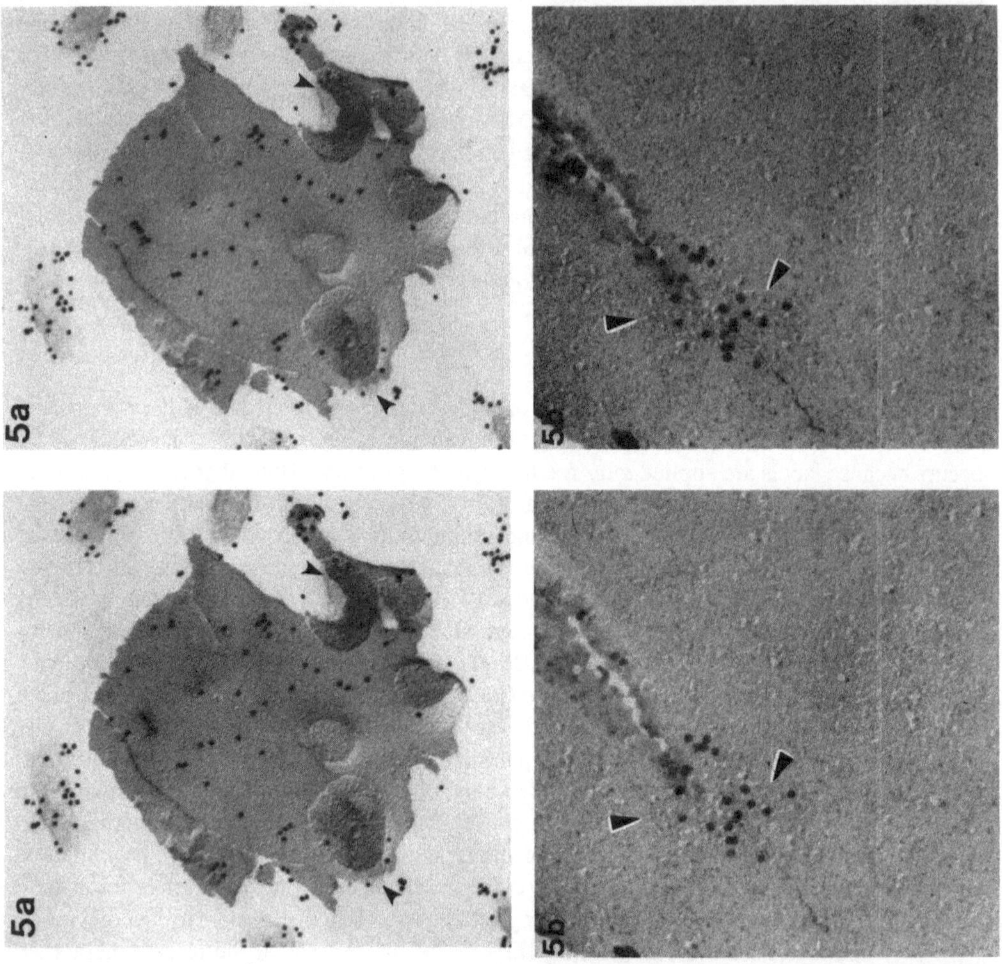
5a
5a
5a
5b
5b
5b

FIGURE 5. Dissociated and washed mouse spleen cells were fixed immediately (''unpatched'', Figure 5a) or were exposed to goat anti-mouse Ig before fixation (to induce ''patching'' of mIg). Cells were then fixed, infiltrated with 30% glycerol, freeze-fractured, replicated with plantinum/carbon, thawed and counter-labeled with goat anti-rabbit immunogold. Because E-surfaces were labeled whereas their attached E-faces were replicated, conventional cleaning techniques would have dissolved the antibodies linking the colloidal gold to the IMPs (see Figure 1b). Thus, these samples were embedded in plastic, thin or thick sectioned, and photographed in the HVEM at two angles of tilt for stereoscopic viewing. In unpatched membranes (Figure 5a), gold labels are ''mono-dispersed'', making it difficult to assign labels to individual particles. However, in cells patched before labeling, clusters of colloidal gold are associated with unusual patches of membrane in which minute (3 to 5 nm IMPs) are closely clustered. Such clusters have not been seen in unpatched membranes but are common[16] in membranes of B lymphocytes that have been patched.[46] The size of the clustered IMPs is consistent with that predicted for the transmembrane domain of mIg, which is thought to be a single α-helical polypeptide sequence.[38] Viewing the images in reversed (or ''intaglio'') perspective is useful in verifying the ''sidedness'' of labels and in interpreting image detail. [These images represent the ''replica-label-section'' technique.] (a; Magnification × 38,000; b; magnification × 100,000.) (Modified from References 16 and 24.)

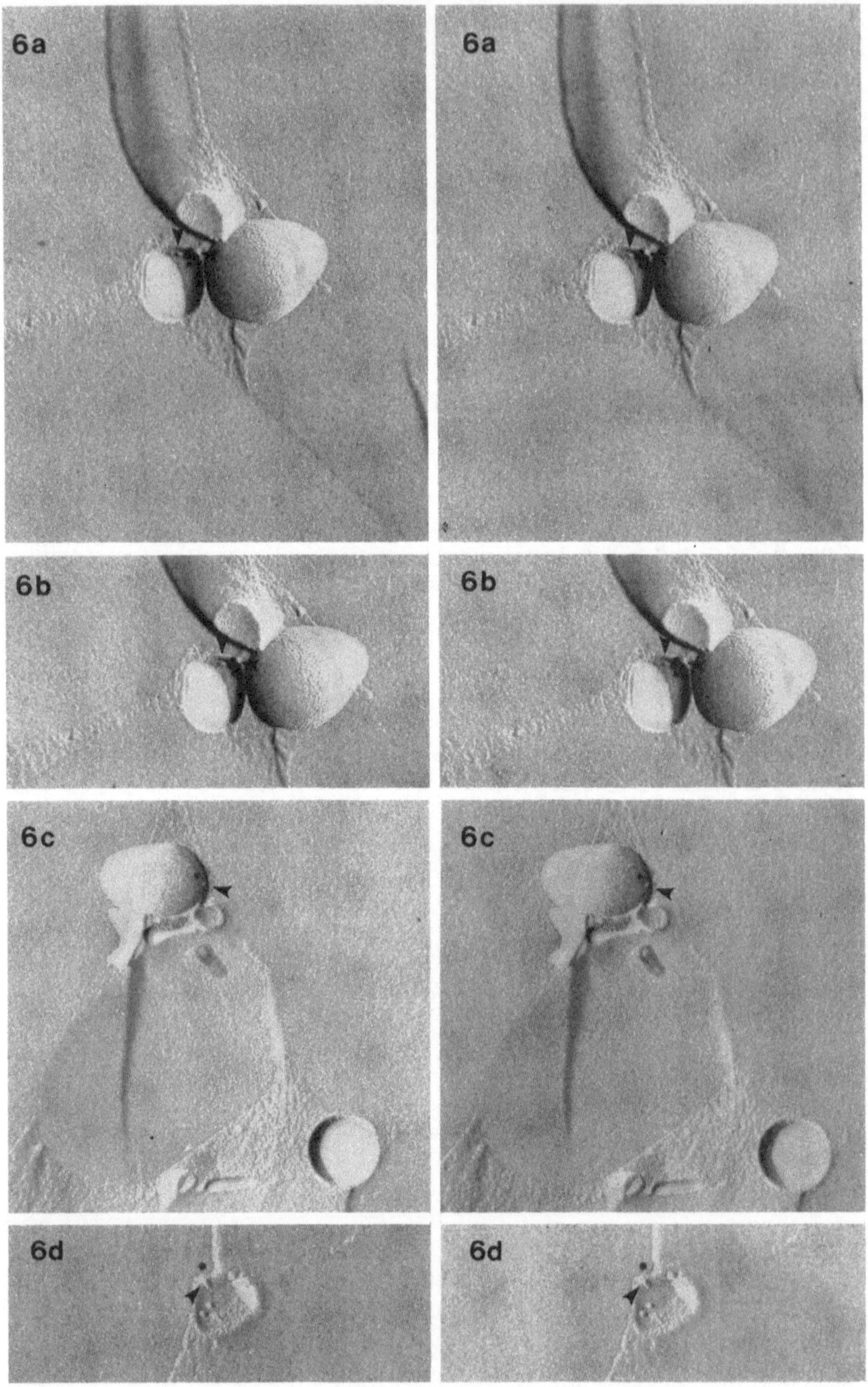

FIGURE 6.

D. "Shadow-Label"

Shadow-label is a new replica labeling technique that provides for high resolution labeling and potential for mapping *intramembrane domains* in larger IMPs (see Figure 1). Fixed and cryoprotected (or unfixed rapidly frozen samples) are shadowed with platinum, but are *not* coated with carbon. Interestingly, the platinum coat (even without a stabilizing secondary coat of carbon) stabilizes the exposed lipid monolayers so that no reorganization of membrane components and no displacement of labels is detected. (Stabilization of lipid leaflets is attributed in part to the high adsorbtive properties of platinum films (also see Figure 3a, E-face). Discontinuities in the platinum film on the "lee" or unshadowed side of large molecules allows shadow-labeling of both *surface* molecules and large diameter IMPs. In Figure 7 (from Reference 42), keyhole limpet hemocyanin molecules were bound to the surfaces of mouse erythrocytes by brief exposure to glutaraldehyde, followed by "quenching" of exposed aldehyde sites by exposure to 50 m*M* glycine.[43] Cell surfaces and attached hemocyanin molecules were exposed by deep-etching, shadowed with platinum (no carbon), thawed, and sandwich-labeled using rabbit antilimpet hemocyanin antibody, followed by immunogold. In this "shadow-label-section" preparation, gold labels were attached to the lee (or uncoated) side of the hemocyanin molecules. In other areas (not shown), labels were present on all sides of unshadowed hemocyanin molecules at a density approximately two to three times that of platinum-coated hemocyanin molecules. These data are consistent with molecule-specific labeling of platinum-shadowed hemocyanin. We conclude that "steric hindrance" from the platinum coat and/or mechanical/thermal damage by vaporized platinum prevented labeling of determinants on the platinum-coated side of the molecule.[28]

In Figure 8 (from Reference 33), bovine erythrocytes were frozen, fractured, and shadowed with platinum (no carbon), thawed and labeled with a monoclonal antibody directed against an uncharacterized minor surface marker. (Mouse anti-bovine erythrocyte plasma membrane monoclonal IgG provided by Dr. Robert E. Lee, College of Veterinary Medicine and Biomedical Sciences, CSU). The shadow-labeled samples were rinsed, freeze-dried, and the labeled P-faces rotary carbon coated. (The resulting carbon "replicas" of residual ice crystals remaining after incomplete freeze-drying attest the sequence of shadow → label → freeze dry → carbon coat.) Two gold granules are suspended approximately 1 and 2 nm above the P-face IMPs. Since it is clear from Figure 7 (above) that gold labels bind only to the lee side of IMPs,[42] the labels may be attached to one or two of the several IMPs located to the left of the gold labels.

Images from shadow-label permit direct correlation of structure and function at the molecular level. An added advantage is that *intramembrane* determinants are exposed for labeling, thereby providing the possibility of simultaneous mapping of both surface and face determinants of intramembrane proteins. Finally, by using techniques that alter the particle partitioning coefficient,[32] *additional* intramembrane determinants not normally exposed by fracturing may be exposed for mapping. This technique may prove especially useful since it has been demonstrated that the polypeptide strands of transmembrane proteins in erythrocyte membranes are not broken during the fracturing process.[44]

VII. LABELING SPECIFICITY

To assign a single gold granule to a single IMP requires not only proximity, but also a

FIGURE 6. Purified sodium channel proteins incorporated at low density in artificially reconstituted membrane vesicles. The samples were rapidly frozen, fractured, replicated (Pt/C), thawed, and labeled with polyclonal antibody directed against surface determinants of the sodium channel. Gold labels are associated with "surface" determinants of replicated IMPs. In an area about four times larger than that shown in Figure 6a, there are three additional gold labels not identifiable as specific, vesicle associated labeling. The nonspecific labeling density in this area is about $2/\mu m^2$. [These images represent the "replica-label-whole mount" technique.] (a; Magnification × 100,000; b; magnification × 100,000; c; magnification × 90,000; d; magnification × 120,000.)

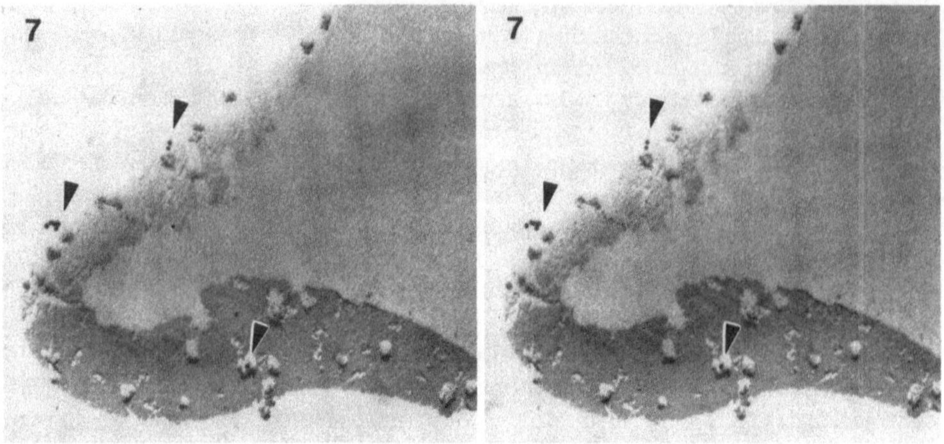

FIGURE 7. Keyhole limpet hemocyanin molecules were attached to mouse erythrocytes by brief glutaraldehyde fixation and "quenched" with glycine.[42] After extensive washes, the cells were infiltrated with 30% ethanol, rapidly frozen, freeze-etched to expose the hemocyanin, "shadowed"[1] with platinum (no carbon coat), thawed in LBB, and immunogold-labeled.[41] Of hundreds of labeled hemocyanin molecules (arrowheads), all labels were attached to the "lee" or uncoated side of the molecules, demonstrating that the platinum film is discontinuous, thereby allowing attachment of colloidal gold labels to the lee side only. Failure of labels to attach to the coated side may be due to steric effects, to molecular damage by bombardment by white hot platinum,[28] or to other factors. [This image represents the "shadow-label-section" technique.] (Magnification × 35,000.)

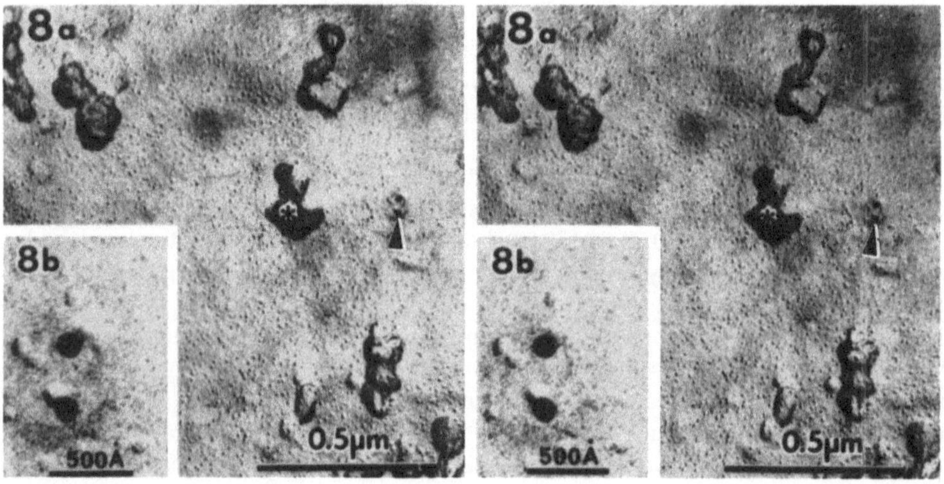

FIGURE 8. Bovine erythrocyte that was fixed, fractured, shadowed (no carbon coat), thawed, and immunogold-labeled using a monoclonal antibody to an uncharacterized bovine erythrocyte antigen. Samples were then freeze-dried and stabilized by rotary deposition of carbon. (Carbon replicas of ice "trees" reveal that freeze-drying was incomplete.) Colloidal gold granules appear to be associated with large diameter IMPs, but that conclusion is premature pending identification of the chemical nature of the bovine antigen. [This image represents the "shadow-label-replica" technique.] (a; Magnification × 48,000; b; magnification × 210,000.)

high confidence that the gold granule represents specificity of labeling. Clearly, a high "signal-to-noise" (S/N) ratio provides for high confidence of specific labeling, whereas a low S/N ratio provides for low confidence. It is also useful to assess the relative extent of nonspecific binding (or N_s). N_s is equal to the number of nonspecifically bound gold granules per unit area (arbitrarily based on the number per square micrometer in a 10 μm^2 area). For example, in Figure 6a, the number of gold granules bound to replicated solvent phase is

2 μm^2. We are also developing models for measuring "labeling efficiency" (E_1). (E_1 is defined as the number of IMPs that are labeled divided by the total number of IMPs that potentially could be labeled.) It is clear that reducing nonspecific labeling and increasing labeling efficiency are essential if individual IMPs are to be identified with confidence.

VIII. SUMMARY

We have reviewed the three methods for simultaneous visualization and specific labeling of IMPs in freeze-fracture replicas. We have documented labeling of individual IMPs or groups of IMPs using each technique. We have demonstrated nonspecific adsorption of labels to the platinum and carbon films and have introduced a "labeling-blocking buffer" to reduce this source of nonspecific labeling. LBB greatly reduces the background "noise" (contributed by nonspecific adsorption of labels) and significantly improves the S/N ratio of these replica labeling techniques. We have used stereoscopic viewing to demonstrate the sidedness of labels and to measure the distance (in three-dimensional space) of labels from associated IMPs. Finally, we have suggested that the simple descriptive labeling terminology of Pinto da Silva be adopted for designating the sequence of labeling steps. That terminology has been modified to include a third term designating the method used to stabilize/retrieve the labeled replica. Examples shown included shadow-label-section, label-fracture-carbon, replica-label-whole mount, replica-label-section, and shadow-label-carbon. This three-part terminological convention is suggested because it avoids ambiguity, designates the IMP surfaces and/or faces that are labeled, and it facilitates interpretation and description of images.

ACKNOWLEDGMENTS

This research was supported by grants from the NIH (NS 14648 and 15991 to JER and TJAJ; and NS-15879 to SRL), the MDA (to JER and R. E. Lee and to SRL), and by a contract from DoD (DAMD-17-84-C-4010 to JER and J. K. Elmund). We thank Dr. Robert E. Lee for monoclonal antibodies to bovine erythrocyte plasma membrane components. Thick sections were photographed at the Laboratory for High Voltage Electron Microscopy, University of Colorado (NIH grant PR 00592 to R. McIntosh). Thanks are extended to Mr. George Wray, University of Colorado, for his excellent technical assistance in the operation of the high-voltage electron microscope.

REFERENCES

1. **Steere, R. L.,** Electron microscopy of structural detail in frozen biological specimens, *J. Biochem. Biophys. Cytol.*, 3, 45, 1957.
2. **Moor, H. and Mühlethaler, K.,** Fine structure in frozen-etched yeast cells, *J. Cell Biol.*, 17, 609, 1963.
3. **Mühlethaler, K., Moor, H., and Szarkowski, J. W.,** The ultrastructure of the chloroplast lamellae, *Planta*, 67, 305, 1965.
4. **Branton, D.,** Fracture faces of frozen membranes, *Proc. Natl. Acad. Sci. U.S.A.*, 55, 1048, 1966.
5. **Deamer, D. W. and Branton, D.,** Fracture planes in an ice bilayer model membrane system, *Science*, 158, 655, 1967.
6. **Pinto da Silva, P., Douglas, S. D., and Branton, D.,** Localization of A antigen sites on human erythrocyteghosts, *Nature (London)*, 232, 194, 1972.
7. **Branton, D.,** Membrane structure, *Annu. Rev. Plant Physiol.*, 20, 209, 1969.
8. **Branton, D.,** Freeze-etching studies of membrane structure, *Philos. Trans. R. Soc. London Ser. B*, 261, 133, 1971.

9. **Hong, K. and Hubbell, W. L.**, Preparation and properties of phospholipid bilayers containing rhodopsin, *Proc. Natl. Acad. Sci. U.S.A.*, 69, 2617, 1972.
10. **Singer, S. J. and Nicolson, G. L.**, The fluid mosaic model of the structure of cell membranes, *Science*, 175, 720, 1972.
11. **Rash, J. E., Copio, D. S., Eldefrawi, M. E., and Hudson, C. S.**, Practical labelling techniques for freeze-fracture, *J. Cell Biol.*, 75, 247a, 1977.
12. **Rash, J. E., Hudson, C. S., and Ellisman, M. H.**, Ultrastructure of acetylcholine receptors at mammalian neuromuscular junctions, in *Cell Membrane Receptors for Drugs and Hormones*, Straub, R. W. and Bolis, L., Eds., Raven Press, New York, 1978, 47.
13. **Rash, J. E.**, The sectioned-replica technique: direct correlation of freeze-fracture replicas and conventional thin-section images, in *Freeze-Fracture: Methods, Artifacts, and Interpretations*, Rash, J. and Hudson, C., Eds., Raven Press, New York, 1979, 153.
14. **Rash, J. E., Johnson, T. J. A., Hudson, C. S., Copio, D. S., Graham, W. F., Eldefrawi, M. E., and Giddings, F. D.**, Identification of intramembrane particles by pre- and post-fracture labelling techniques: a progress report, *Proc. Electron Microsc. Soc. Am.*, 38, 692, 1980.
15. **Rash, J. E., Johnson, T. J. A., Hudson, C. S., Giddings, F. D., Graham, W. F., and Eldefrawi, M. E.**, Labelled-replica techniques: post-shadow labelling of intramembrane particles in freeze-fracture replicas, *J. Microsc.*, 128, 121, 1982.
16. **Dinchuk, J. E., Johnson, T. J. A., and Rash, J, E.**, Post-replication labeling of E-leaflet molecules: membrane immunoglobulins localized in sectioned labeled replicas examined by TEM and HVEM, *J. Electron Microsc. Tech.*, 7, 1, 1987.
17. **Rash, J. E., Hudson, C. S., Albuquerque, E. X., Eldefrawi, M. E., Mayer, R. F., Graham, W. F., Johnson, T. J. A., and Giddings, F. D.**, Freeze-fracture, labelled-replica, and electrophysiological studies of junctional fold destruction in myasthenia gravis and experimental autoimmune myasthenia gravis, *Ann. N.Y. Acad. Sci.*, 377, 38, 1981.
18. **Turkevich, J., Stevenson, P. C., and Hilier, J.**, The formation of colloidal gold, *J. Phys. Chem.*, 57, 670, 1953.
19. **Slot, J. W. and Geuze, H. J.**, A new method of preparing gold probes for multiple-labeling cytochemistry, *Eur. J. Cell Biol.*, 38, 87, 1985.
20. **Pinto da Silva, P., Parkison, C., and Dwyer, N.**, Fracture-label: cytochemistry of freeze-fracture faces in the erythrocyte membrane, *Proc. Natl. Acad. Sci. U.S.A.*, 78, 343, 1981.
21. **Pinto da Silva, P., Torrisi, M. R., and Kachar, B.**, Freeze-fracture cytochemistry: localization of wheat-germ agglutinin and concanavalin A binding sites on freeze-fractured pancreatic cells, *J. Cell Biol.*, 91, 361, 1981.
22. **Pinto da Silva, P., Kachar, B., Torrisi, M. R., Brown, C., and Parkison, C.**, Freeze-fracture cytochemistry: replicas of critical point-dried cells and tissues after fracture-label, *Science*, 213, 230, 1981.
23. **Pinto da Silva, P. and Kan, F. K.**, Label-fracture: a method for high resolution labeling of cell surfaces, *J. Cell Biol.*, 99, 1156, 1984.
24. **Dinchuk, J. E.**, Refinement of molecular labeling techniques for membranes and their application to the study of B-lymphocyte biology, Ph.D. Dissertation, Colorado State University, Fort Collins, 1987, 198.
25. **Dinchuk, J. E., Johnson, T. J. A., and Rash, J. E.**, A rapid assay and method for minimizing non-specific binding of colloidal gold labels to platinum/carbon replicas, in preparation.
26. **Fisher, K. A. and Branton, D.**, Freeze-fracture autoradiography: feasibility, *J. Cell Biol.*, 70, 53, 1976.
27. **Rix, E., Schiller, A., and Taugner, R.**, Freeze-fracture autoradiography *Histochemistry*, 50, 91, 1976.
28. **Peters, K.-R.**, Continuous ultrathin metal films, in *Science of Biological Specimen Preparation for Microscopy and Microanalysis*, Revel, J.-P., Barnard, T., and Haggis, G. H., Eds., SEM Inc., AMF O'Hare, IL, 1983, 221.
29. **Linsen, B. G., Ed.**, *Physical and Chemical Aspects of Adsorbants and Catalysts*, Academic Press, New York, 1970, 427.
30. **Pinto Da Silva, P.**, Molecular cytochemistry of fractured cells: freeze-etching, fracture-label, fracture-permeation, and label-fracture, *Adv. Cell Biol.*, 1, 157, 1987.
31. **Branton, D., Bullivant, S., Gilula, N. B., Karnovsky, M. J., Moor, H., Muhlethaler, K., Northcote, D. H., Packer, L., Satir, B., Satir, P., Speth, V., Staehelin, L. A., Steere, R. L., and Weinstein, R. S.**, Freeze-etching nomenclature, *Science*, 190, 54, 1975.
32. **Satir, B. H. and Satir, P.**, Partitioning of intramembrane particles during the freeze-fracture procedure, in *Freeze-Fracture: Methods, Artifacts, and Interpretations*, Rash, J. E. and Hudson, C. S., Eds., Raven Press, New York, 1979, 43.
33. **Rash, J. E., Johnson, T. J. A., Dinchuk, J. E., Giddings, F. D., and Kuenning, W.**, Labelling intramembrane particles and surface markers in freeze-etch replicas, in *Frontiers in Membrane Research in Agriculture*, 9, 7, 1986.
34. **Tillack, T. W., Scott, S. E., and Marchesi, V. T.**, The structure of erythrocyte membrane studied by freeze-etching. II. Localization of receptors for phytohemaglutinin and influenza virus to the intramembrane particles, *J. Exp. Med.*, 35, 1209, 1972.

35. **Harding, C., Heuser, J., and Stahl, P.,** Receptor mediated endocytosis of transferrin and recycling of the transferrin receptor in rat reticulocytes, *J. Cell Biol.*, 97, 329, 1983.
36. **Strader, C. B. D., Revel, J. P., and Raftery, M. A.,** Demonstration of the transmembrane nature of the acetylcholine receptor by labelling with anti-receptor antibodies, *J. Cell Biol.*, 83, 499, 1979.
37. **Gruijters, W. T. M., Kistler, J., Sullivant, S., and Goodenough, D. A.,** Immunolocalization of MP70 in lens fiber 16 to 17 -nm intercellular junctions, *J. Cell Biol.*, 104, 565, 1987.
38. **Taylor, R. B., Duffus, P. H., Raff, M. C., and de Petris, S.,** (1971) Redistribution and pinocytosis of lymphocyte surface immunoglobulin molecules induced by anti-immunoglobulin antibody, *Nature (London), New Biol.*, 233, 225, 1971.
39. **Roger, J., Early, P., Calame, K., Bond, M., Hood, L., and Wall, R.,** Two mRNAs with different 3′ ends encode membrane-bound and secreted forms of immunoglobin μ chain, *Cell*, 20, 303, 1980.
40. **Duch, S. D. and Levinson, S. R.,** Neurotoxin-modulated uptake of soduim by highly purified preparations of the electroplax tetrodotoxin-binding glycopeptide reconstituted into lipid vesicles, *J. Membr. Biol.*, 98, 43, 1987.
41. **Johnson, T. J. A., Levinson, S. R., Dinchuk, J. E., and Rash, J. E.,** Labeling individual IMPs corresponding to sodium channels in reconstituted vesicles, in preparation.
42. **Dinchuk, J. E., Rash, J. E., and Johnson, T. J. A.,** Post-shadow labeling of individual macromolecules: Immunogold binds to the lee side only, *J. Cell biol.*, 103, 337a, 1986.
43. **Johnson, T. J. A.,** Glutaraldehyde fixation chemistry: a scheme for rapid crosslinking and evidence for rapid oxygen consumption, in *The Science of Biological Specimen Preparation 1985*, Mueller, M., Becker, R. P., Boyde, A., Wolosewick, J. J., Eds., SEM Inc., AMF O'Hare, IL, 1986, 51.
44. **Fisher, K. A. and Yanagimoto, K. C.,** Effect of membrane splitting on transmembrane polypeptides, *J. Cell Biol.*, 102, 551, 1986.
45. **Schiller, A. and Taugner, R.,** Freeze-fracturing and deep etching with the volatile cryoprotectant ethanol reveals true membrane surfaces of kidney structures, *Cell Tissue Res.*, 210, 57, 1980.
46. **de Groot, C., Kapsenberg, M. L., and Leene, W.,** Observations on transmembrane structures of surface immunoglobins in the plasma membrane of B lymphocytes, *Biochim. Biophys. Acta*, 689, 275, 1982.

Chapter 4

DISTRIBUTION AND MOBILITY OF RECEPTORS IN THE PLASMA MEMBRANE*

Horst Robenek

TABLE OF CONTENTS

I. Introduction 62

II. Appraisal of Methods 62
- A. Platinum-Carbon Surface Replication Techniques 62
  - 1. Exoplasmic Plasma Membrane Surface 62
  - 2. Protoplasmic Plasma Membrane Surface 63
- B. Choice of Electron-Dense Markers 65

III. Results and Discussion 66
- A. Receptor Domains in the Plasma Membrane of Cultured Cells 66
  - 1. Mouse Peritoneal Macrophages 66
    - a. Single Exposure to Ligand-Gold Conjugates 66
    - b. Sequential Exposure to Different Ligand-Gold Conjugates 67
  - 2. LDL Receptors in the Plasma Membrane of Human Skin Fibroblasts 68
    - a. Distribution of Coated Pits 68
    - b. Distribution of LDL Receptors 69
  - 3. Distribution of Immunoglobulin A Receptors in the Plasma Membrane of Rat Hepatocytes 73
- B. Immunocytochemical Studies on the Protoplasmic Plasma Membrane Surface 77

IV. Conclusions 78

Acknowledgements 80

References 84

* Dedicated to Professor Wohlfarth-Bottermann on the occasion of his 65th birthday.

I. INTRODUCTION

Living cells communicate with their environment via cell surface receptor molecules that bind specific extracellular macromolecules. Over the past few years, important advances in understanding the biology of receptors have been made.

There are two classes of cell surface receptors: those whose primary function is to transmit signals, stimulating or suppressing some cellular process, and those whose primary function is to transport high molecular weight ligands across the cell membrane.[1] Examples of the first class of receptors include the immunoglobulins borne by various cells of the immune systems and receptors for polypeptide hormones (e.g., insulin and epidermal growth factor) found on the surfaces of a variety of divergent cell types. The second class comprises receptors for lipoproteins, which carry cholesterol, transferrin, which transports iron, and lysosomal enzymes.

The principle pathway for the selective transport of extracellular macromolecules across the membranes of mammalian cells is receptor-mediated endocytosis. Cell surface receptors for specific ligands aggregate in specialized structures called coated pits prior to being internalized in intracellular vesicles.[2,3] Some receptors, such as secretory component[4] or the α_2-macroglobulin receptor,[5] require ligand binding for aggregation in coated pits, whereas others, such as the low density lipoprotein (LDL) receptor,[6] do not. Many receptors recycle, and this has raised questions regarding their sites of reinsertion into the plasma membrane and their mode of transport to coated pits.

Understanding the details of these processes has been the focus of an enormous experimental effort over the past 10 years.[2,6,7] In this review, we discuss the model systems and techniques that my colleagues and I have been employing during the last few years to study the distribution, insertion, movement, and recycling of plasma membrane receptors and the underlying mechanism.

II. APPRAISAL OF METHODS

A. Platinum-Carbon Surface Replication Techniques

1. Exoplasmic Plasma Membrane Surface

Studies on plasma membrane receptors are usually performed *in vitro* on cultured cells. Biochemical methods have been the mainstay for the assessment of the presence of receptors in a given cell population. However, biochemical and other averaging techniques lack the resolution for visual localization of individual receptors on the plasma membrane of a single cell. Therefore, attempts have been made to develop methods that allow the direct visualization of membrane receptors. Many such investigations have been carried out using light microscopy and fluorescent probes.[8,9]

For electron microscopical localization of receptors, ultrathin sectioning,[10,11] scanning electron microscopy,[12] platinum-carbon replication of the exoplasmic and protoplasmic membrane surface following deep-etching, freeze-drying, or critical-point drying, as well as freeze-fracture combined with cytochemical and immunocytochemical techniques, have been employed.[13,14]

Light microscopical and scanning electron microscopal techniques unfortunately suffer from low resolution. They also do not yield sufficient information about regional differences in the distribution of receptors in the plasma membrane of individual cells. Furthermore, neither thin sectioning nor deep-etching can provide detailed information regarding the topographical distribution of receptors on the surface of cell membranes. In this regard, thin sectioning techniques are severely limited because the relationships of the individual markers or the topography of structures lying outside the plane of section are difficult to ascertain. Additionally, the labeled areas cannot be located selectively for study. Thus it is virtually impossible to obtain quantitative information.

Although deep-etching has proved highly effective for visualizing labeled plasma membrane surfaces, its application is restricted to cells in suspension and monolayer culture because marker substances do not usually penetrate tissues adequately. The sample is usually incubated with the label before freezing and the surface of the plasma membrane (with attached label) is then exposed by deep-etching or freeze-drying. The deep-etching step is normally preceded by freeze fracture, although this is not obligatory unless the aim is to correlate the distribution of surface label with the occurrence of unlabeled intramembrane components. Even when sufficient penetration of label into tissue samples can be attained, the surfaces of cells within the tissue usually cannot be satisfactorily exposed owing to the presence of nonsublimable extracellular constituents and cryoprotectants (normally required for adequate cryopreservation). Fast-freezing techniques ameliorate this problem somewhat by making at least the surface cells of a tissue sample amenable to the deep-etch labeling approach. In addition, experience has shown that deep-etching is only of limited value for studying cell surfaces of monolayer cultures because of the relatively small surface areas that are normally exposed.

Surface replicas of critical-point dried or freeze-dried specimens allow extensive areas of cell surfaces to be viewed at higher resolution than can be achieved by scanning electron microscopy.[13,15] We have used platinum-carbon surface replicas of cell monolayers in conjunction with colloidal gold probes in our experiments to analyze the distribution of receptors on the cell surface.[16-19] This is a highly effective approach to understanding the biology of receptors. The critical-point drying method, whether used to study membrane surfaces or fracture faces, renders many fine details (e. g., capillary fenestrae, nuclear pores, tight junctions, coated pits, and other small specializations of the membrane in the range 70 to 100 nm in size) sufficiently faithfully, but does not resolve ultrafine structural details (e. g., intramembrane particles) preserved by conventional freeze-fracture or freeze-etching techniques.

Surface replicas of critical-point dried cell monolayers are prepared in a manner similar to that used for freeze-fractured tissue. In fact, these surface labeling methods have evolved in parallel to freeze-fracture procedures, sharing similar technology and objectives. Briefly, for surface replication of cells in monolayers, 1 × 1 cm pieces of plastic are removed from the bottom of the culture flasks using a soldering iron. The cells are dehydrated and dried in a critical-point drying device. Surface replicas are made by shadowing the dried cells with platinum-carbon at an angle of 38°, followed by carbon at 90°. We use a Balzers BA 300 freeze-etching unit (Balzers AG, Liechtenstein) equipped with electron gun evaporators and a quartz crystal thickness monitor. The replicas are cleaned overnight in household bleach, washed in distilled water, and picked up on 200 mesh copper grids.

With the aid of the platinum-carbon replication technique combined with colloidal gold-labeled ligands and antibodies we were able to demonstrate characteristic microdomains in the distribution pattern of receptors in the plasma membrane of cultured cells. We have mapped the distributions of receptors for LDL in human skin fibroblasts, of receptors for acetylated LDL (ac LDL), high density lipoprotein (HDL), iron-saturated transferrin (Tf), and bovine serum albumin (BSA) in mouse peritoneal macrophages, and of receptors for immunoglobulin A (IgA) in primary rat hepatocytes.

2. *Protoplasmic Plasma Membrane Surface*

Our finding that a special type of receptor is present in very high densities in a differentiated area of the membrane, while being absent in other areas, suggests that the cell exerts some sort of control over the topographical distribution of its receptors.[16] Since both the receptor-rich and receptor-poor regions of the plasma membrane are expected to consist of common fluid bilayer, it is of interest to identify the means by which the cell maintains these receptor domains and prevents the randomization of the receptors. The association of cytoskeletal elements with the plasma membrane and with certain of its constituents (e. g., receptors) is thought to be functionally important in the control of receptor distribution and dynamics.

To extend our understanding of receptor dynamics and mobility, information on the relationships between receptors and cytoskeletal elements is required. The biochemical identification of the molecules involved in direct linkage between cytoskeletal structures, such as actin filaments, and intramembranous components is still confined to a few examples.[20-23] On the other hand, more detailed topographical information on the association of structural elements with the plasma membrane has been obtained by the recent platinum-carbon replica technique in conjunction with deep-etching or freeze-drying of exposed cytoplasmic plasma membrane surfaces of tissue culture cells or of cells in suspension. Evidence for the interaction of cytoskeletal structures and coated regions of the plasma membrane as well as of cytoskeletal elements and the ventral portion of the plasma membrane has been published.[14,24-33] If such replicas are to yield true images of the order inside living cells, a rapid method for opening the cells, for removing their cytoplasmic contents, and for conserving the topography of their ventral or dorsal membrane is of utmost importance. To preserve membrane structure adequately, methods which open the cells by physical rupture are to be preferred over techniques involving detergent extraction. Of the various methods already described, two have proved most successful for breaking open adherent cells: (1) tearing the dorsal portions away[23,34] and (2) shearing off the nonadherent parts with a stream of buffer from a syringe or a Pasteur pipette (squirting technique).[33,35,36] In the latter case, a brief incubation in hypotonic buffers to weaken or lyse the plasma membrane may be necessary. Both techniques are suitable for exposing large areas of ventral or dorsal plasma membrane. With the lysis-squirting technique, most of the cytoplasmic components are removed in the squirting step, and the remaining membranes can be fixed within seconds of opening the cells. Thus, distortion of the molecular arrangement of the adherent cell surface should be minimal. The quick-freeze, deep-etch replica as well as the lysis-squirting technique have disclosed a rich variety of intracelluar ultrastructural detail, leading to new perspectives in our appreciation of membrane-cytoskeleton and structure-function relationships.

Unfortunately, the quick-freeze, deep-etch technique is primarily a morphological tool, and because application of direct techniques to the cytochemical and immunocytochemical localization of cellular components has been hindered by the practical difficulty of applying labels at the low temperature and high vacuum conditions inherent in deep-etching, the method continues to be used mainly for purely morphological purposes. Consequently, there remains a substantial deficit in our understanding of the significance and function of much of the fine structural specializations discovered by this technique. Nevertheless, direct chemical identification of the components revealed by deep-etching is germane to achieving a unified understanding of membrane-cytoskeleton structure and function.

In our laboratory we have already addressed this goal by combining the lysis-squirting technique with gold immunolabeling of selected structural components associated with the plasma membrane. In general, the combination of lysis-squirting with critical-point drying and subsequent platinum-carbon replication yields preparations with a high degree of structural preservation of both the plasma membrane and cytoskeletal elements; the preservation is comparable to that observed after deep-etching. In addition, this method has the obvious advantage that it is amenable to immunolabeling procedures. In our experiments, ventral membranes were prepared by using a slight modification of the technique described by Nermut.[33] Fibroblasts grown on coverslips were rinsed for 10 s in squirting buffer (20 m*M* PIPES, 100 m*M* KC1, 5 m*M* $MgC1_2$, 3 m*M* EGTA, pH 6.1). The cells were then transferred to the same buffer diluted to 20% of the original concentration for 4 min. This caused swelling of the cells and an overall destabilization of their plasma membranes. The dorsal membranes and the bulk of the cellular contents were then swept away with a stream of the buffer from a hypodermic needle. The coverslips were quickly rinsed in PIPES buffer (20 m*M* PIPES, 100 m*M* KC1, 5 m*M* $MgC1_2$, 3 m*M* EGTA, pH 7.0), and then fixed with 0.1

to 1% glutaraldehyde in PIPES buffer. For the immunocytochemical procedures, ventral membranes were prefixed with 0.1 to 1% glutaraldehyde for 10 to 15 min, rinsed in PIPES, treated with 0.1 *M* ethanolamine in PIPES for 10 to 15 min and rinsed in PIPES. The preparations were then placed in TBS-BSA (20 m*M* tris(hydroxymethyl)-aminomethane, 150 m*M* NaCl, 0.1% bovine serum albumin, pH 8.2) for 5 to 15 min, incubated at room temperature for 30 to 60 min with the primary antibody, rinsed three times in TBS-BSA followed by incubation in either protein A-gold diluted 1:10 with TBS-BSA, but with 1% BSA, or a one- to fivefold dilution of IgG-gold (Janssen Biochimica) for 30 to 60 min. The labeled cells were then rinsed three times in TBS-BSA and twice in PIPES. They were postfixed with 1% glutaraldehyde in PIPES for 30 min, washed in PIPES, fixed in 1 to 2% OsO_4 for 30 min, washed in distilled water, and treated with 1% uranyl acetate (aqueous) for 5 min. In control experiments the primary antibody was omitted. After immunocytochemical labeling, specimens were dehydrated, critical-point dried and replicated with platinum-carbon using standard techniques.

Although fine structural details are well preserved using the lysis-squirting replication method outlined here, the light prefixation necessary for successful immunolabeling is inevitably a compromise between adequate ultrastructural preservation and antigenic destruction. Thus, the degree of fine structural preservation achievable is, to some extent, limited by the resistance of the particular antigen to fixation. For illustrative purposes, we will presently show the immunocytochemical labeling of cytoskeletal elements with specific antibodies detected with either gold-labeled secondary antibodies or protein A-gold.

The experimental applications of gold-labeled antibodies in surface replication of the protoplasmic plasma membrane face carried out to date in our laboratory indicate that there are no insurmountable technical problems in the preparation of specimens. Although most of the cytoplasmic components are removed during the lysis-squirting step, distortion of fine structure at the adherent plasma membrane seems to be minimal. Furthermore, subsequent processing of the specimens for platinum-carbon replication does not dislodge the gold marker particles, thus permitting unambiguous identification and localization of antigens in the replica.

B. Choice of Electron-Dense Markers

In principle, many existing cytochemical techniques can be combined with surface replication. The traditional approach to labeling membranes for face-on viewing in replicas is the application of an ultrastructurally visible ligand or ligand conjugated with a visible marker. However, markers, such as ferritin, hemocyanin, silica spheres, latex particles, and peroxidase are of limited value in conjunction with platinum-carbon replicas. First, these markers can be difficult to identify after replication since they often cannot be clearly distinguished from the rough background material or from small membrane structures that cast similar shadows. Thus, labeling can be difficult to assess even by comparison with unlabeled control preparations. A second problem is that the marker particles located in the shadow of larger surface structures are likely to remain undetected. Third, it is difficult to determine accurately the extent of spuriously adsorbed label. Hence, precise mapping and quantification of cell surface components may not be feasible. These difficulties can be circumvented by using colloidal gold as the marker. This marker has become increasingly popular in recent years for studying qualitatively and quantitatively the distribution of cellular macromolecules in the electron microscope. One important feature of colloidal gold is that the gold particles can be coupled to a range of macromolecules. In this way, a macromolecule-gold conjugate is produced in which each gold particle is buried within a sheath of the desired macromolecules. This has the obvious advantage that only the macromolecule moiety and not the gold marker itself is exposed for binding to any potential receptor or antigen at the cell surface. Gold particles can be prepared in a variety of sizes suitable for use with different microscopical techniques. At the smallest diameters used, an extremely high topographical resolution of label distribution is achieved. Furthermore, unlike the markers mentioned above, the high intrinsic contrast of the gold particles allows their direct visualization against the replica background, thus entirely avoiding the possibility of confusing real membrane structure with the replicated marker.

The experimental applications of gold-labeled ligands in surface replication undertaken so far indicate that the complexes are easy to prepare and use. The results obtained to date affirm that high specificity and sensitivity may be had and that quantitative analysis is a feasible proposition.[4,40]

III. RESULTS AND DISCUSSION

A. Receptor Domains in the Plasma Membrane of Cultured Cells

We have exploited the selectivity of gold-labeled lipoproteins, transferrin, bovine serum albumin, and immunoglobulin A for their respective receptors to detect receptor domains on the surface of the plasma membrane of cultured macrophages, fibroblasts, and hepatocytes. Control experiments were carried out to confirm the specificity of the method used by exposing the cells first to unlabeled ligands and then to the same ligands conjugated to gold. In these experiments the cell surface was always completely devoid of gold label; the unlabeled ligand competes with its gold conjugate, thus preventing subsequent binding of the labeled ligand. This indicates that the ligand-gold conjugates interact specifically with their receptors and that there are no binding sites for the gold itself under the experimental conditions used. The reason that the ligand-gold conjugates display binding properties similar to those of their unlabeled counterparts becomes clear when negatively stained preparations are examined. The examples in Figure 1 depict gold conjugates with LDL, iron-saturated transferrin (Tf), high density lipoproteins (HDL) and show that each gold particle is completely buried within a covering of ligand molecules. Thus, only the ligand and not the gold itself is exposed and available for binding to any potential receptor at the cell surface.

1. Mouse Peritoneal Macrophages

a. Single Exposure to Ligand-Gold Conjugates

We used two colloidal gold-labeled lipoprotein species, ac LDL and HDL, Tf and BSA to map the distribution of receptors in the plasma membrane of cultured mouse peritoneal macrophages. The entire cell surface and even the most minute details of the surface of individual macrophages are visualizable with the aid of the surface-replication technique. The cells are well spread and flat. The plasma membrane surface can be subdivided into three morphologically distinct areas: the cell periphery, which is smooth and characterized by numerous marginal processes of variable length; an intermediate zone between the cell periphery and the central region; and a cental region which overlays the nucleus and is covered by numerous microvilli. The distinction between the peripheral and the intermediate region is very obvious, the boundary between these two areas being more sharply delineated than that between the intermediate and the central region. After incubation of the macrophages with the gold-labeled ligands, the gold particles can be easily identified against the replica background due to their high intrinsic electron density. Thus, cultured mouse peritoneal macrophages are ideally suited for studying receptor distribution.

Figures 2 and 3 illustrate the distribution of binding sites on the macrophage cell surface as revealed after different cultures were treated with one of the four different types of ligand-gold conjugates. In these examples Tf, HDL, and BSA coupled to 16-nm diameter gold particles and ac LDL coupled to 40-nm diameter gold particles were used. The experiments show that the plasma membrane in the peripheral domain appears to have the highest concentration of Tf receptors (Figure 2a), whereas the intermediate region contains most of the ac LDL (Figure 2b) and HDL receptors (Figure 3a). Receptors for BSA are concentrated at the cell periphery and on the peripheral processes (Figure 3b). The distribution of BSA

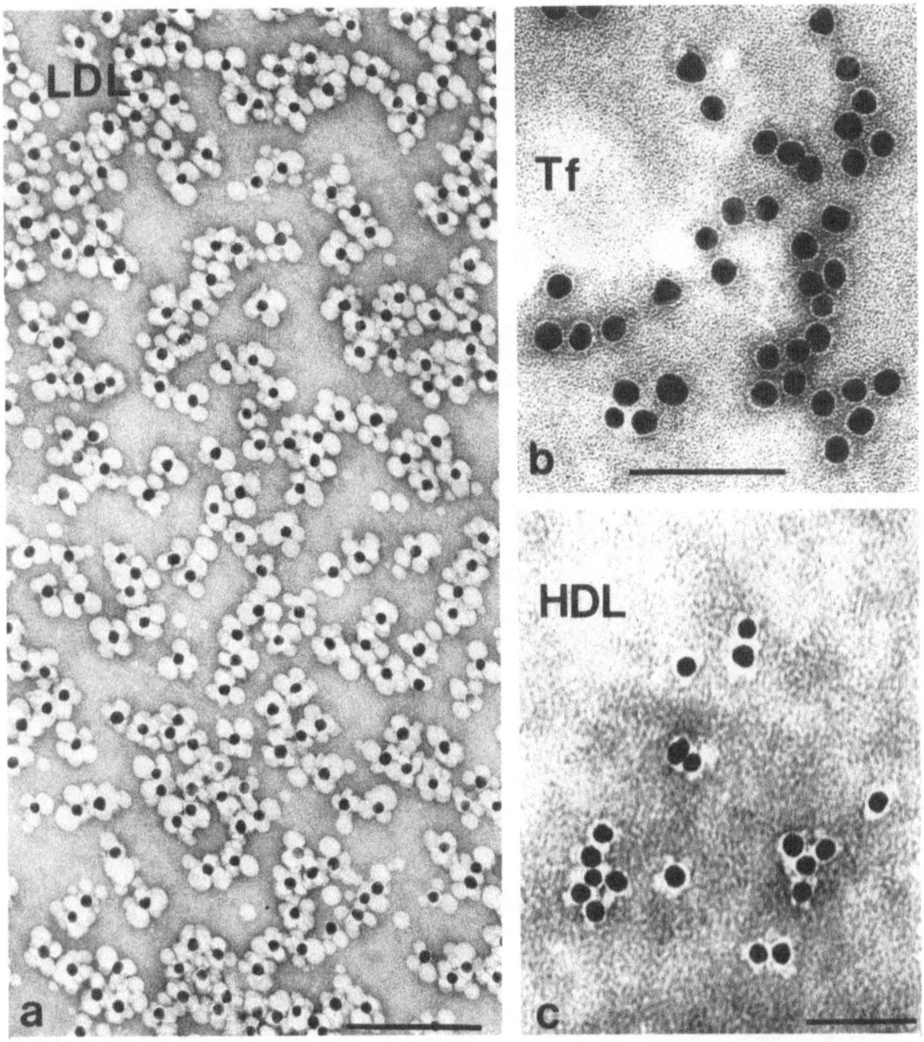

FIGURE 1. Negatively stained LDL-gold conjugates (LDL) (a), transferrin-gold conjugates (Tf) (b), and high density lipoprotein-gold conjugates (HDL) (c). The gold particles have a diameter of about 16 nm. Bars: (a) 0.2 μm; (b,c) 0.1 μm.

receptors on the remainder of the cell surface is more uniform than for the other receptor types.

b. Sequential Exposure to Different Ligand-Gold Conjugates

To extend our understanding of receptor function, information on the relationship between the specificities of different classes of receptors is required. The acquisition of such information would be facilitated if it were possible to visualize and discriminate between the binding sites of two or more ligands in the same cell. In principle, this could be achieved by using two readily distinguishable markers. We show here that a double-labeling technique employing colloidal gold particles of distinctly different diameters can be used successfully in conjunction with surface replicas to differentiate two types of lipoprotein binding sites on a single membrane. Apart from gaining insight into the specificity of ligand-receptor interaction, this appraoch facilitates studying the spatial relationships of two classes of receptors. Labeling of the two binding sites may be carried out sequentially and simulta-

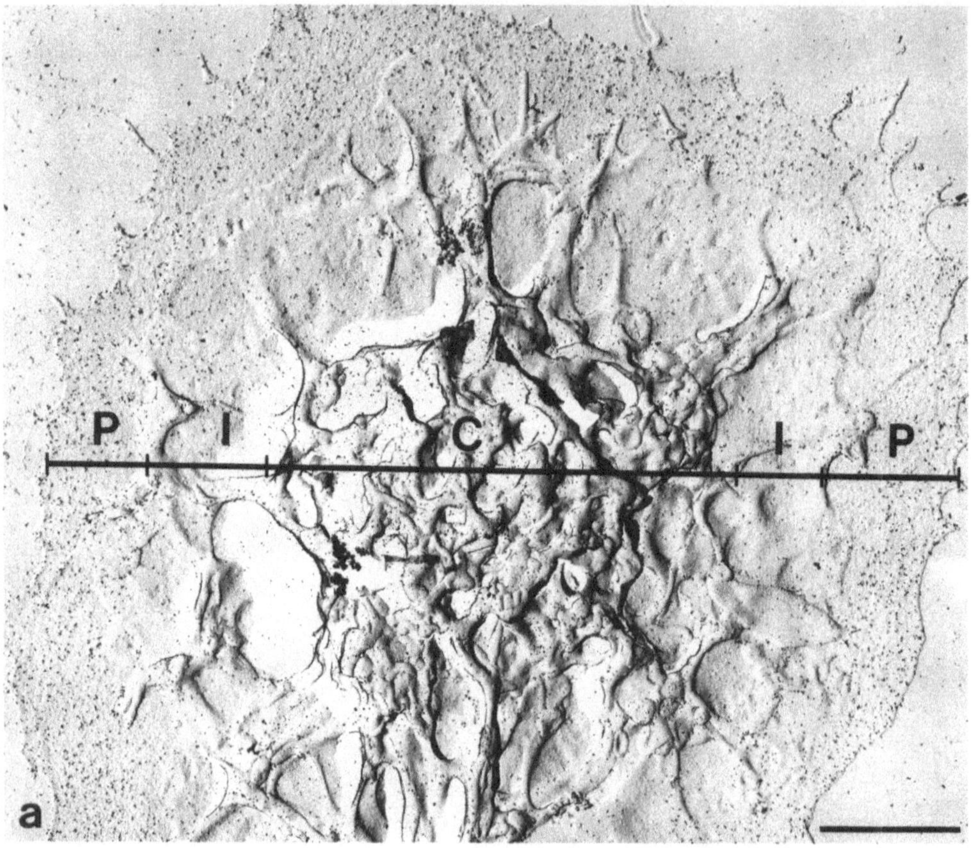

FIGURE 2. Low magnification surface replica of the plasma membrane of macrophages showing the distribution of 16 nm gold-transferrin (a) and 40 nm gold-ac LDL complexes (b) after incubation with the respective conjugates for 1 h at 4°C. P = cell periphery, I = intermediate zone, C = central region. Bars: 3 μm.

neously. Sequential treatment of macrophages with ac LDL conjugated to 16-nm gold particles (1 h at 4°C) followed by HDL conjugated to 40-nm gold particles (1 h at 4°C) produced a labeling pattern almost identical to that obtained when ac LDL or HDL conjugates were applied separately (Figure 4). This demonstrates that HDL binding sites are not blocked by prior exposure to ac LDL. Thus two distinct classes of receptors exist (i.e., discrete receptors for each lipoprotein species). Similar results were obtained when the same two labeled lipoproteins were administered in the reversed order. Ac LDL- and HDL-gold conjugates were dispersed more or less at random in the intermediate region and only in rare instances was a small amount of clustering of HDL-gold conjugates observed. Gold particles are virtually absent from the periphery and from the marginal processes of the cells. These distributions were similar to those obtained when the conjugates were applied in single labeling experiments.

2. *LDL Receptors in the Plasma Membrane of Human Skin Fibroblasts*

a. Distribution of Coated Pits

The initial step in the metabolism of LDL is the binding of this molecule to specific areas of the cell surface designated as coated pits. LDL binding to coated pits has been shown to occur in several types of cells. After the LDL has bound to the cell surface, the coated pits invaginate and the LDL molecules are internalized via endocytotic vesicles in a process called receptor-mediated endocytosis.[2,11,13,37,38]

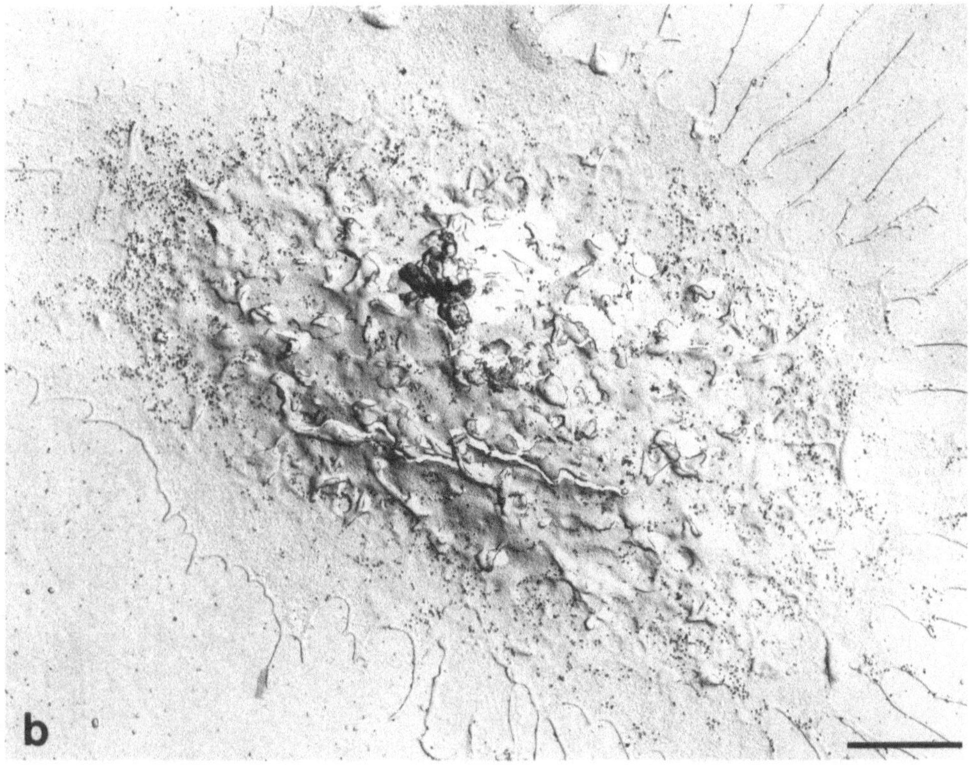

FIGURE 2b.

Coated pits appear in electron micrographs of thin sections as indentations in the plasma membrane about 200 nm in diameter. The cytoplasmic side of the membrane of the coated vesicles is differentiated as a lattice composed primarily of a protein called clathrin.[39] Freeze-fracture replicas of cell membranes of cultured fibroblasts affirm that the coated pits are specialized regions of the membrane. The intramembranous particles of coated pits differ markedly in number and size from those of the remaining membrane. Since coated pits are difficult to identify in overviews of normal membrane fracture faces, we employed the polyene antibiotic filipin to aid in differentiating these structures. Typical filipin-sterol complexes are abundant on the plasma membrane fracture faces of filipin-treated fibroblasts (for details see Severs and Robenek[40] and references therein). The complexes are completely absent in coated pits.[41] The freeze-fracture technique is well suited for exposing large membrane areas, and the use of filipin for differentiating coated pits permits the study of the distribution of the coated pits on fracture faces of the plasma membrane, even at low magnification.[19,41] With the aid of this technique, it is possible to show that coated pits are scattered randomly over the surface of the plasma membrane. The number of coated pits differs from cell to cell. They usually occur singly, but may also be found in groups of up to ten closely packed units (Figure 5a).

b. Distribution of LDL Receptors

Clusters of gold particles can be observed under low magnification (Figure 5b) on the surface of cells incubated for 1 h with gold-labeled LDL molecules (Figure 1a) at 4°C. The distribution of the clusters on the replicas correlates principally with the distribution of coated pits on the plasma membrane fracture faces (cf. Figure 5a). The clusters generally occur singly or in groups. Appreciable numbers of diffusely distributed solitary gold particles were

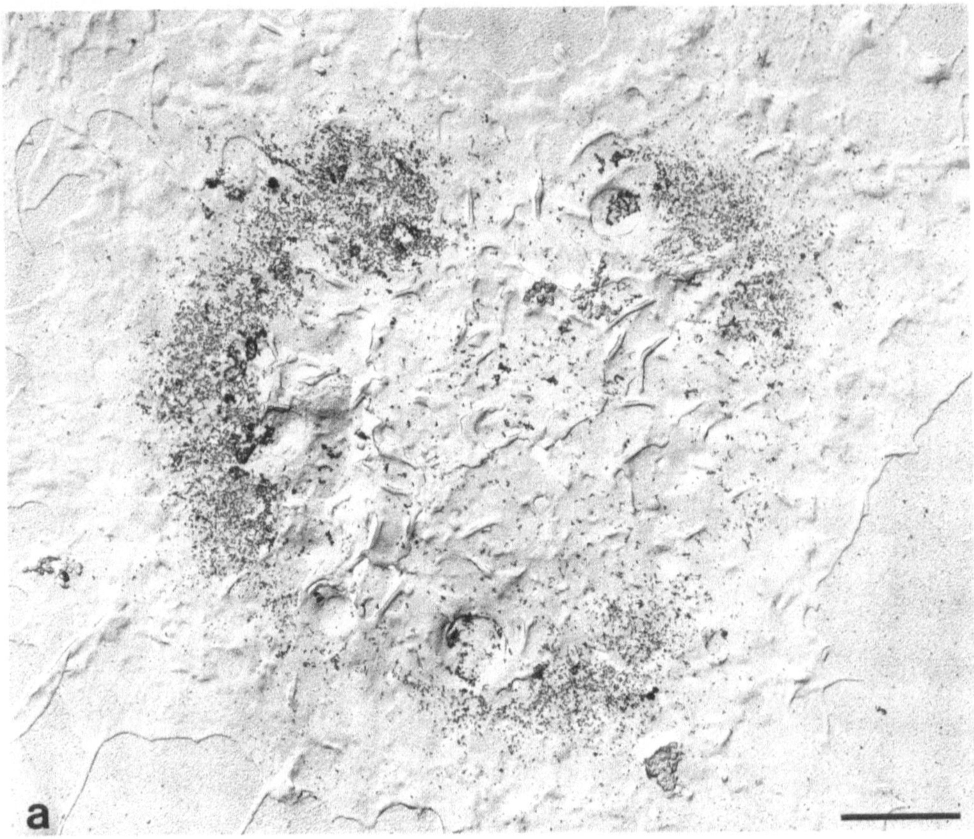

FIGURE 3. Survey view of a surface replica of the plasma membrane of macrophages showing the distribution of 16 nm gold-HDL complexes (a) and 16 nm gold-BSA complexes (b) after incubation with the conjugate for 1 h at 4°C. Bars: 3 μm.

never detected. Since the same labeling pattern was observed when glutaraldehyde-fixed cells were incubated with the LDL-gold conjugate the distribution of the gold particles seems to reflect that of unoccupied LDL receptors. The arrangement of gold particles in the clusters varies considerably. Such variation may represent different stages of the internalization process. The pattern of gold particles within different clusters ranges from widely scattered to so closely packed that it is difficult to distinguish individual particles. This observation suggests that LDL receptors are loosely arranged in nascent coated pits, whereas they aggregate as endocytosis takes place. Figure 6 may represent the sequence of events from the insertion of the receptors into the membrane (Figure 6a), their lateral movement (Figure 6b to d) and clustering in coated pits (Figure 6e), to their subsequent internalization (Figure 6f). In this sequence the first step in the process, the insertion of receptors into the lipid bilayer of the membrane, is represented by the diffusely distributed gold particles within these clusters. It appears that the new LDL receptors are not inserted at random into the membrane, but rather in clusters. In an experiment designed to detect only newly inserted receptors, fibroblasts were first incubated at 4°C with excess concentrations of unlabeled LDL. The unbound LDL was then removed by exchanging the medium, the cells were incubated with LDL-gold and warmed to 37°C and samples were fixed at brief intervals (every 2 min). Labeled receptors began to appear at the cell surface after 4 min at 37°C. The marker particles were in clusters. After 6 to 8 min, almost all the gold particles were observed in clearly delineated membrane regions representing coated pits.

FIGURE 3b.

A diagrammatic representation of the possible sequence of events leading to internalization, recycling and insertion of LDL receptors into the plasma membrane is shown in Figure 7. Although this scheme seems to be complete, it is based on ultrastructural observations, and one should bear in mind the basic fact that the results and data collected are static by nature; they represent time points and not a dynamic sequence of events.

Conflicting views concerning LDL receptor insertion into the plasma membrane and the migration of the receptors to coated pits have been proffered; (1) receptors are inserted into the cell membrane at random locations and diffuse to coated pits[2] and (2) receptors are inserted at random and active flow drives them to coated pits.[6] Our data differ, to some extent, from those of other authors. Under our experimental conditions the insertion of LDL receptors occurred in clusters of different size and shape and into unspecialized regions of the plasma membrane. Newly incorporated receptors are characterized by a random and uniform distribution within these clusters. We found but few randomly distributed receptors on the plasma membrane outside of these clusters. The different kinds of gold particle patterns observed in the plaque areas suggest that they are intermediates in the pathway of receptor clustering into coated pits and receptor endocytosis (Figure 6).

It must be emphasized that although we observed that the degree of clustering of LDL receptors at coated pits increases, we did not find evidence for the migration of individual receptors over large distances. Our results are not consistent with the model that LDL receptors migrate appreciably in the plane of the membrane before being incorporated into coated pits. Rather, our studies provide convincing morphological evidence that the site of replacement and incorporation of LDL receptors into the plasma membrane corresponds to the site at which new coated pits are formed and internalized. Indeed, coated pits appear to form where the insertion of receptors into the plasma membrane takes place. However, it

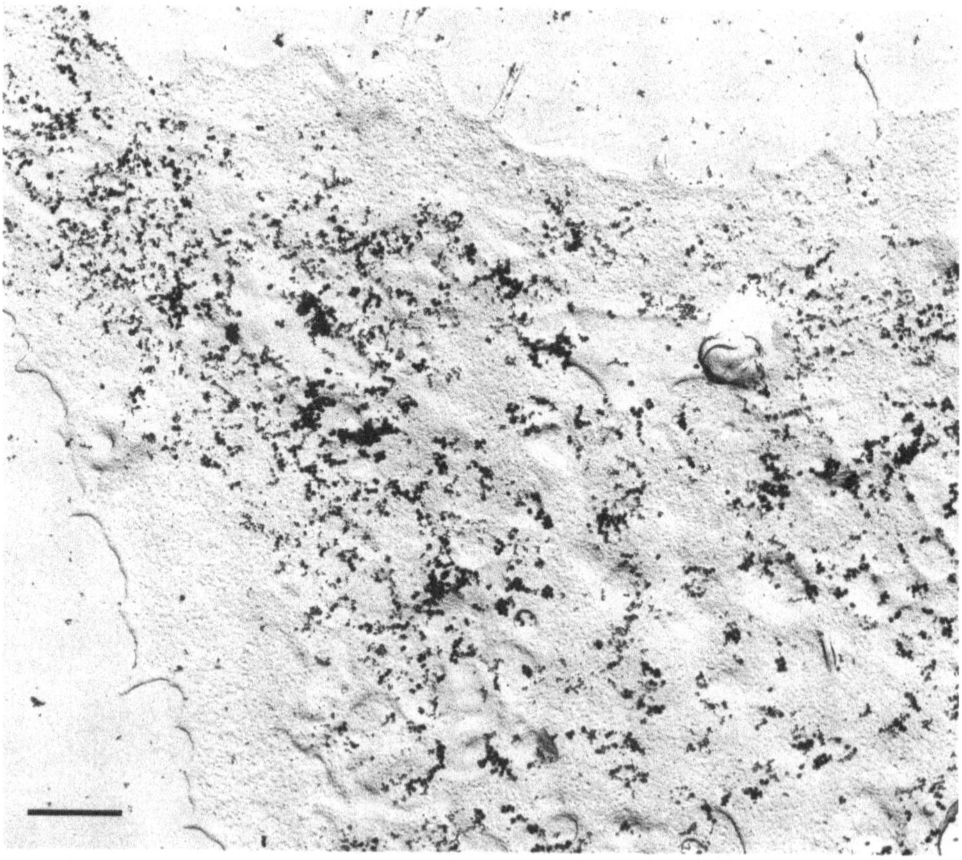

FIGURE 4. Surface replica of a macrophage showing labeling pattern after sequential exposure to 16 nm gold-ac LDL (1h, 4°C) followed by 40 nm gold-HDL (1h, 4°C). The macrophage was pretreated with unlabeled ac LDL for 24 h at 37°C. Bar: 1 μm.

cannot be excluded that conjugation of LDL to gold particles may influence the configuration and function of the LDL molecules. It might be argued that the LDL-gold conjugates used are multivalent and thus may have bound more efficiently to aggregated than to solitary receptors. If so, the LDL-gold binding pattern may not reflect the true receptor distribution.

In order to demonstrate that LDL receptor clusters do exist, we carried out further experiments. In one series of experiments, cells were incubated for 1 h at 4°C with a mixture containing some LDL-gold particles but many unbound LDL molecules. In favorable surface replicas both gold particles and free LDL molecules were observed exclusively in clusters; neither gold particles nor LDL molecules were seen outside of the clusters (Figure 8a and b).

In a second set of experiments, immunolabeling of the LDL receptor itself was accomplished using monoclonal antibodies directed against LDL receptors or immunolabeling of surface-bound LDL molecules was affected using polyclonal antibodies directed against apolipoprotein B. The antibodies were detected with protein A-gold conjugates. Visualization of LDL receptors (Figure 8c) and LDL molecules (Figure 8d) using these immuocytochemical methods yielded results comparable to those obtained with gold-labeled LDL or with the mixture of gold-labeled and free LDL molecules. These studies showed that no alteration of the binding of LDL is incurred due to conjugation of the lipoprotein molecules with colloidal gold particles. These findings indicate that unoccupied LDL receptors in the native membrane are arranged in the form of clusters.

3. Distribution of Immunoglobulin A Receptors in the Plasma Membrane of Rat Hepatocytes

To obtain more accurate information about clustering of receptors, we investigated the topography and dynamics of immunoglobulin A (IgA) receptors on the plasma membrane of rat hepatocytes in primary culture. These cells have been shown to be competent for IgA binding,[42] uptake, and intracellular processing,[46,47] as well as biliary secretion.[42,48] The IgA receptor, also known as secretory component (SC), is well characterized[49] as a glycoprotein synthesized by the hepatocytes[44,45] and expressed at their sinusoidal surfaces[43,44] at a density of about 200,000 molecules per cell.[44] Binding to the receptor is restricted to the polymeric forms of IgA.[50] Binding is required for regular endocytosis and biliary secretion of this immunoglobulin.[42,50,51]

We analyzed quantitatively the topography of SC and of IgA-SC complexes on the cell surface of normal hepatocytes.[4] In addition, we characterized the influence of cytochalasin B (CB) on this topography and demonstrated that the drug induces patch formation of occupied but not of free IgA receptors. These results suggest that microfilaments perform an important role in the clustering of IgA-SC complexes at coated pits. Binding of IgA-gold conjugates to the plasma membrane of cultured hepatocytes at 4°C showed the same features found previously for unlabeled IgA.[42] As measured with radiolabeled IgA-gold conjugates, binding was strongly reduced when the cells were incubated with an excess of unlabeled IgA, but the binding was not inhibited by asialofetuin. Preincubation with anti-SC antiserum also blocked binding. These results indicate that IgA binding was dependent on a specific interaction between the gold-labeled IgA and the IgA receptor. Pretreatment of the cells with CB did not change the characteristics of IgA binding.

We also demonstrated that at 4°C the distribution of the IgA-gold conjugates in surface replicas of rat hepatocytes is of a mixed random type (Figure 9a). This distribution is intermediate between the exclusively clustered type characteristic of LDL receptors and random scattering. Since direct decoration of SC by anti-SC antibodies or by gold-labeled IgA after fixation yielded virtually the same results, this labeling pattern evidently reflects the distribution of unoccupied receptors. Incubation of hepatocytes at 37°C subsequent to binding of IgA-gold conjugates at 4°C initiated a time-dependent redistribution characterized by progressive coalescence of gold particles into larger clusters (Figure 9b and c). A direct quantitative analysis revealed that small clusters predominate at 4°C; more than 80% of the gold conjugates were found in clusters containing less than 50 particles and about 20% of the particles appeared in groups of less than 4. After 40 min at 37°C, the percentage of aggregates containing more than 50 particles increased considerably. The proportion of aggregates consisting of more than 100 particles increased tenfold. Conversely, the percentage of clusters with less than 15 particles was significantly diminished. In the course of this redistribution, the larger clusters became increasingly associated with coated pits as ascertained from thin sections. When gold-labeled IgA was replaced by free dimeric IgA during the incubation, virtually the same distribution was observed after fixation and subsequent decoration of the IgA-SC complexes by gold-labeled IgA antibodies.

In cultures exposed to CB, the distribution of IgA-gold conjugates in surface replicas after incubation at 4°C was almost the same as that of controls. In contrast, incubation of CB-treated hepatocytes at 37°C changed the distribution of IgA-gold conjugates on the surface markedly. Within 20 to 40 min, huge patches comprising several hundred gold particles were formed preferentially at the periphery of most cells (Figure 9d). Although this patch formation did not completely block the transfer of conjugates from coated pits to coated vesicles, endocytosis was reduced compared to controls. Cells treated with CB and incubated in the absence of IgA for 20 min at 4°C and then for 40 min at 37°C before exposure to gold-labeled IgA showed no patch formation. These results indicate that both redistribution as well as patch formation do not occur spontaneously, but are triggered by binding of the ligand to SC. In order to rule out the possibility of a different patch-inducing potential of gold-labeled (multivalent) and unlabeled (monovalent) IgA, cultured hapatocytes were also

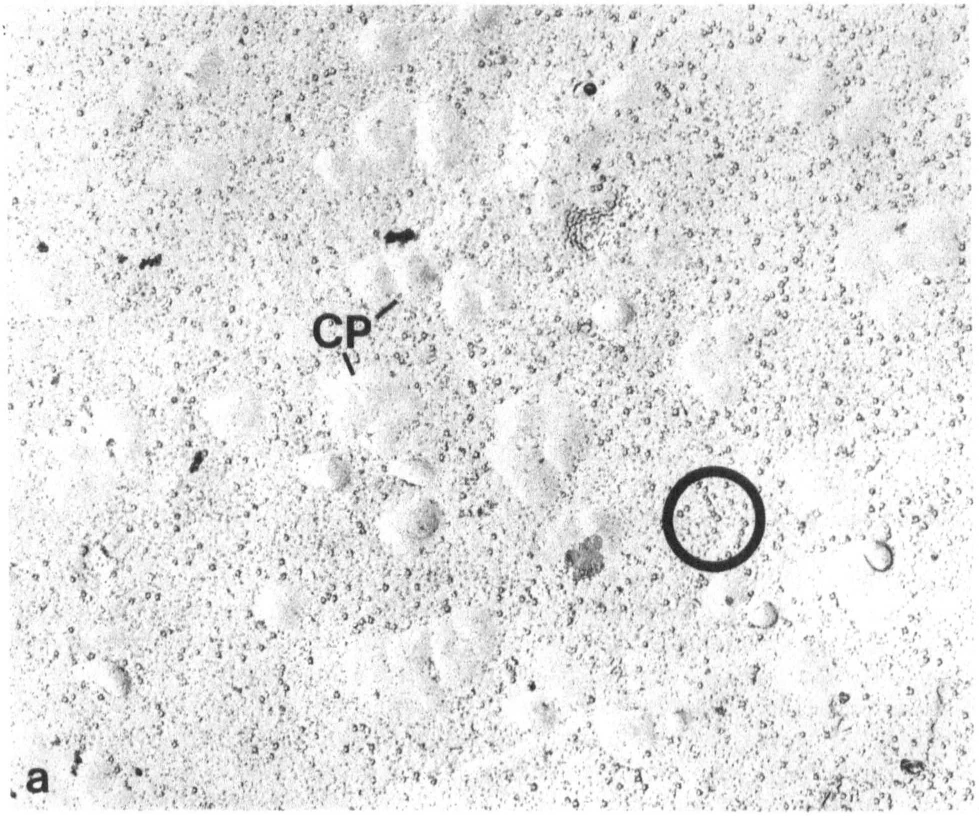

FIGURE 5. (a) Fracture face of the plasma membrane of a cultured fibroblast after filipin treatment showing the distribution of coated pits (CP). The membrane fracture face is studded with filipin-cholesterol complexes (circle). Note that coated pits are free of filipin-cholesterol complexes. (b) Distribution of gold-LDL complexes on the plasma membrane surface of a fibroblast incubated for 1h at 4°C with gold-LDL conjugate. The gold particles are present in clusters, their distribution largely correponding to that of coated pits. Bar: 1 μm.

incubated with free unlabeled IgA at 4°C. After incubation at 37°C and fixation, the topography of IgA-SC complexes was determined by labeling with gold-conjugated anti-IgA antibodies. No patch formation was observed in controls, whereas in CB-treated cultures patch formation was prominenent after 20 to 40 min. These findings indicate that the unlabeled native ligand is capable of triggering patch formation.

The crucial question for the correct interpretation of the data presented is whether the changes subsequent to binding of the ligand are attributable to the fact that IgA-gold conjugates are relatively large multivalent ligands compared to native IgA; multivalent ligands might cause artificial changes in receptor distribution. The main line of evidence against artificially induced changes comes from experiments where gold-conjugated IgA was replaced by free dimeric IgA which is known to bind to SC in a one-to-one ratio preserved in secretory IgA. In order to ensure that the native distribution of receptors was retained, fixation of the cells was carried out prior to decorating the ligand-receptor complexes with gold labeled anti-IgA. Similar results to those with unfixed cells were obtained under all incubation protocols tested. In addition, it should be noted that by using the same technique completely different results were obtained for other multivalent ligands, for example, LDL. Therefore, the findings described here reflect bona fide interactions of IgA with SC on the hepatocyte plasma membrane.

Binding of IgA to SC induced a time-dependent redistribution of the resulting complexes

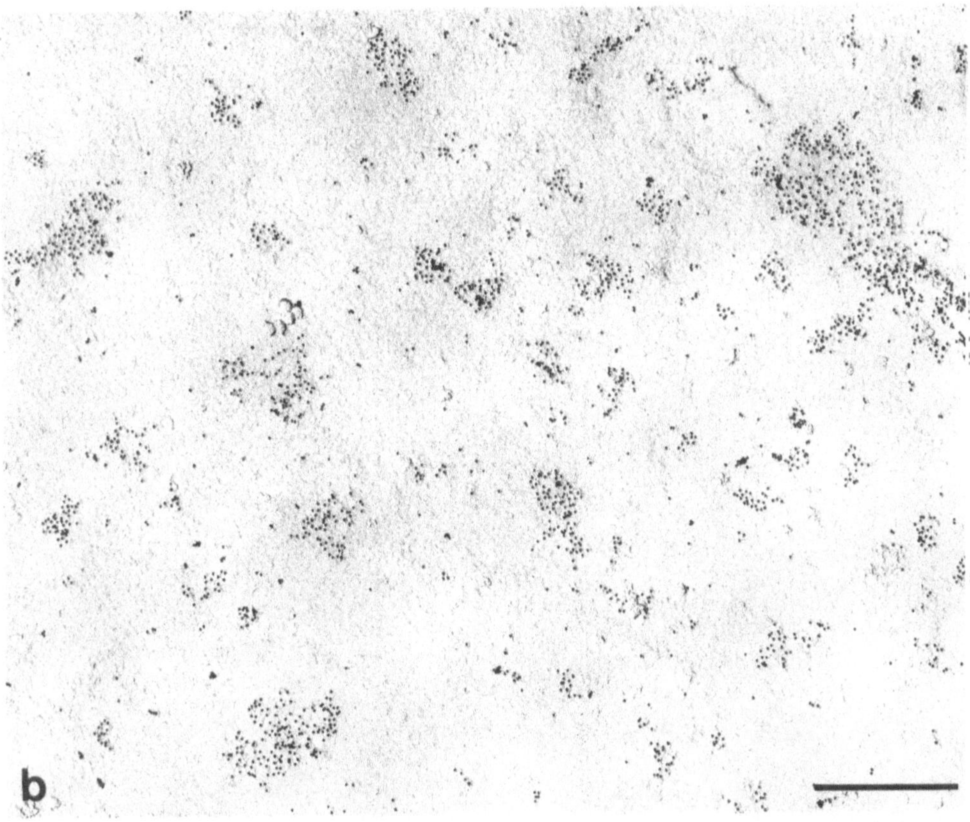

FIGURE 5b.

after incubation at 37°C. Although this is a ligand-induced phenomenon, it is unlikely that the ligand itself is responsible for the aggregation. Instead, we suggest that a conformational change in the receptor molecule increases the mutual affinity of the receptors themselves. This can be assumed to be an intrinsic property of this immunoglobulin-like molecule.

In the light of our results the ligand-induced clustering is a tightly regulated process with respect to cluster size and topography. The observed impairment of this control by CB provides support for the previously tendered hypothesis[48] that microfilaments play an important role in the transcellular biliary transport of IgA. Obviously, CB affects the redistribution of IgA-SC complexes in at least two different ways: (1) single ligand-receptor complexes and those arranged in small groups redistribute much more quickly and (2) the aggregation process is no longer restricted with respect to the final size of the cluster, but rather seems to continue uncontrolled leading to patch formation. These effects are specific for antimicrofilament drugs; antimicrotubular agents do not cause a similar disturbance of receptor redistribution.

In view of recent results from our[48] as well as other laboratories,[52] it seems reasonable to ascribe the effect of CB to the disruption of the association of the microfilamentous web and the plasma membrane. This interpretation supports the anchorage model of receptor redistribution[53] which implicates a linkage between microfilaments and clustered ligand-receptor complexes. The long cytoplasmic tail[49] of the IgA receptor (103 residues) could provide the basis for such an interaction.

In conclusion, our results on redistribution and patch formation of the IgA receptor can be interpreted by assuming the following series of events:

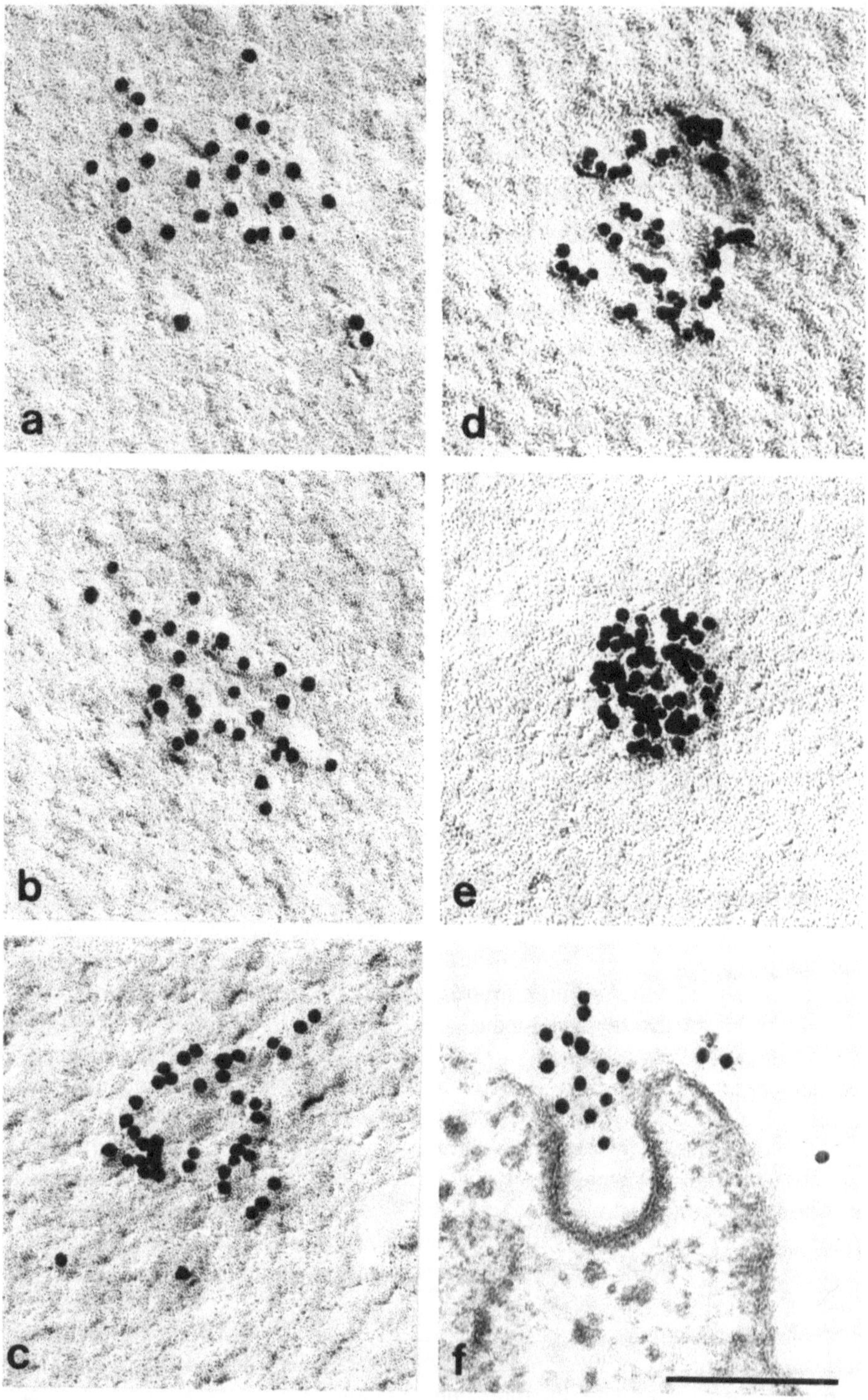

FIGURE 6. Sequence of micrographs (a to e), depicting the progressive concentration of receptors in coated pits following gold-LDL binding and culminating in the internalization of gold-labeled receptors in a coated vesicle (f). Bar: 0.2 μm

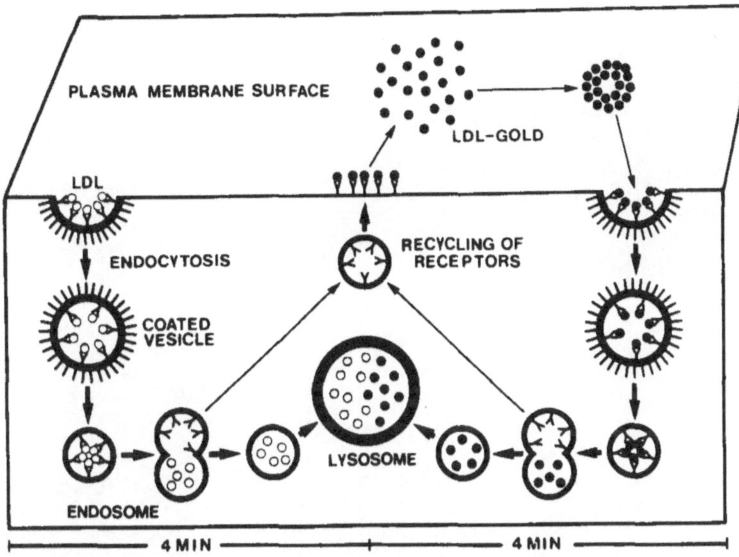

FIGURE 7. Digrammatic representation of putative sequence of events from the insertion of LDL receptors into the plasma membrane to the internalization of LDL-receptor complexes.

1. In contrast to the LDL receptor, the IgA receptor is randomly distributed or present in the form of clusters in the sinusoidal plasma membrane.
2. As a result of binding, the natural tendency of IgA receptor complexes to aggregate is strikingly reinforced due to a conformational change which increases the mutual affinity of these receptor molecules.
3. At this stage, microfilaments may be involved in the control of the redistribution process; in particular, they may limit the extent of clustering.
4. The accelerated, ligand-induced aggregation in turn may enhance the recruitment of clathrin to the clustered ligand-receptor complexes to form coated pits as postulated for LDL receptors.
5. Enhanced endocytosis would then be a consequence of the increased number of coated structures.
6. When the control of aggregation by the filamentous system is perturbed experimentally by CB, the redistribution process is no longer limited leading to patch formation by unconstrained diffusion within the plasma membrane.

B. Immunocytochemical Studies on the Protoplasmic Plasma Membrane Surface

The combination of lysis-squirting and immunocytochemical labeling procedures allows the visualization of the protoplasmic plasma membrane surface and its constituents. Using this new method, which still is at an early stage of refinement, we were able to localize cytoskeletal structures at the ventral plasma membrane surface of cultured fibroblasts. We utilized specific antibodies in conjunction with either gold-labeled secondary antibodies or protein A-gold.

The most prominent structures seen were bundles of microfilaments adjacent to the ventral membrane and running largely parallel to the long axis of the cell. In microfilament-free areas of the ventral membrane clathrin sheets, which displayed the characteristic "honeycomb" pattern, coated pits or coated vesicles were found. Smooth areas and areas studded with small particles were also observed. Regions with "honeycomb" patterns were specifically labeled with an antibody directed against a minor component of coated vesicles isolated

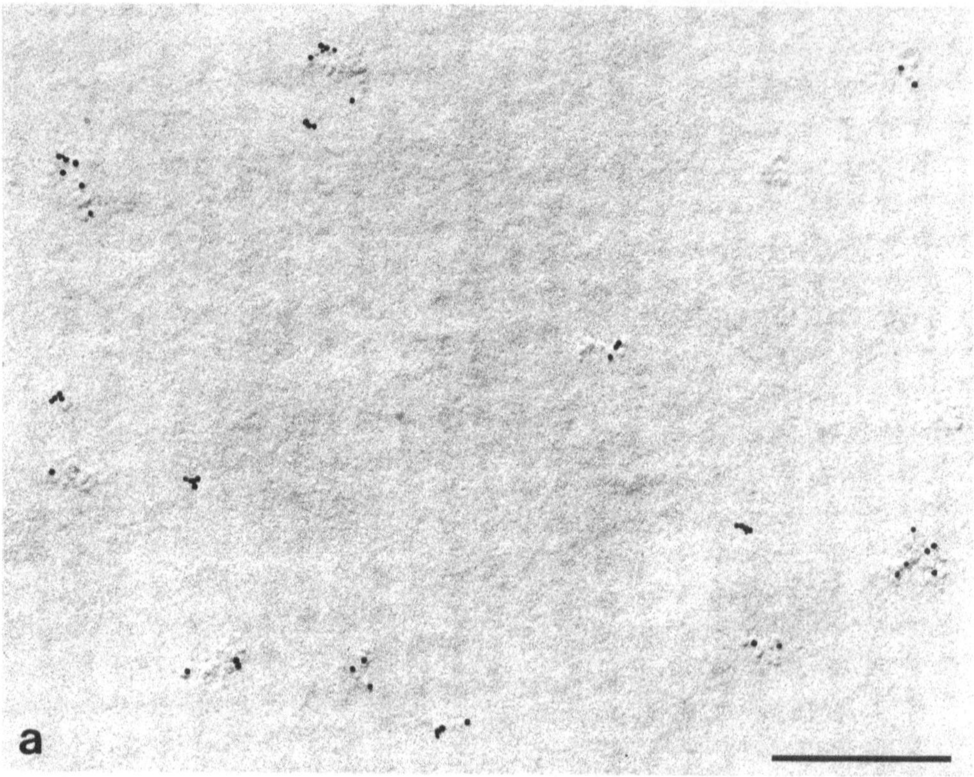

FIGURE 8. Localization of LDL receptors in the plasma membrane of fibroblasts with the aid of a mixture of gold-labeled (LDL-Au) and unlabeled LDL molecules (LDL) (a, b), by gold-immunolabeling of the LDL receptor using monoclonal antibodies directed against the LDL receptor (c), and by gold-immunolabeling of surface-bound LDL molecules using polyclonal antibodies directed against apolipoprotein B (d). Bars: (a,c) 0.5 μm; (b,d) 0.2 μm.

from bovine brain.[14] The colloidal gold label appeared attached to coated pits or coated vesicles found within irregular webs of microfilaments (Figure 10a), laterally attached to microfilament bundles (Figure 10b), or free on the protoplasmic surface of the plasma membrane. When the antibody followed by the protein A-gold conjugates was applied, virtually no label was associated with the microfilaments and very little was found outside coated pits, at the plasma membrane or in control preparations.

When preparations were incubated with anti-actin antibodies microfilaments were densely labeled while coated regions of the plasma membrane were completely free of colloidal gold (Figure 10c). Areas of plasma membrane devoid of discernible structural elements showed but little background label.

Using a monoclonal antibody directed against vimentin, we were able to visualize vimentin-containing intermediate filaments (Figure 11). We could show that intermediate filaments are rarely connected to the plasma membrane, whereas they are strongly associated with stress fibers. Intermediate filaments appear to function as cross-linkages between parallel orientated stress fibers.

IV. CONCLUSIONS

Until recently, morphological data regarding the topographical distribution of receptors and the mobility of receptors in the plasma membrane have been sparse. This paucity has been caused mainly by the lack of a suitable marker and a convincing method to trace the

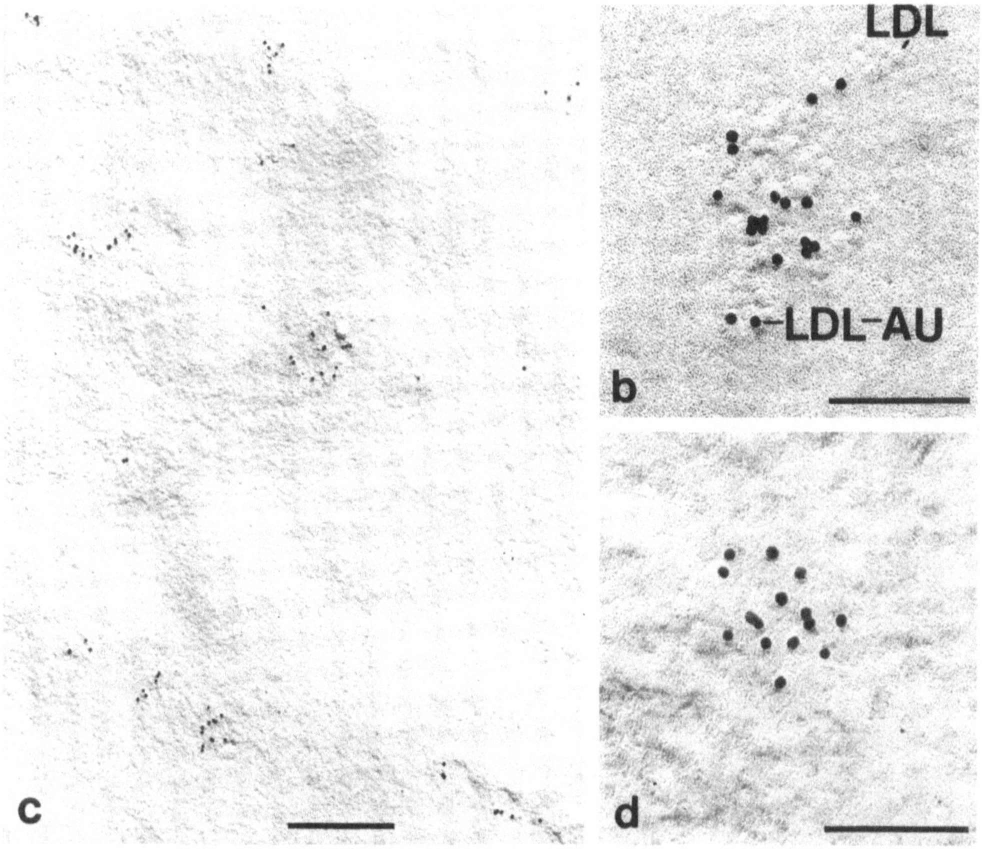

Figures 8b, c, and d.

movement of receptors within the plane of the lipid bilayer at high resolution. However, techniques have been developed that permit chemical labeling and examination of both the external and protoplasmic surfaces of the plasma membrane. These methods, combined with the unique advantages of colloidal gold markers, represent important advances for the investigation of the identity, distribution and dynamics of plasma membrane receptors. The labeling of ligands with colloidal gold, combined with standard platinum-carbon surface replica techniques, provides a highly effective approach for visualizing the planar distribution of cell surface receptors.

Using this new methodology, we demonstrated the existence of characteristic microdomains in the distribution pattern of receptors on the plasma membrane surface of various cultured cells. We also utilized this procedure to investigate the sequence of events from the insertion of receptors into the plasma membrane to the internalization of the ligand-receptor complexes.

Employing gold marker particles of distinctly different sizes, we could visualize two types of receptors on a single membrane simultaneously. Apart from investigating receptor specificity, the double labeling approach offers fascinating insights into how different receptors are displayed and organized in relation to one another, as well as into the dynamic alterations and interactions that may result when two ligands are presented to the cell.

We have investigated the topography and dynamics of IgA-SC complexes on the plasma membrane of rat hepatocytes in primary culture. In addition, we have characterized the influence of cytochalasin B on this topography and demonstrated that this drug induces patch

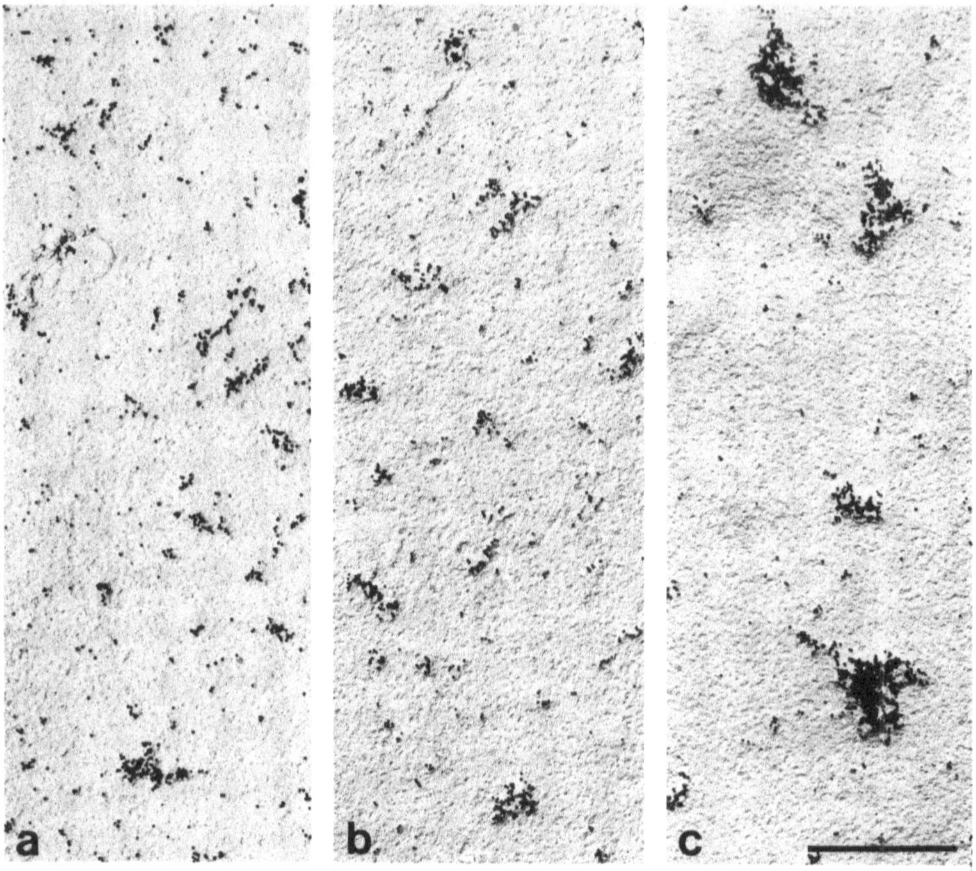

FIGURE 9. Localization of gold-labeled IgA in surface replicas of normal hepatocytes. Cultured hepatocytes were incubated with IgA-gold conjugates at 4°C for 20 min (a) and warmed to 37°C for 30 min (b) and 60 min (c). Surface replicas show the progressive clustering of IgA-gold conjugates. Hepatocytes treated with CB and prelabeled with IgA-gold conjugates at 4°C and then warmed to 37°C show extensive patches of gold conjugate in the cell periphery within 40 min of incubation (d). Bars: (a to c) 1μm; (d) 2 μm.

formation of the occupied, but not of the free IgA receptor. These results suggest an important role of microfilaments in the clustering of IgA-receptor complexes at coated pits. Finally, we studied the organization of cytoskeletal elements at the ventral plasma membrane surface in fibroblasts. The methods employed included exposure of the protoplasmic face of the ventral plasma membrane by lysis-squirting, labeling the exposed structures with specific antibodies conjugated to colloidal gold, and platinum-carbon replication. Although this technique is at an early stage of refinement, the results obtained so far encourage further studies with multiple labeling of different structures in the same cellular fragment. Our future experiments are aimed at localizing receptor molecules at the protoplasmic plasma membrane surface using specific antibodies directed against the cytoplasmic tail of these receptors.

ACKNOWLEDGMENTS

I wish to thank my collaborators B. Harrach and R. Semich, who generously contributed their data and their ideas to this review. I also gratefully acknowledge the excellent technical assistance of K. Schlattmann and S. Passmann. The studies on lipoprotein receptors were carried out with Dr. G. Schmitz (Münster). Dr. R. Gebhardt and his group (Tübingen)

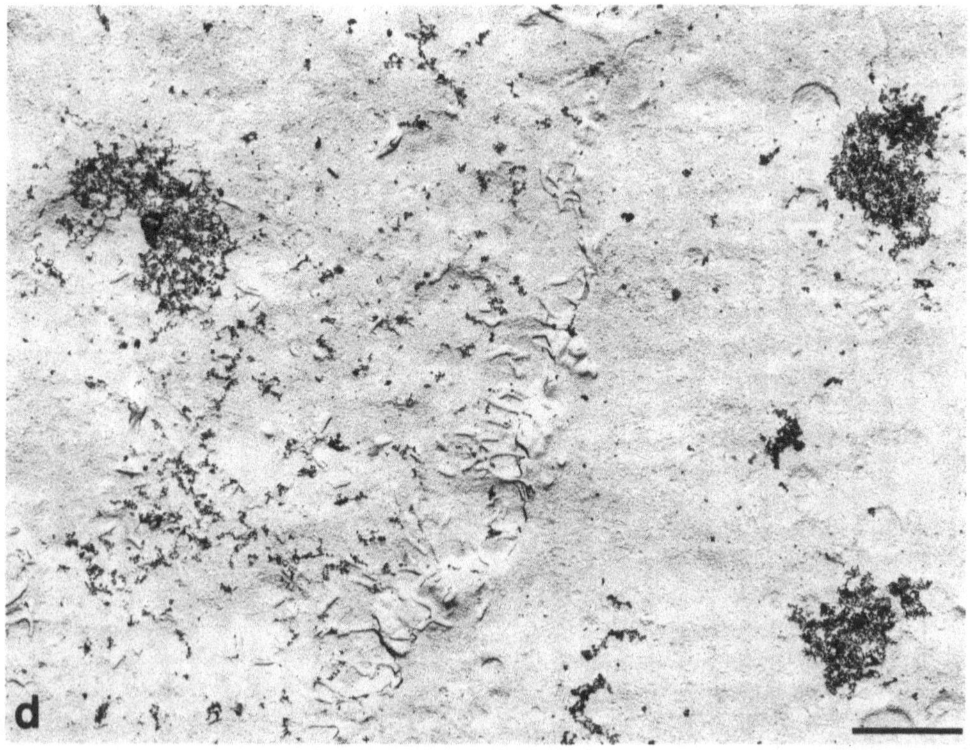

FIGURE 9d.

collaborated constructively on the work concerning the distribution of IgA receptors in rat hepatocytes. Dr. B. M. Jockusch (Bielefeld) and Dr. M. V. Nermut (London) contributed to the studies on the protoplasmic surface of fibroblast plasma membranes. I am grateful to Dr. D. Troyer for his indispensable aid in preparing the English manuscript. Much of the work presented here was supported by the "Minister für Wissenschaft und Forschung des Landes NRW" and by the "Deutsche Forschungsgemeinschaft" (SFB 223 and 310).

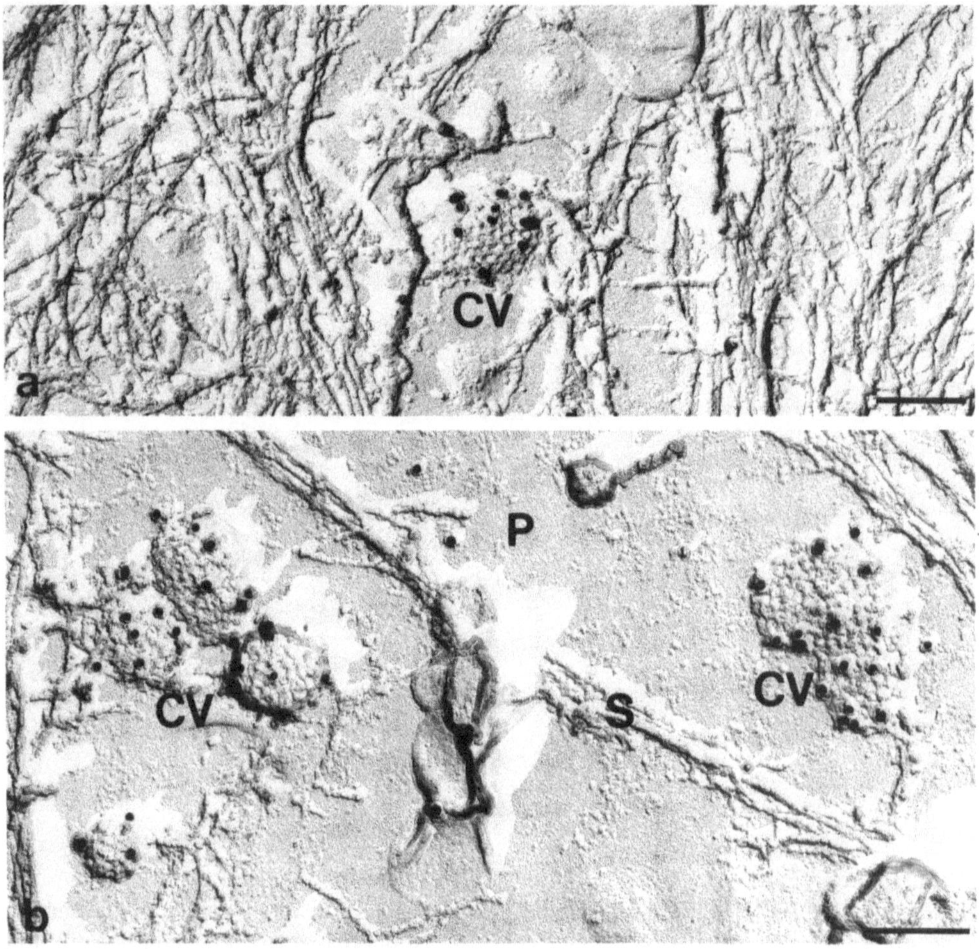

FIGURE 10. Replicas of the ventral protoplasmic plasma membrane surface of human fibroblasts after labeling with antibodies directed towards coated vesicles followed by protein A-gold (17 nm) (a, b). Replica of the ventral protoplasmic plasma membrane surface of normal rat kidney (NRK) cell after labeling with antibodies against actin followed by protein A-gold (17 nm) (c). CV = coated pit, coated vesicle, S = stress fiber, P = protoplasmic plasma membrane surface. Bars: (a, b) 0.2 μ; (c) 1 μm.

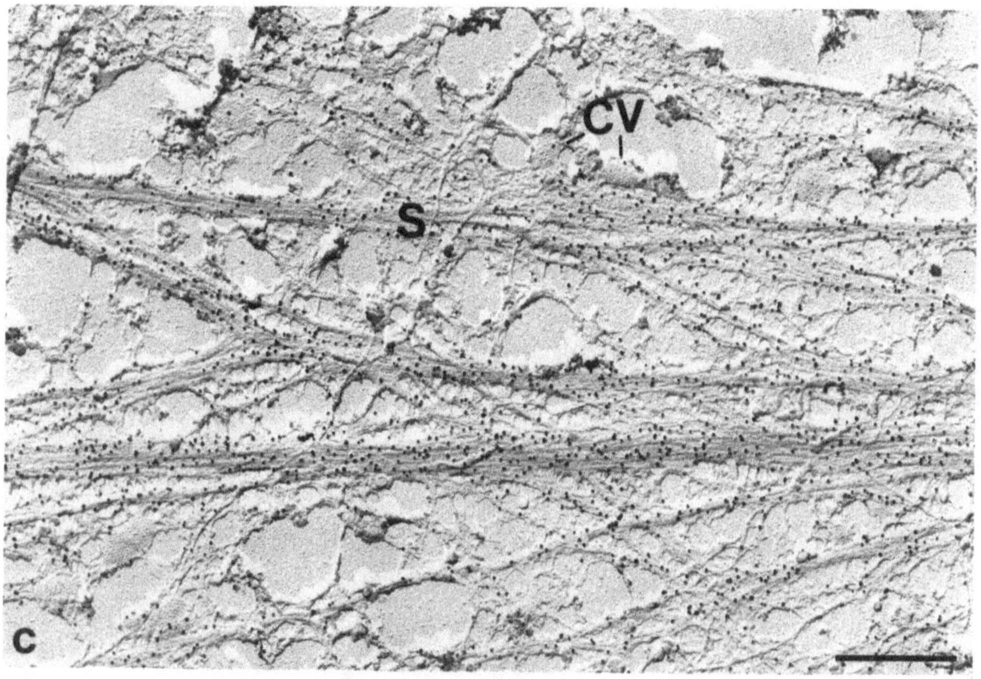

FIGURE 10c.

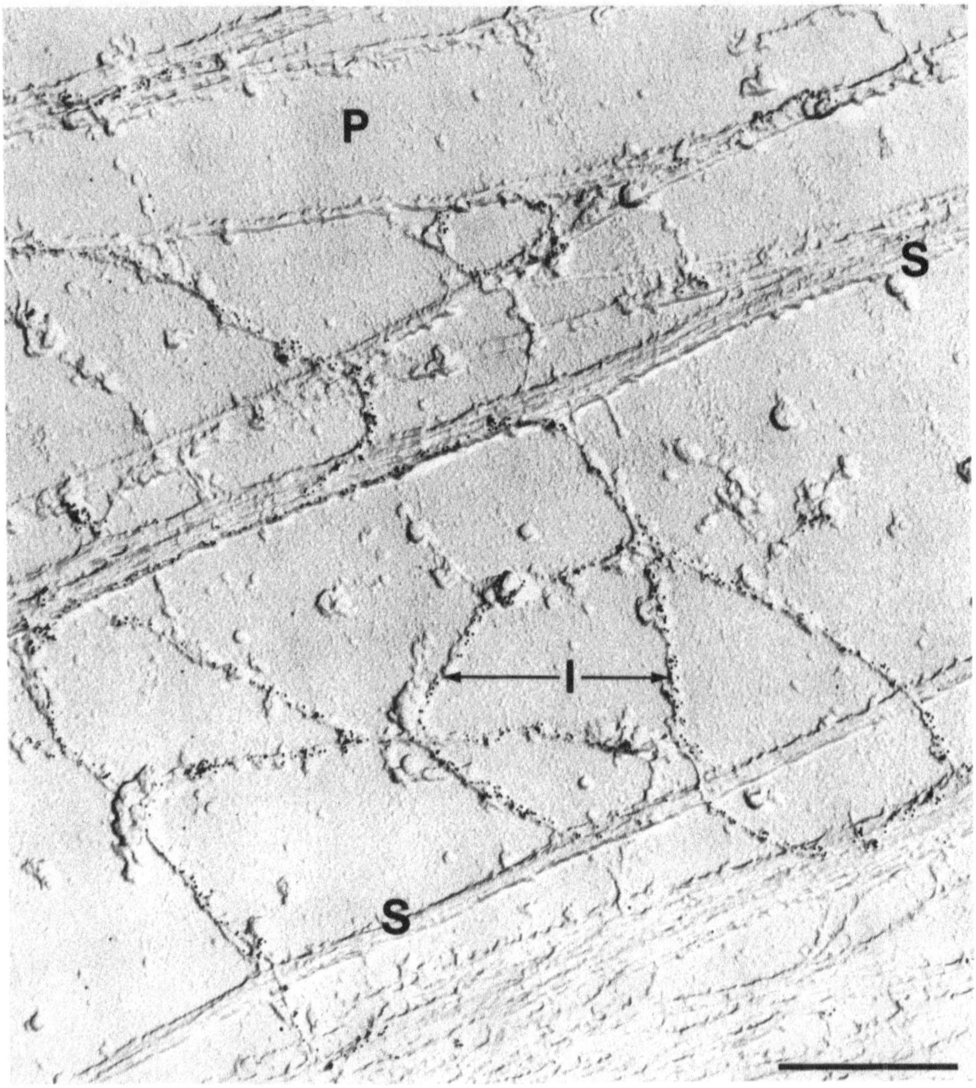

FIGURE 11. Replicas of the ventral protoplasmic plasma membrane surface (P) of a human fibroblast after labeling with monoclonal antibodies against vimentin followed by gold-labeled goat anti-mouse IgG. I = vimentin-containing intermediate filament, S = stress fiber. Bar: 1 μm.

REFERENCES

1. **Kaplan, J.,** Polypeptide-binding membrane receptors: analysis and classificaiton, *Science,* 212, 14, 1981.
2. **Anderson, R. G. W., Brown, M. S., and Goldstein, J. L.,** Role of the coated endocytotic vesicle in the uptake of receptor-bound low density lipoprotein in human fibroblasts, *Cell,* 10, 351, 1977.
3. **Maxfield, F. R., Schlessinger, J., Shechter, Y., Pastan, I. H., and Willingham, M. C.,** Collection of insulin, EGF, and alpha$_2$-macroglobulin in the same patches on the surface of cultured fibroblasts and common internalization, *Cell,* 14, 805, 1978.
4. **Gebhardt, R. and Robenek, H.,** Ligand-dependent redistribution of the IgA receptor on cultured rat hepatocytes and its disturbance by cytochalasin B, *J. Histochem. Cytochem.,* 35, 301, 1987.

5. **Willingham, M. C., Maxfield, F. R., and Pastan, I. H.,** Alpha$_2$-macroglobulin binding to the plasma membrane of cultured cells: diffuse binding followed by clustering in coated regions, *J. Cell Biol.*, 82, 614, 1979.
6. **Bretscher, M. S.,** Directed lipid flow in cell membranes, *Nature (London)*, 260, 21, 1976.
7. **Robenek, H. and Hesz, A.,** Dynamics of low density lipoprotein receptors in the plasma membrane of cultured human skin fibroblasts as visualized by colloidal gold in conjunction with surface replicas, *Eur. J. Cell Biol.*, 31, 275, 1983.
8. **Anderson, R. G. W., Brown, M. S., Beisiegel, U., and Goldstein, J. L.,** Surface distribution and recycling of the low density lipoprotein receptor as visualized with antireceptor antibodies, *J. Cell Biol.*, 93, 523, 1982.
9. **Bourguignon, L. Y. W. and Singer, S. J.,** Transmembrane interactions and the mechanism of capping of surface receptors by their specific ligands, *Proc. Natl. Acad. Sci. U.S.A.*, 74, 5031, 1977.
10. **Bretscher, M. S. and Thomson, J. N.,** Distribution of ferritin receptors and coated pits on giant HeLa cells, *EMBO J.*, 2, 599, 1983.
11. **Handley, A. C., Arbeeny, C. M., Witte, L. D., and Chien, S.,** Colloidal gold-low density lipoprotein conjugates as membrane receptor probes, *Proc. Natl. Acad. Sci. U.S.A.*, 78, 368, 1981.
12. **Horisberger, M. and Rosset, J.,** Colloidal gold, a useful marker for transmission and scanning electron microscopy, *Biol. Cell.*, 36, 253, 1979.
13. **Robenek, H., Rassat, J., Hesz, A., and Grünwald, J.,** A correlative study on the topographical distribution of the receptors for low density lipoprotein (LDL) conjugated to colloidal gold in cultured human skin fibroblasts employing thin section, freeze-fracture, deep-etching and surface replication techniques, *Eur. J. Cell Biol.*, 27, 242, 1982.
14. **Nicol, A., Nermut, M. V., Doeinck, A., Robenek, H., Wiegand, Ch., and Jockusch, B. M.,** Visualization of cytoskeletal elements at the ventral plasma membrane of fibroblasts by gold immunolabelling, *J. Histochem. Cytochem.*, 35, 499, 1987.
15. **Robenek, H. and Schmitz, G.,** Receptor domains in the plasma membrane of cultured mouse peritoneal macrophages, *Eur. J. Cell Biol.*, 39, 77, 1985.
16. **Robenek, H. and Severs, N. J.,** Double labeling of lipoprotein receptors in fibroblast cell surface replicas, *J. Ultra-struct. Res.*, 87, 149, 1984.
17. **Robenek, H., Schmitz, G., and Assmann, G.,** Topography and dynamics of receptors for acetylated and malondialdehyde-modified low density lipoprotein in the plasma membrane of mouse peritoneal macrophages as visualized by colloidal gold in conjunction with surface replicas, *J. Histochem. Cytochem.*, 32, 1017, 1984.
18. **Schmitz, G., Robenek, H., Lohmann, U., and Assmann, G.,** Interaction of high density lipoproteins with cholesteryl ester-laden macrophages: biochemical and morphological characterization of cell surface receptor binding, endocytosis and resecretion of high density lipoproteins by macrophages, *EMBO J.*, 4, 613, 1985.
19. **Robenek, H., Hesz, A., and Rassat, J.,** The variability of the topography of low density lipoprotein (LDL) receptors in the plasma membrane of cultured human skin fibroblasts as revealed by gold-LDL conjugates in conjunction with the surface replication technique, *J. Ultrastruct. Res.*, 82, 143, 1983.
20. **Branton, D., Cohen, C. M., and Tyler, J.,** Interaction of cytoskeletal proteins on the erythrocyte membrane, *Cell*, 24, 24, 1981.
21. **Burridge, K. and Connel, L.,** Talin: a cytoskeletal component concentrated in adhesion plaques and other sites of actin membrane interaction, *Cell Motility*, 3, 405, 1983.
22. **Coudrier, F., Reggio, H., and Louvard, D.,** Characterization of an integral membrane glycoprotein associated with the microfilaments of pig intestinal microvilli, *EMBO J.*, 2, 469, 1983.
23. **Rogalski, A. A. and Singer, S. J.,** An integral glycoprotein associated with the membrane attachment sites of actin microfilaments, *J. Cell Biol.*, 101, 785, 1985.
24. **Aggeler, J. and Werb, Z.,** Initial events during phagocytosis by macrophages viewed from outside and inside the cell: membrane-particle interactions and clathrin, *J. Cell Biol.*, 94, 613, 1982.
25. **Aggeler, J., Takemura, R., and Werb, Z.,** High resolution three-dimensional views of membrane-associated clathrin and cytoskeleton in critical-point-dried macrophages, *J. Cell Biol.*, 97, 1552, 1983.
26. **Chandler, D. E.,** Rotary shadowing with platinum-carbon in biological electron microscopy: a review of methods and applications, *J. Electron Microsc. Tech.*, 3, 305, 1986.
27. **Heuser, J. E.,** Three-dimensional visualization of coated vesicle formation in fibroblasts, *J. Cell Biol.*, 84, 560, 1980.
28. **Heuser, J. E. and Kirschner, M. W.,** Filament organization revealed in platinum replicas of freeze-dried cytoskeletons, *J. Cell Biol.*, 86, 212, 1980.
29. **Hirokawa, N. and Heuser, J. E.,** Quick-freeze, deep-etch visualization of the cytoskeleton beneath surface differentiations of intestinal epithelial cells, *J. Cell Biol.*, 91, 399, 1981.
30. **Hirokawa, N. and Tilney, L. G.,** Interactions between actin filaments and between actin filaments and membranes in quick-frozen and deeply etched hair cells of the chick ear, *J. Cell Biol.*, 95, 249, 1982.

31. **Lang, R. D. A., Nermut, M. V., and Williams, L. D.,** Ultrastructure of sheep erythrocyte plasma membranes bound to solid supports — effects of selective removal of membrane components, *J. Cell Sci.*, 49, 383, 1981.
32. **Nermut, M. V.,** Visualization of the "membrane cytoskeleton" in human erythrocytes by freeze-etching, *Eur. J. Cell Biol.*, 25, 265, 1981.
33. **Nermut, M. V.,** The "cell monolayer technique" in membrane research, *Eur. J. Cell Biol.*, 28, 160, 1982.
34. **Roos, E., Spiele, H., Feltkamp, C. A., Huisman, H., Wiegant, F. A. C., Traas, J., and Mesland, D. A. M.,** Localization of cell surface glycoproteins in membrane domains associated with the underlying filament network, *J. Cell Biol.*, 101, 1817, 1985.
35. **Clarke, M., Schatten, G., Mazia, D., and Spudich, J. A.,** Visualization of actin fibers associated with the cell membrane in amoebae of *Dictyostelium discoideum*, *Proc. Natl. Acad. Sci. U.S.A.*, 72, 1758, 1975.
36. **Avnur, Z. and Geiger, B.,** Substrate-attached membranes of cultured cells. Isolation and characterization of ventral cell membranes and the associated cytoskeleton, *J. Mol. Biol.*, 153, 361, 1981.
37. **Goldstein, J. L., Anderson, R. G. W., and Brown, M. S.,** Coated pits, coated vesicles, and receptor-mediated endocytosis, *Nature (London)*, 279, 679, 1979.
38. **Handley, D. A., Arbeeny, C. M., Eder, H. A., and Chien, S.,** Hepatic binding and internalization of low-density lipoprotein-gold conjugates in rats treated with 17(alpha)-ethinyl estradiol, *J. Cell Biol.*, 90, 778, 1981.
39. **Pearse, B. M. F.,** Clathrin: a unique protein associated with intracellular transfer of membrane by coated vesicles, *Proc. Natl. Acad. Sci. U.S.A.*, 73, 1255, 1976.
40. **Severs, N. J. and Robenek, H.,** Detection of microdomains in biomembranes. An appraisal of recent developments in freeze-fracture cytochemistry, *Biochim. Biophys. Acta*, 737, 373, 1983.
41. **Robenek, H., Jung, W., and Gebhardt, R.,** The topography of filipin-cholesterol complexes in the plasma membrane of cultured hepatocytes and their relation to cell junction formation, *J. Ultrastruct. Res.*, 78, 95, 1982.
42. **Gebhardt, R.,** Primary cultures of rat hepatocytes as a model system of canalicular development, biliary secretion, and intrahepatic cholestasis. III. Properties of the biliary transport of IgA revealed by immunofluorescence, *Gastroenterology*, 84, 1462, 1983.
43. **Orlans, E., Peppard, J., Fry, J. F., Hinton, R. H., and Mullock, B. M.,** Secretory component as the receptor for polymeric IgA on rat hepatocytes, *J. Exp. Med.*, 150, 1577, 1979.
44. **Socken, D. J., Jeejeebhoy, K. N., Bazin, H., and Underdown, B. J.,** Identification of secretory component as an IgA receptor on rat hepatocytes, *J. Exp. Med.*, 50, 1538, 1979.
45. **Zevenbergen, L. L., May, C., Wanson, J. C., and Vaerman, J. P.,** Synthesis of secretory component by rat hepatocytes in culture, *Scand. J. Immunol.*, 11, 93, 1980.
46. **Jones, A. L., Husling, S., Hraelek, G. T., Gaines, H. S., Christiansen, W. D., and Underdown, B. J., Uptake and intracellular disposition of IgA by rat hepatocytes in monolayer culture, *Hepatology*, 2, 769, 1982.**
47. **Limet, J. N., Schneider, Y. J., Vaerman, J. P., and Trouet, A.,** Binding, uptake and intracellular processing of polymeric rat immunoglobulin A by cultured rat hepatocytes, *Eur. J. Biochem.*, 125, 437, 1982.
48. **Gebhardt, R.,** Participation of microtubules and microfilaments in the transcellular biliary secretion of immunoglobulin A in primary cultures of rat hepatocytes, *Experientia*, 40, 269, 1984.
49. **Mostov, K. E., Friedlander, M., and Blobel, G.,** The receptor for transepithelial transport of IgA and IgM contains multiple immunoglobuln-like domains, *Nature (London)*, 308, 37, 1984.
50. **Fisher, M. M., Nagy, B., Bazin, H., and Underdown, B. J.,** Biliary transport of IgA: role of secretory component, *Proc. Natl. Acad. Sci. U.S.A.*, 76, 2008, 1979.
51. **Kuhn, L. C. and Kraehenbuhl, J. P.,** The sacrificial receptor — translocation of polymeric IgA across epithelia, *Trends Biochem. Sci.*, 7, 299, 1982.
52. **Mesland, D. A. M., Los, G., and Spiele, H.,** Cytochalasin B disrupts the association of filamentous web and plasma membrane in hepatocytes, *Exp. Cell Res.*, 135, 431, 1981.
53. **Koch, G. L. E.,** The anchorage of cell surface receptors to the cytoskeleton, in *Cell Adhesion and Motility*, Curtis, A. S. G. and Pitts, J. D., Eds., Cambridge University Press, New York, 1980, 425.
54. **Nicolson, G. L.,** Transmembrane control of the receptors on normal and tumor cells. I. Cytoplasmic influence over cell surface components, *Biochim. Biophys. Acta*, 457, 57, 1976.
55. **Singer, S. J., Ash, J. F., Bourguignon, L. Y. W., Heggeness, M. H., and Louvard, D.,** Transmembrane interactions and the mechanism of transport of proteins across membranes, *J. Supramol. Struct.*, 9, 373, 1978.

Chapter 5

FREEZE-FRACTURE STEROL LOCALIZATION WITH FILIPIN

Rodman G. Miller

TABLE OF CONTENTS

I. Introduction 88

II. An Overview of the Use of Filipin in Freeze Fracture 88

III. Sterol, Sterol, Where is the Sterol: Lateral Diffusion in Biological Membranes 89

IV. Three Bags Full: How Much Cholesterol? 91

V. Topologic Localizations 97

VI. Other Sterol Binding Drugs Used in Freeze Fracture 99

VII. Conclusion 100

Note Added in Proof 100

Acknowledgments 100

References 100

As long as there are foolish men, a goose that lays golden eggs will be of value, even though the eggs are of fool's gold.

Suffi Proverb

I. INTRODUCTION

Biological membranes may be considered as of a pair apposed lipid monolayers with proteins intercalated into the hydrophobic core of the monlayers.[1] The freeze-fracture technique owes much to the lipid bilayer arrangement of membranes. Conversely, a major evidence that the structure of biological membranes is a bilayer is the freeze-fracture technique and its interpretation of the fracture plane being in the center of the membrane.[2] The major paradigm of both the structure of biological membranes and the freeze-fracture technique grew together.

In freeze fracture, membrane proteins are visualized as particles on a smooth background representing the acyl tails of the membrane lipids. Much has been said, written, and hypothesized about these lipids. Nonmorphological techniques offer many means of definition of the biophysical properties of the lipids, but there are few morphological techniques which promise the possibility of better defining lipids in membrane. What makes protocols using filipin and other cytochemicals which interact with lipids so seductive is that heretofore there has been no means by which these ubiquitous cell components have been able to be localized. There are some sound reasons, which will be dealt with in the next section, for a relative paucity of acceptable morphologic probes for membrane lipids. For now, I wish to continue to look into how and why filipin could rise to the status of a virtual panacea for many morphologists in spite of rather unstable physical foundations.

In the past few years, the existence of "microdomains" within the membrane has been postulated and demonstrated to varying confidences by some.[4,5] It matters not that some of the original portrayers of microdomains in biological membranes defined neither the size nor lifetimes of these microdomains. Although microdomains have been ill defined on physical terms, pointing to a cartoon of a membrane that looks like a patchwork quilt usually is sufficient for many to get a feeling for what a microdomain might be.

Membranes have such diversity of functions, that it seems reasonable that if specialized domains with different physicochemical properties were available, the cell would make valid use of them. And what are the crucial determinants which might functionally and experimentally differentiate different domains? Certainly, cholesterol concentration must be one important variable. Filipin, a "marker" for cholesterol concentration became a highly valued tool to describe and define microdomains. That filipin labeling is simple to perform added adherants to the technique. And finally, nearly every complex cell showed some sort of inhomogeneity with respect to filipin labeling, providing the researcher something to write up upon completion of the study.

Given this backdrop, it is understandable that the introduction of this probe to define the location of cholesterol should be met with enthusiasm and be ardently courted by many freeze-fracture morphologists.

II. AN OVERVIEW OF THE USE OF FILIPIN IN FREEZE FRACTURE

Filipin is a group of polyene antibiotics produced by *Streptomyces filipinesis* which collectively have the property of binding to sterols with a planar ring structure and a free 3-OH group. Since the most common of such sterols is cholesterol, throughout this text I will speak of cholesterol when I really mean any sterol with the free 3-OH which binds to filipin. Filipin was introduced into morphological studies by Kinsky et al.,[6] who found that filipin

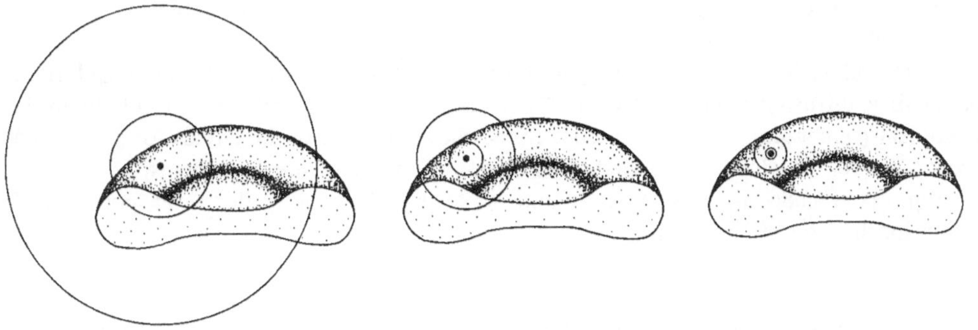

FIGURE 1. Mean diffusion of a molecule in 1 and 10 s (small and large circles, respectively) assuming $D = 10^{-8}$, 10^{-9}, and 10^{-10} cm²/s corresponding to D for cholesterol, phospholipid and mobile proteins in the left, center and right pannels, respectively. In each pannel, a sketch of a red blood cell is included for scale (7 μm diameter). See Edidin[60] for a more complete discussion on lateral mobility and diffusion coefficients for membrane components and text for further details of figure.

complexed with cholesterol to form a 30-nm deformation or filipin-cholesterol complex (FCC) in the contour of the membrane which was visible through thin section and negative stain electron microscopy. Subsequently, Verkleij and co-workers[7] showed that these membrane deformations could be visualized through freeze fracture as pits and protrusions. Since that time many have used filipin to label cellular and liposomal membranes in attempts to determine the location of cholesterol in these membranes. In general, filipin incubation is performed after glutaraldehyde fixation in a solution of up to 200 ± mg/ml which contains 1% DMSO which is necessary to disperse filipin. Incubation times vary betweeen 10 min and overnight. Subsequently, the tissue is prepared for freeze fracture with standard protocols.

There have been several pertinent overviews concerning filipin and microdomains determined through lipid binding drugs.[5,9] Some of the arguments in the present communication closely parallel arguments of one of the reviews.[13]

III. STEROL, STEROL, WHERE IS THE STEROL: LATERAL DIFFUSION IN BIOLOGICAL MEMBRANES

The most significant problematic area for the use of filipin to localize cholesterol in the plane of the membrane is the dynamics of lipids in monolayers. The diffusion coefficient is a constant which describes the speed of movement of a diffusing element. Cholesterol has the largest diffusion coefficient of the common membrane components, with $D = 10^{-8}$ cm²/sec.

Since the diffusion coefficient can be defined by the equation:[14]

$$D = \frac{X^2}{2t}$$

where X = the root-mean squared distance a molecule travels in a time, t, it is possible to describe the probable distance a membrane component will travel in a given time. Applying this equation to membrane cholesterol dynamics, we find that the location of a cholesterol molecule on a 7-μm red blood cell at one time is not correlated with its location after about 6 s. This is represented pictorally in Figure 1; on the left panel, a cartoon of a red blood cell is depicted. Overlaid are flat circles representing the mean distance that a cholesterol molecule will have moved in 1 s (small circle) and 10 s (large circle). Clearly, since the 10-s circle is larger than the diameter of the red blood cell, a localization technique in which the time to develop a signal is greater than 10 s is incompatible with free diffusion of cholesterol.

In the middle frame, the circles again represent average displacements for 1 and 10 s, but with a diffusion coefficient of $10^{-9}cm^2/sec$, similar to membrane phospholipids in biological membranes. Finally, the third frame shows the size of the circles for a diffusion coefficient 10^{-10}, which corresponds to diffusion of the most mobile proteins.

Filipin aficionados may use three arguments to partially mollify this bleak portrayal of filipin localization of cholesterol:

1. These diffusion coefficients were measured in membranes above the phase transition temperature, while filipin labeling may be done below the phase transition temperature. At such lowered temperatures the problem of cholesterol diffusion may not be as acute as described. Working below the phase transition temperature introduces additional problems: formation of FCCs require a temperature of over 12°C.[13] Overnight incubations in filipin probably do not add to the development of visualizable FCCs. Below the phase transition temperature of membrane lipids (typically in the region of 15°C), some of the lipids may be crystalized out into microdomains which were not present *ab initio*.
2. Fixation of tissue results in decreased mobility of the tissue (the tissue becomes rubbery) and membranes. Perhaps this decrease in membrane mobility will be reflected as a reduction in cholesterol mobility. However, aldehyde fixation results in cross-linking of amine groups, particularly abundant on proteins, but absent on cholesterol. In addition, it may be argued that amino lipids (e.g., phosphatydtil ethanolamine) may be cross-linked. We have no *a priori* reason to believe that aldehyde fixation should alter cholesterol diffusion. Fixation with osmium, as for thin section electron microscopy, might alter mobility of lipid components in the membrane, but is not compatible with normal freeze-fracture technique.
3. Finally, it may be argued that the diffusion coefficient for cholesterol is measured on the basis of a ''microviscosity'' wherein a rotational, or local, movement of cholesterol is converted to a macroscopic diffusional coefficient. Such a microviscosity does not specify the long-range diffusion of cholesterol, and the above formula is not applicable. This strong argument depends upon the possibility that microdomains of lipids exist which are effectively isolated from other domains by either proteins, proteins with relatively stable annular lipid, stable rafts of lipids, or any combination of the three (percolation theory — for a more complete discussion of this problem, see Reference 15). Although this argument is difficult to rule out with presently available data, (see Reference 13), it rests the validity of the filipin-labeling technique upon the presumption that microdomains are significant in biological membranes. Since one of the goals in the use of filipin to localize cholesterol is to better define microdomains, having the technique ride upon an assumption of the existence of microdomains is logically unsound.

The foregoing arguments concerning the mobility of membrane components apply equally well to the mobility of filipin within the membrane. An additional complication with filipin is its insertion into the membrane. Filipin is not very soluble in aqueous media. The standard protocol for filipin labeling calls for solubilization of filipin in DMSO (up to 20 mg/ml) and subsequent 100-fold dilution in the labeling media. Opalescence occurs upon 1:10 dilution, indicating that the filipin is not monodisperse (Miller, unpublished observation). Behnke et al.[10] found that filipin labeling activity may be spun out of concentrated ''solutions'' while dilute solutions (below 30 μg/ml) retain filipin-labeling activity in the supernate. Thus, it is likely that much incorporation of filipin into the membranes being studied is done in

FIGURE 2. Lack of FCC labeling of the nuclear envelope (lower fractured membrane) in the face of significant labeling of the plasma membrane (upper portion of micrograph) is thought by some both to reflect the relative lack of cholesterol in the former and to be a vindication of localization of cholesterol with filipin. Bar = 0.2 μm.

packets (see Reference 10 for a discussion of this). We do not know what effects this delivery system might have. Preferred sites of incorporation of monomeric and/or aggregated filipin into the membrane would have profound effects on the results of filipin labeling.

Although the final labeling solution usually contains 1% DMSO, the preferential partition of DMSO into the membrane may result in a membrane concentration much higher than 1% DMSO. The morphological effects of DMSO upon membranes, and more specifically membrane lipid domains, is unknown. Some have used fixatives containing DMSO to reveal structures not normally seen with fixatives lacking DMSO.[16]

Let us assess where we are; the ultrastructural localization of cholesterol in biological membranes is an important objective, but its investigation through the use of filipin includes implicit assumptions which appear unsound at present. We have not yet studied the possibility that filipin may be of great use in determination of cholesterol concentrations (not lateral localization) within membranes. The next section covers this possibility.

IV. THREE BAGS FULL: HOW MUCH CHOLESTEROL?

Although FCCs may not accurately pinpoint the location of cholesterol, surely filipin may still be used to indicate the presence and relative abundance of cholesterol. Vindication of this approach has been in the form of pointing out that nuclear and rough endoplasmic reticulum (RER) membrane is known to be low in cholesterol concentration and, in general, develops less FCCs (see Figures 2 and 3) whereas plasma membrane, secretory granule membrane, and lysosomes are known to be high in cholesterol concentration and develop high densities of FCCs (see Figures 3 and 4). There are few studies which attempt to quantitate the dose response relationship between cholesterol concentration and FCC density. Friend and Bearer[17] incorporated various levels of cholesterol into *Drosophilia* larval cells, measured

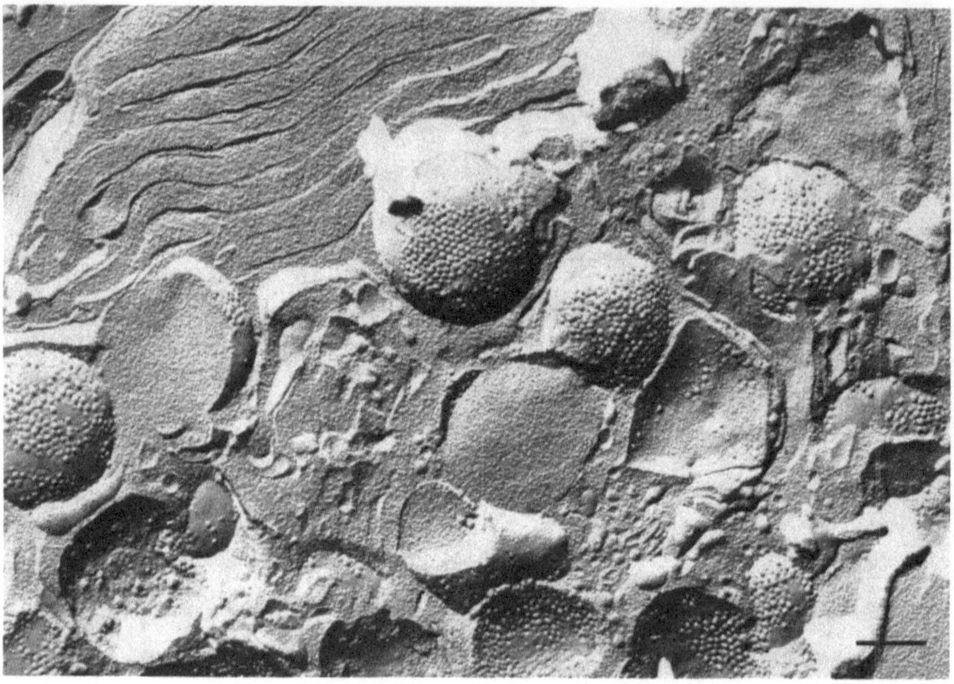

FIGURE 3. Rat parotid gland, acinar cell, labeled with filipin. Note the profiles of the RER are smooth, lacking FCCs, and secretory granules (spherical in shape) are heavily labeled with the polarity of the FCCs the same as the curvature of the granule. Condensing vacuoles demonstrate less FCCs which have a mixture of polarities. Bar = 0.2 μm.

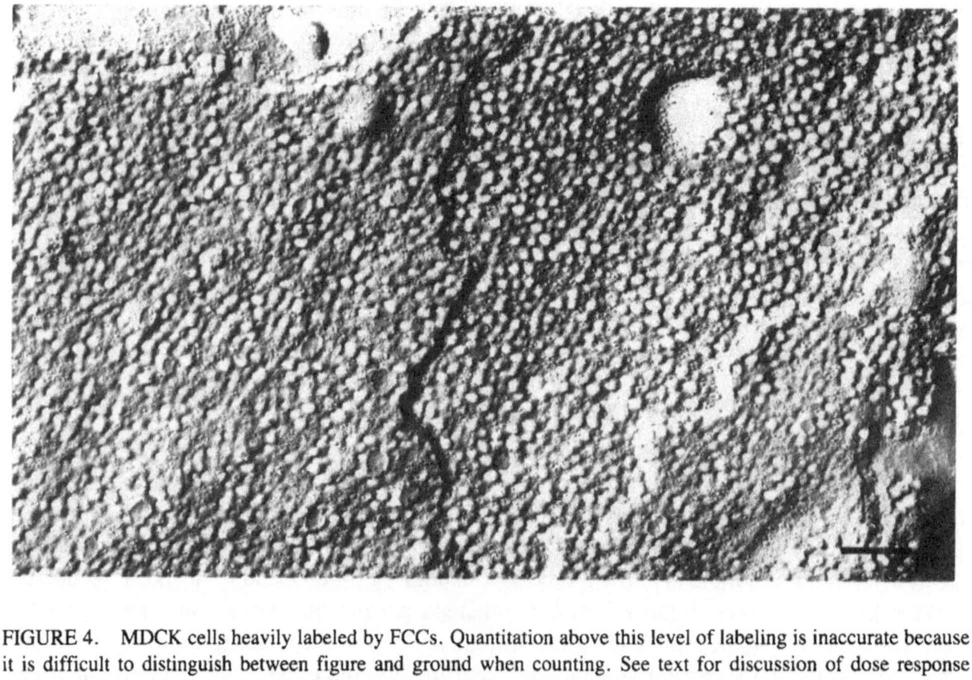

FIGURE 4. MDCK cells heavily labeled by FCCs. Quantitation above this level of labeling is inaccurate because it is difficult to distinguish between figure and ground when counting. See text for discussion of dose response curves in this region. Density of FCCs (protrusions only) in the left region this micrograph is 280/μm^2 as counted by one observer and 350 μm^2 as counted by another observer. FCC pits are seen with even less reliability. Bar = 0.2 μm.

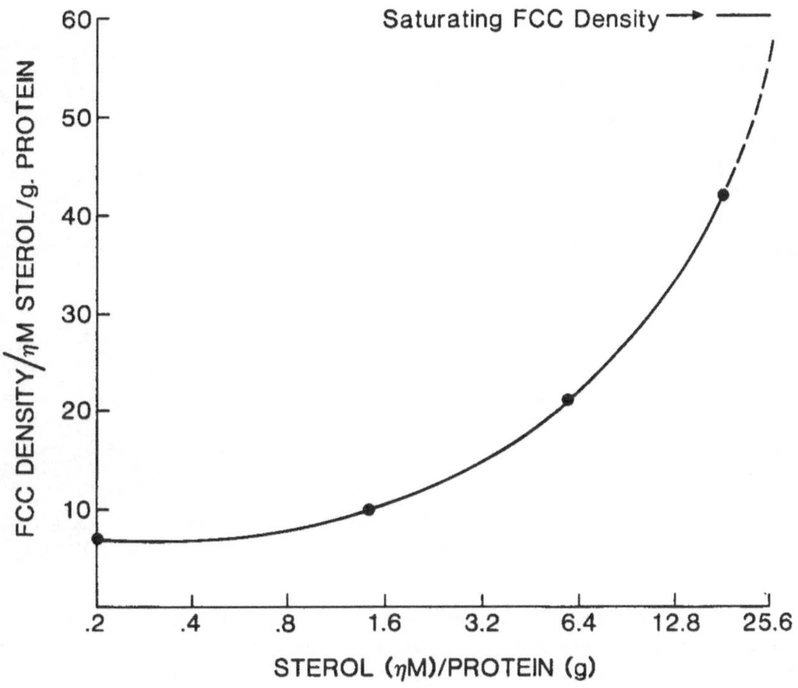

GRAPH 1. A dose response curve relating quantity of cholesterol incorporation (adjusted to protein content) and FCC density. At low concentrations, development of FCCs is much more efficient. The maximal FCC density corresponds to the point where FCCs are created "wall to wall"; higher densities are not possible without additional buckling of the membrane. In practice, accurate counting of FCC densities above 100 to 200/μ^2 is inaccurate and depends greatly upon the observers' counting criteria.

final cholesterol concentrations, and correlated these with FCC density. The dose response was monotonic but far from comforting; Graph 1 shows a rearrangement of their data from a representative experiment. It would appear that at low cholesterol concentrations, spuriously low densities of FCCs are observed, while at high cholesterol concentrations, increasing densities per amount of cholesterol are observed. The highest concentration observed is approaching maximal possible FCC density* and is at the limit of reliable quantitation of FCCs.[18] Figure 4 shows a heavily labeled pair of membranes in which the accuracy of quantitation suffers due to high FCC density. The reader is encouraged to try to count pits or protusions in any area of this micrograph. The possibility of correlating an FCC density with a sterol concentration led to the hope that histochemical quantitation of cholesterol might be possible with filipin. Subsequently, it has become evident that a dose response curve for one type of membrane does not apply to others; there are other factors involved in the development of a visualizable FCC.

Within the membrane proper, other chemicals, sphingomyelins in particular, have been shown to effect the density of FCC labeling. In addition, sphingomylins alter the size of the deformation, with high concentrations of sphingomyelin resulting in a doubling of the diameter of the "FCC".[9] We do not know whether there are other chemicals which can directly modify the development of FCCs. We do know, however, that there are other elements both within and exterior to the membrane that are also crucial in determination of whether an FCC is developed; these have in common and are most easily understood in terms of their ability to stabilize the membrane against deformations.

* Freeze-fracture FCC protrusions measure 20 to 25 nm, therefore the outer dimensions including membrane thickness of a deformation must be on the order of 30 nm diameter. Packing these into a square array with 30-nm spacing suggests a maximum packing density of about 1100/μm^2.

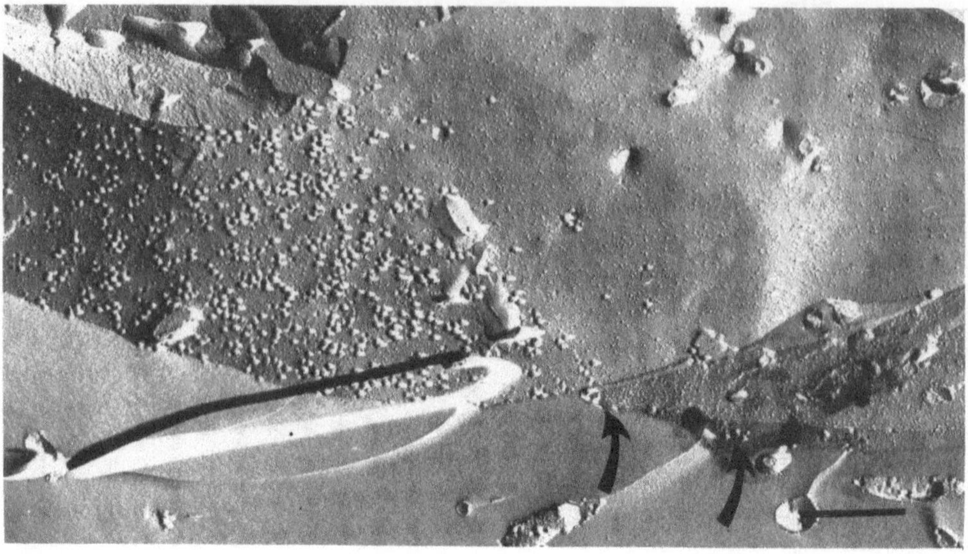

FIGURE 5. MDCK cells labeled with filipin. Note that there is an inhomogeneous distribution of FCCs on the plasma membrane although there appears to be no membrane specialization to maintain a steep concentration gradient which would be suggested if the FCCs are indicative of *ab initio* cholesterol concentration. Arrow points to the probable cause of scanty labeling; another cell is in close proximity to the region with few FCCs. Bar = 0.5 μm.

An FCC as seen through freeze fracture is a local deformation in the contour of the membrane. The creation of such a deformation will push the surface of the membrane into the cytoplasm or into the extracellular space. Any constraints upon the membrane, either within the membrane, in the extracellular space, or within the cytoplasm could have an effect upon creation of a deformation. Recently, it has been shown that such constraints do play a role in FCC density. As a rule, crystalline or paracrystalline arrays within the membrane are associated with exclusion of FCCs. Extracellular constraints can appear as regions of the plasma membrane which are resistant to FCC formation due to the presence of extracellular material. Figure 5 documents an instance where the close proximity of a pair of plasma membranes appears to block FCC formation. Cytoskeletal attachments are more cryptic in that they are frequently not observed in standard freeze fracture protocols. Although their presence may not be directly detected through freeze fracture, frequently, the existence of cytoskeletal attachments to the membrane are documented through thin section electron microscopy and the lack of FCC labeling can be surmised to be one of their effects. In Figure 6a, spectrin on the red blood cell may be responsible for anomalously low FCC labeling in the face of high cholesterol content. In Figure 6b, putative tears within the spectrin network permit high labeling of the plasma membrane with filipin.

Reexamination of the results of some recent studies of filipin labeling reveal that cytoskeletal attachments and constraints within the membrane proper are particularly relevant in the inhibition of FCC development (Figure 7). The following are given as examples of the spectrum of studies which have been done in the past few years.[18-43] It is not intended to be all inclusive. A discussion of many of the earlier studies of filipin labelling may be found in Reference 13.

Gap junctions, an array of particles which might be expected to constrain the membrane, contain significant cholesterol when isolated and analyzed biochemically,[44] but do not demonstrate FCCs.[20]

Several reports have been made concerning the distribution of FCCs on membranes of peripheral myelinated nerve. Blanchard et al.[21] found that the node of Ranvier labeled while

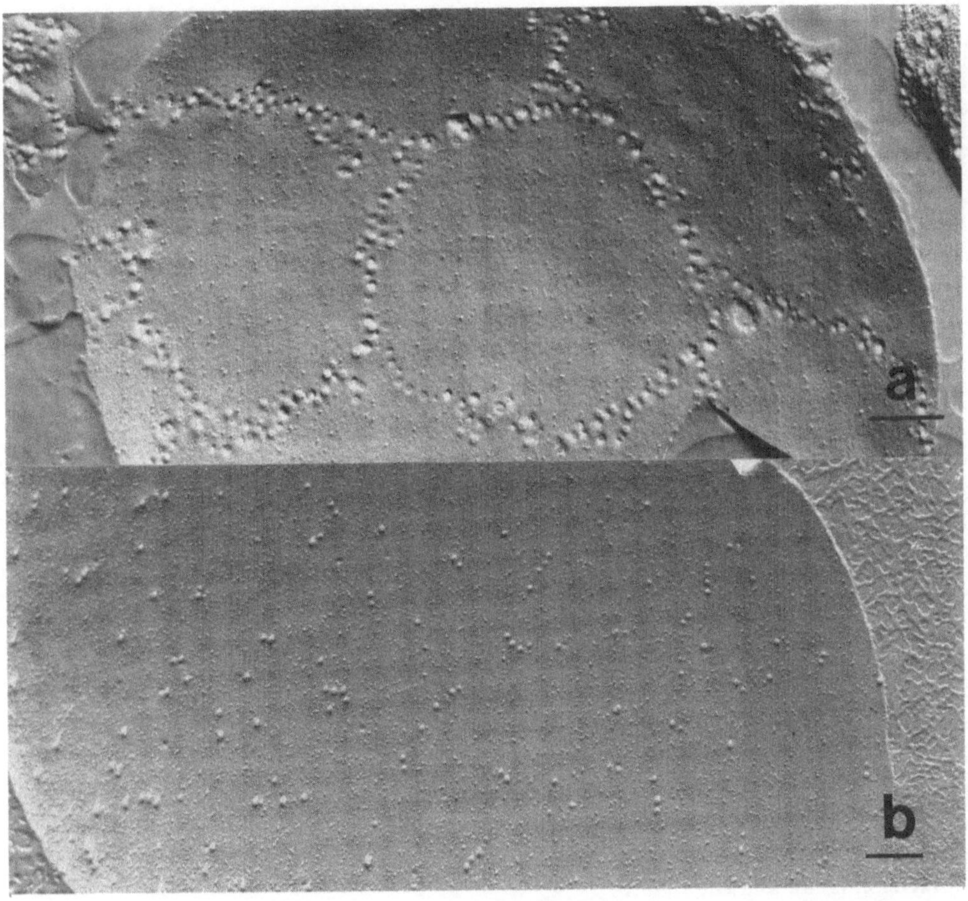

FIGURE 6. Typically, red blood cells do not develop many FCCs in spite of a high concentration of cholesterol. In a, the E-face of a cell with a higher density of FCCs which are distributed in a network, probably corresponding to "tears" in the spectrin network, and in b, the E-face of a red blood cell with low FCC density. Bar = 0.2 μm.

the paranodal region of the axolemma resisted FCCs. Both the nodal and paranodal region of the axolemma are highly specialized membrane domains with the nodal region being high in channels and the paranodal region being covered with an array of particles (impressions). It should come as no surprise that FCCs would have difficulty deforming a membrane made of a paracrystal. That the nodal region might be easily enough deformed by FCCs is noteworthy in light of the cytoskeletal network which has been shown to be associated with this membrane domain.[45]

Lamellar myelin is a spiral extension of the Schwann cell or oligodendrocytic, in which the cytoplasm has been removed by close apposition of the cytoplasm. This flap wraps the axon creating another close apposition between the extracellular surfaces of the plasma membrane. Early studies of the freeze-fracture morphology of myelin revealed that there were many particles on the myelin membrane and that under most conditions, these particles were susceptible to aggregation creating artifactually particle-free regions which are propagated through many lamellae[46] which is the freeze-fracture correlate of neurokeratin formation observed in light microscopy.[47] Thus, although there is no place for the more widespread types of cytoskeletal proteins, myelin contains constraining contacts across both the cytoplasmic and extracellular surfaces which might be expected to inhibit the formation of FCCs in lamellar myelin[19] even though the cholesterol concentration in myelin is high.

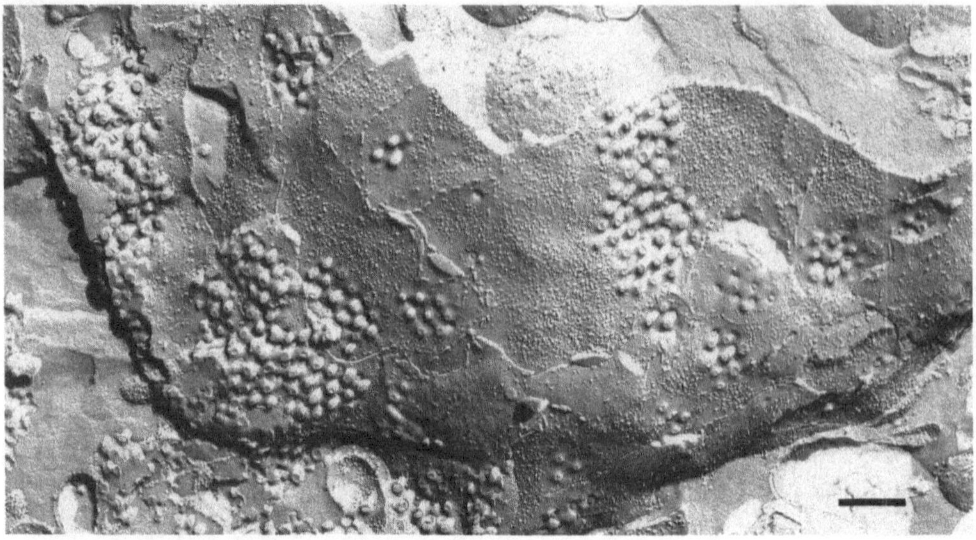

FIGURE 7. Junctional region of the rat parotid gland acinar cells showing the tight and desmosomal junctions inhibit the development of FCCs. Bar = 0.2 μm.

Recently, Blanchard et al.[22] have demonstrated that compact lamellar myelin is free of FCCs while they are abundant on Schmidt-Lanterman cleft membrane, which may be presumed to be less constrained.

Fields and co-workers[26] have shown that prior to myelination, FCCS readily form on the axolemma reflecting an inhomogeniety of the membrane which is not apparent on the basis of freeze-fracture particle density. Whether this reflects an inhomogeneity in cholesterol or cytoskeleton is not known.

The pattern of FCC labeling of Purkinje dendrites is dependent upon the degree of input from climbing fibers:[30] harmaline, which induces rhythmic activity of climbing fibers to the Purkinje dendrites, increases FCC susceptibility of the dendrites. This effect is inhibited if the climbing fibers are destroyed. Although these experiments could be interpreted in terms of alterations in the cholesterol content of dendrites, alterations in the cytoskeleton induced by activity is equally likely and fits with some models of neuronal plasticity.

Early studies of inhomogenieties of FCC labeling demonstrated that in smooth muscle cells FCCs only develop in the longitudinal regions which contain caveolae whereas the bands of caveolar-free membrane were resistant to filipin labeling.[36] Although the reason for these prominant major domains on the smooth muscle membrane is not known, similar vesicular sites are seen in cultured fibroblasts and their distribution appears to be closely linked to the presence of cytoskeletal elements which effectively prohibit vesicle attachment sites in regions where fibers are running close to the surface of the membrane.[48] Severs and Simon[39] investigated the differential labeling of caveolar bands in smooth muscle through the sequential labeling with two cholesterol binding probes, filipin and tomatin (see below). They found that while the caveolar bands labeled densely with filipin, the interbands were susceptible to labeling with tomatin. Which probe is telling the truth about cholesterol localization? They conclude that cytoskeletal constraints are responsible for false negative results of sterol probes. In other studies, plasmalemmal vesicle rims in endothelium have been shown by Simonescu et al.[40] to be susceptible to filipin. Simons and Severs,[41] using a slightly different system, find that preferential localization of FCCs did not occur at vesicular rims: both filipin and tomatin label more or less homogeneously in rabbit aortic endothelium.

Membranes stabilized through the cytoskeleton associated with pseudopods of platelets,[34] microvilli in epithelia,[23] cell-to-matrix attachment plaques (hemidesmosomes)[24] and cell-to-cell desmosomes[20] contain far fewer FCCs than surrounding plasma membrane. High densities of membrane proteins might be expected to stabilize the membrane and are associated with low FCC densities.[28]

Active zones at the neuromuscular junction, regions where we expect that there may be considerable membrane associated apparatus responsible for release of synaptic vesicles, do not label with either filipin, "saponin", or digitonin.[32] On the other side of the junction, the clusters of acetylcholine receptors on the muscle membrane exclude the formation of FCCs.[37]

At this time, there has been an increased recognition that cytoskeleton presence may have a profound effect upon the demonstration of FCCs and several groups have approached this question directly. Noting that the red blood cell normally develops low FCC density[13] in spite of a known high concentration of cholesterol, Clark et al.[25] compared filipin labeling of normal and spectrin deficient mouse red blood cells. They found that the spectrin deficient membrane demonstrated a fivefold increase in FCCs relative to normal, spectrin-containing cells. Steer et al.[42] have also approached this problem directly through labeling of clatherin-coated vesicles, which normally do not show FCCs. Removal of the clatherin cage results in a strong FCC response.

From what has been said thus far, it would seem that the ideal membranes to deal with would be ones which did not have any constraints in terms of membrane-associated proteins. Liposomal membranes fit this description; it is disconcerting to see a filipin-labeled liposome with all of the FCCs bunched tightly at one pole while other liposomes with the same concentration of cholesterol either contain no FCCs or are an apparent knot of pure FCCs.[13] It has been demonstrated that filipin labeling of liposomal membrane results in a capping and sloughing-off of FCCs.[9] It would seem that just the right amount of constraint is necessary for appropriate labeling by filipin. Both too much and too little constraint result in misleading conclusions concerning cholesterol in membranes.

V. TOPOLOGIC LOCALIZATIONS

There are several other possible uses of filipin which are not as dependent upon the diffusion coefficient of cholesterol and, if we are dealing with membranes with similar cytoskeletal constraining properties, may yet be valid. Although both cholesterol and filipin and even the complex of the two are potentially mobile within the membrane, once within the membrane, they are likely to remain within the membrane. Thus, it may be possible to compare cholesterol levels between two similar membranes using filipin.

Some membrane specializations define domains within a single membrane. A question which might be approached through a lipid localizing label is whether the domains formed differ with respect to lipids. In this laboratory, we were interested in whether the tight junction acted as a barrier to membrane lipids as could be predicted by a lipidic model of the tight junction.[49,50] A study of filipin-labeled MDCK cells revealed that tight junctions were a boundary for different densities of FCCs.[18] Neglecting the possibility that adjacent cells had significantly different cytoskeletal properties, the tight junction appeared to inhibit the free exchange of cholesterol both between adjacent cells and between the apical and basolateral membrane. It was unclear whether it was valid to neglect cell-to-cell differences in cytoskeletal-membrane attachment. Concerns about equivalent access of filipin to different regions of the plasma membrane (which will be raised in the next paragraph) were met by scraping cells from the substrate prior to labeling. It is known that there are major differences between apical and basolateral membrane proteins and lipids[51,52] which could contribute to differences in FCC density seen on the two regions of the plasma membrane. More recently the conclusion concerning the free exchange of lipid from one apical cell membrane to an adjacent one has been confirmed using fluorescently labeled dyes in MDCK cells.[53]

Filipin has been used for a different sort of topological problem concerning tight junctions. The tight junction has long been known to represent a barrier to passage of small molecules and ions between the luminal and basal extracellular regions in epithelia, both on the basis of electrophysiological studies and on the basis of tracers (such as lanthanum) using thin sections. Pelletier and Friend[54] reasoned that since filipin is a small molecule and that both apical and basolateral membranes connected to a tight junction should contain some cholesterol, filipin may be used as a tracer for freeze fracture in much the same manner that lanthanum has been used in thin section electron microscope in the past. At present, we have little idea of whether filipin is "small" in aqueous media (see Behnke et al. and above); labeling may be performed with aggregates of filipin rather than with monodispersed filipin. Nontheless, this study showed that sertoli cells with intact tight junctions are labeled only on the basolateral side of the tight junction when the filipin is presented through the bloodstream. The logical hazards associated with this use of filipin suggest that the use of thin section markers may be used with better confidence.

Although the lateral diffusion of the lipids within either leaflet is rapid in comparison to the normal time required for incubation in filipin solutions, the diffusion (translocation or flip or flip-flop) of a lipid from one leaflet to the apposed leaflet has a high enough energy barrier as to allow residence in one leaflet for a relatively long time.[55] Since the deformation created by the FCC is asymmetric with respect to the membrane, the possibility exists that markers such as filipin may be able to detect asymmetries of cholesterol within the membrane. There are various conditions which must be met for the use of filipin in this manner to be valid. *A priori,* there is no reason to believe that these conditions should be met, but the experimental fact is that many membranes do produce FCCs which are consistently biased such that usually the protrusions are produced on one side of the membrane and pits on the other or vice versa. This writer has been intrigued by the prospect that the FCC asymmetries may represent cholesterol asymmetries.

The density of FCCs is low on the *cis* side of the Golgi apparatus, increasing on the *trans* Golgi region and condensing vacuoles and high in the secretory granules of pancreas.[38,56] In addition, the polarity of the FCCs reverses in the transition from condensing vacuole to secretory granule.[57] Subsequent work by this writer has shown that the parotid and submandibular glands are similar to the pancreas with respect to the polarities of FCCs in the Golgi apparatus, condensing vacuole and secretory granule membranes (Miller, unpublished). Maturation of a condensing vacuole into a secretory granule requires concentration of the vacuolar contents and concomitant removal of over half of the vacuolar membrane through the removal of small vesicles. These vesicles appear in thin sections to be of a size such that the outer surface of the membrane is significantly (up to 50%) larger than the inner surface. Hence, removal of many such vesicles might result in a membrane with a highly asymmetric surface pressure, resulting in the asymmetry in the flip-flop rate of cholesterol (a more complete rendition of this model is in Reference 58). If this scenario is accurate, the FCC polarities suggest that the fracture face which contains pits represents the leaflet which contains more cholesterol. Friend[29] has reported experiments in which liposomes were incubated with exogeneous cholesterol to produce liposomes which were presumed to have cholesterol in the outer leaflet and none in the inner leaflet. Under this interpretation, FCCs were seen as protrusions on the leaflet with cholesterol. An alternative interpretation would suggest that although cholesterol may enter the liposome from the outside, flip-flop is rapid enough such that it should equilibrate between the two leaflets more rapidly than it takes to develop FCCs and FCC asymmetries reflect asymmetries in the surface pressure created at the time of creation of the liposome. Since both of these interpretations may be faulted, we may have learned better the problem, but have no better idea of the answer to the question. This potential use of filipin may be valid, but requires much more experimental work to determine what FCC polarity corresponds to a given cholesterol polarity.

FIGURE 8. Digitonin cholesterol complexes are in the form of hemitubules which corregate the membrane. Note that all of the corregations on the P-face are present convex curvature while all on the E-face are concave. See text and Reference 59 for details. Bar = 0.2 μm.

VI. OTHER STEROL BINDING DRUGS USED IN FREEZE FRACTURE

There have been several other lipid binding drugs which have been called upon to report position of sterols through freeze fracture. Saponins, such as tomatin and digitonin, bind to cholesterol creating deformations in membrane contour which can be observed through freeze fracture (Figure 8). In the case of commercial saponins (i.e., Sigma: "Saponin (pfs) white powder") these deformations are very pleomorphic elevations (and depressions) on the order of 0.1 μm (see Reference 37). Tomatin and digitonin create corrugations, or hemitubular arrays, in the membrane and appear to be less sensitive to cytoskeletal effects (e.g., see Reference 39). The corrugations frequently contain particles, indicating that the corrugations may be formed with significant impurities. If this is the case, quantitation is out of the question. Pure digitonin suspensions have tubular morphology making it unclear whether any particular hemitubular structure is digitonin, digitonin-cholesterol complex or a digitonin-cholesterol complex with impurities.[59] As mentioned above, tomatin has been used efficaciously in combination with filipin to demonstrate that filipin does not reveal all of the cholesterol in smooth muscle cells.[41] All of the saponin alternatives to filipin bear sugar moieties (by definition of saponin), which hinder their ability to translocate across the membrane. It has been reported for digitonin[59] that nearly all deformations of plasma membranes are of the same polarity, indicating that translocation is not usual. *Vida infra*, similar conclusions can be made concerning tomatin on the basis of published micrographs. In short, except for special applications, the saponins such as digitonin and tomatin offer no particular advantages over filipin.

VII. CONCLUSION

Many of us have been tantalized and infatuated by the apparant attainability of localization of membrane lipids through freeze-fracture techniques. Courted by many, filipin has shown promise of lateral localization of membrane cholesterol, of localization of permeability barriers (e.g., tight junctions) and determination of which leaflet of the bilayer contains more cholesterol. But filipin has not delivered upon her promises and her glamour is fading fast. Indeed, filipin has been a fickle mistress, servicing journal publishers more faithfully than the dissemination of knowledge of cholesterol disposition in membranes. But, if we strive for wisdom, filipin will have taught us what we should be looking for in a true marker for membrane lipids. In this age of designer genes, jeans, and drugs, surely a faithful marker of membrane lipids for freeze fracture will be forthcoming.

NOTE ADDED IN PROOF

Since submission of this chapter, there have appeared numerous new studies using filipin to localize cholesterol. It seems beyond the scope of this note to belabour the ideas presented herein upon these studies. There are several studies which do not concern filipin labeling directly, but have a direct bearing upon cytochemical probes for lipids; Saxton[61] has considered percolation theory as it applies to lateral diffusion of proteins in the red blood cell membrane with attached spectrin network. He concludes that proteins which are not attached to spectrin may be significantly diffusionally hindered by proteins which are attached to spectrin. Several reports[62-66] present evidence supporting membrane lipid asymmetries being maintained by specific association of the lipids to cytoskeletal (or, presumably, membrane) proteins. If the cytoskeletal associations are extensive, it is possible that rafts of anchored lipid may act as barriers to free diffusion to other lipids, such as cholesterol. If this were the case, the conversion from rotational movement (microviscosity) to a diffusion coefficient may not be valid. I have previously suggested [13] that lipid association with membrane proteins (annular lipid) is insufficient to create diffusional barriers to cholesterol.

ACKNOWLEDGMENTS

I thank Caroline Collins for preparation of the manuscript, Louise Cassar for technical assistance, and Wolfram Tetzlaff for critical reading of the manuscript. I also acknowledge and thank support from MRC and NSERC (Canada).

REFERENCES

1. **Singer, S. J. and Nicolson, G. L.,** The fluid mosaic model of the structure of cell membranes, *Science*, 175, 720, 1972.
2. **Branton, D.,** Fracture faces of frozen membranes, *Proc. Natl. Acad. Sci. U.S.A.*, 55, 1048, 1966.
3. **Pinto da Silva, P. and Branton, D.,** Membrane splitting in freeze-etching, *J. Cell Biol.*, 45, 598, 1970.
4. **Karnovsky, M. J., Kleinfeld, A. M., Hoover, R. L., and Kalusner, R. D.,** The concept of lipid domains in membranes, *J. Cell Biol.*, 94, 1, 1982.
5. **Severs, N. J. and Robenek, H.,** Detection of microdomains in biomembranes. An appraisal of recent developments in freeze-fracture cytochemistry, *Biochim. Biophys. Acta*, 737, 373, 1983.
6. **Kinsky, S. C., Luse, S. A., and van Deenen, L. L. M.,** Interaction of polyene antibiotics with natural and artificial membrane systems, *Fed. Proc. Fed. Am. Soc. Exp. Biol.*, 25, 1503, 1966.
7. **Verkleij, A. J., de Kruijf, B., Gerritsin, W. F., Demel, R. A., van Deenen, L. L. M., and Ververgaert, P. H. J.,** Freeze-etch electron microscopy of erythrocytes, Acholeplasma laidlawii cells and liposomal membranes after the action of filipin and amphotericin B, *Biochim. Biophys. Acta*, 291, 577, 1973.

8. **Elias, P. M., Friend, D. S., and Goerke, J.**, Membrane sterol heterogeneity. Freeze-fracture detection with saponins and filipin, *J. Histochem. Cytochem.*, 27, 1247, 1979.
9. **Behnke, O., Tranum-Jensen, J., and van Deurs, B.**, Filipin as a cholesterol probe. I. Morphology of filipin-cholesterol interaction in lipid model systems, *Eur. J. Cell Biol.*, 35, 189, 1984.
10. **Behnke, O., Tranum-Jensen, J., and van Deurs, B.**, Filipin as a cholesterol probe. II. Filipin-cholesterol interaction in red blood cell membranes, *Eur. J. Cell Biol.*, 35, 200, 1984.
11. **Bolard, J.**, How do the polyene macrolide antibiotics affect the cellular membrane properties?, *Biochim. Biophys. Acta*, 864, 257, 1986.
12. **Robinson, J. M. and Karnovsky, M.**, Evaluation of the polyene antibiotic filipin as a cytochemical probe for membrane cholesterol, *J. Histochem. Cytochem.*, 28, 161, 1980.
13. **Miller, R. G.**, The use and abuse of filipin to localize cholesterol in membranes, *Cell Biol. Int. Rep.*, 8, 519, 1984.
14. **Jost, W.**, in *Diffusion in Solids, Liquids, and Gases*, Academic Press, New York, 1952, 652.
15. **Saxton, M. J.**, Lateral diffusion in an archipelago, *Biophys. J.*, 39, 165, 1982.
16. **Sandborn, E. B., Stephens, H., and Bendayan, M.**, The influence of dimethyl sulfoxide on cellular ultrastructure and cytochemistry, *Ann. N.Y. Acad. Sci.*, 243, 122, 1975.
17. **Friend, D. S. and Bearer, E. L.**, beta-Hydroxysterol distribution as determined by freeze-fracture cytochemistry, *Histochem. J.*, 13, 535, 1981.
18. **Miller, R. G. and Baldridge, W. H.**, The tight junction as a barrier to cholesterol in canine epithelial cells, *J. Ultrastruct. Res.*, 90, 275, 1985.
19. **Allt, G., Blanchard, C. E., MacKenzie, M. L., and Sikri, K.**, Distribution of filipin-sterol complexes in the myelinated nerve fiber, *J. Ulstrastruct. Res.*, 91, 104, 1985.
20. **Berdan, R. C. and Shivers, R. R.**, Filipin-cholesterol complexes in plasma membranes and cell junctions of tenebrio molitor epidermis, *Tissue Cell*, 17, 177, 1985.
21. **Blanchard, C. E., Mackenzie, M. L., Sikri, K., and Allt, G.**, Filipin-sterol complexes at nodes of Ranvier, *J. Neurocytol.*, 14, 1053, 1985.
22. **Blanchard, C. E., Sikri, K., and Allt, G.**, Filipin-sterol complexes at Schmidt-Lanterman incisures, *Acta Neuropathol.*, 72, 355, 1987.
23. **Chailley, B. and Boisvieux-Ulrich, E.**, Detection of plasma membrane cholesterol by filipin during microvillogenesis and ciliogenesis in quail oviduct, *J. Histochem. Cytochem.*, 33, 1, 1985.
24. **Cho, M.-I. and Garant, P. R.**, A freeze-fracture cytochemical study of filipin-sterol complexes in pulp fibroblasts of the rat incisor, *Aarchs Oral Biol.*, 31, 549, 1986.
25. **Clark, M. R., Mel, S., Lupu, F., and Friend, D. S.**, Influence of the membrane undercoat on filipin perturbation of the red blood cell membrane, *Exp. Cell Res.*, 171, 321, 1987.
26. **Fields, R. D., Black, J. A., and Waxman, S. G.**, Filipin-cholesterol binding in CNS axons prior to myelination: evidence for microheterogeneity in premyelinated axolemma, *Brain Res.*, 404, 21, 1987.
27. **Forge, A.**, Cholesterol distribution in cells of the stria vascularis of the mammalian cochlea and some effects of ototoxic diuraetics, *J. Cell Sci.*, 79, 181, 1985.
28. **Forsman, C. A.**, Freeze-fracture cytochemistry of sympathetic ganglia, *Histochemistry*, 82, 209, 1985.
29. **Friend, D. S.**, Plasma-membrane diversity in a highly polarized cell, *J. Cell Biol.*, 93, 243, 1982.
30. **Garcia-Segura, L. M.**, Harmaline-induced changes in plasma membrane of Purkinje cells: a trans-synaptic effect mediated by climbing fibers, *Brain Res.*, 372, 390, 1986.
31. **Ginsbach, C. and Fahimi, H. D.**, Labeling of cholesterol with filipin in cellular membranes of parenchymatous organs, *Histochemistry*, 86, 241, 1987.
32. **Ko, C.-P. and Propst, J. W.**, Absence of sterol-specific complexes at active zones of degenerating and regenerating frog neuromuscular junctions, *J. Neurocytol.*, 15, 231, 1986.
33. **Koehler, J. K., Clark, J. M., and Smith, D.**, Freeze-fracture observations on mammalian oocytes, *Am. J. Anat.*, 174, 317, 1985.
34. **Lupu, F., Calb, M., Scurei, C., and Simionescu, N.**, Changes in the organization of membrane lipids during human platelet activation, *Lab. Invest.*, 54, 136, 1986.
35. **McGookey, D. J., Fagerberg, K., and Anderson, R. G. W.**, Filipin-cholesterol complexes form in uncoated vesicle membrane derived from coated vesicles during receptor-mediated endocytosis of low density lipoprotein, *J. Cell Biol.*, 96, 1273, 1983.
36. **Montesano, R.**, Inhomogeneous distribution of filipin-sterol complexes in smooth muscle cell plasma membrane, *Nature (London)*, 280, 328, 1979.
37. **Pumplin, D. W. and Bloch, R. J.**, Lipid domains of acetylcholine receptor clusters detected with saponin and filipin, *J. Cell Biol.*, 97, 1043, 1983.
38. **Severs, N. J.**, Localization of cholesterol in the Golgi apparatus of cardiac muscle cells, *Experientia*, 37, 1195, 1981.
39. **Severs, N. J. and Simons, H. L.**, Caveolar bands and the effects of sterol-binding agents in vascular smooth muscle plasma membrane, *Lab. Invest.*, 55, 295, 1986.
40. **Simionescu, N., Lupu, F., and Simionescu, M.**, Rings of membrane sterols surround the openings of vesicles and fenestrae in capillary endothelium, *J. Cell Biol.*, 97, 1592, 1983.

41. **Simons, H. L. and Severs, N. J.,** Plasmalemmal vesicles and the effects of sterol-binding agents in rabbit aortic endothelium, *J. Cell Sci.*, 83, 141, 1986.
42. **Steer, C. J., Bisher, M., Blumenthal, R., and Steven, A. C.,** Detection of membrane cholesterol by filipin in isolated rat liver coated vesicles is dependent upon removal of the clathrin coat, *J. Cell Biol.*, 99, 315, 1984.
43. **Toshimori, K., Higashi, R., and Oura, C.,** Distribution of intramembranous particles and filipin-sterol complexes in mouse sperm membranes: polyene antibiotic filipin treatment, *Am. J. Anat.*, 174, 455, 1985.
44. **Henderson, D., Eibl, H., and Weber, K.,** Structure and biochemistry of mouse hepatic gap junctions, *J. Mol. Biol.*, 132, 193, 1979.
45. **Ellisman, M. H.,** Molecular specializations of the axon membrane at nodes of Ranvier are not dependent upon myelination, *J. Neurocytol.*, 8, 719, 1979.
46. **Pinto da Silva, P. and Miller, R. G.,** Membrane particles on fracture faces of frozen myelin, *Proc. Natl. Acad. Sci. U.S.A.*, 72, 4046, 1975.
47. **Nageotte, J.,** Note sur le mécanisme de la formation des réseaux artificiels dans la gaine de myéline, *C. R. Seances Soc. Biol. Paris*, 69, 628, 1910.
48. **Bretscher, M. S. and Whytock, S.** Membrane associated vesicles in fibroblasts, *J. Ultrastruct. Res.*, 61, 215, 1977.
49. **Kachar, B. and Pinto Da Silva, P.,** Rapid massive assembly of tight junction strands, *Science*, 213, 541, 1981.
50. **Kachar, B. and Reese, T. S.,** Evidence for the lipidic nature of tight junction strands, *Nature (London)*, 296, 464, 1982.
51. **Brasitus, T. A. and Schachter, D.,** Lipid dynamics and lipid protein interactions in rat enterocyte basolateral and microvillus membranes, *Biochemistry*, 19, 2763, 1980.
52. **Hansson, G. C., Simons, K., and van Meer, G.,** Two strains of the Madon-Darby Canine kidney (MDCK) cell line have distinct glycosphiongolipid compositions, *EMBO J.*, 5, 483, 1986.
53. **van Meer, G. and Simons, K.,** The function of tight junctions in maintaining differences in lipid composition between the apical and the basolateral cell surface domains of MDCK cells, *EMBO J.*, 5, 1455, 1986.
54. **Pelletier, R.-M. and Friend, D. S.,** Sertoli cell junctional complexes in gossypol-treated neonatal and adult guinea pigs, *J. Androl.*, 7, 127, 1986.
55. **Lange, Y., Cohen, C. M., and Poznansky, M. J.,** Transmembrane movement of cholesterol in human erythrocytes, *Proc. Natl. Acad. Sci. U.S.A.*, 74, 1538, 1977.
56. **Orci, L., Montesano, R., Meda, P., Malaisse-Lagae, F., and Brown, D.,** Heterogeneous distribution of filipin-cholesterol complexes across the cisterna of the Golgi apparatus, *Proc. Natl. Acad. Sci. U.S.A.*, 78, 293, 1981.
57. **Orci, L., Miller, R. G., Montesano, R., Perrelet, A., Amherdt, M., and Vassalli, P.,** Opposite polarity of filipin-induced deformations in the membrane of condensing vacuoles and zymogen granules, *Science*, 210, 1019, 1980.
58. **Miller, R. G.,** Asymmetry of surface tension in membranes, Act of the 7th Congress of the Italian Biophysical Society, September 1986.
59. **Miller, R. G.,** Interactions between digitonin and bilayer membranes, *Biochim. Biophys. Acta*, 774, 1151, 1984.
60. **Edidin, M.,** Rotational and translational diffusion in membranes, *Annu. Rev. Biophys. Bioeng.*, 3, 179, 1974.
61. **Saxton, M. J.,** The spectrin network as a barrier to lateral diffusion in erythrocytes. A percolation analysis, *Biophys. J.*, 55, 21, 1989.
62. **Bevers, E. M., Verhallen, P. F., Linskens, W. M., Comfurius, P., and Zwaal, R. F.,** Loss of phospholipid asymmetry in dilauroylphosphatidylcholine induced plasma membrane vesicles from human platelets, *Biochim. Biophys. Acta*, 903, 197, 1987.
63. **Del-Buono, B. J., Williamson, P. L., and Schlegel, R. A.,** Relation between the organization of spectrin and of membrane lipids in lymphocytes, *J. Cell. Biol.*, 106, 697, 1988.
64. **Middelkoop, E., Lubin, B. H., Bevers, E. M., Op-den-Kamp, J. A., Comfurius, P., Chiu, D. T., Zwaal, R. F., van-Deenen, L. L., and Roelofsen, B.,** Studies on sickled erythrocytes provide evidence that the asymmetric distribution of phosphatidylserine in the red cell membrane is maintained by both ATP-dependent translocation and interaction with membrane skeletal proteins, *Biochim. Biophys. Acta*, 937, 281, 1988.
65. **Verhallen, P. F., Bevers, E. M., Comfurius, P., and Zwaal, R. F.,** Correlation between calpain-mediated cytoskeletal degradation and expression of platelet procoagulant activity. A role for the platelet membrane-skeleton in the regulation of membrane lipid asymmetry?, *Biochim. Biophys. Acta*, 903, 206, 1987.
66. **Williamson, P., Antia, R., and Schlegel, R. A.,** Maintenance of membrane phospholipid asymmetry. Lipid-cytoskeletal interactions or lipid pump?, *FEBS Lett.*, 219, 316, 1987.

Chapter 6

POLYMORPHIC PHASE BEHAVIOUR OF MEMBRANE LIPIDS

Arindam Sen

TABLE OF CONTENTS

I. Introduction 104

II. Single Lipid Systems 104
A. Bilayer Phases 104
B. Nonbilayer Phases 104

III. Mixed Lipid Systems 109
A. Bilayer Defect Structures 109
B. Lipidic Particles 109
C. Effect of Environmental Factors on Lipid Polymorphism 109

IV. Total Membrane Lipids 115
A. Chloroplast Total Lipid Extract 118
B. Retinal Disk Lipid Extract 118

V. Destabilization of Membrane Bilayers 119
A. Effect of Membrane Proteins 119
B. Destabilization by Environmental Factors 120

VI. Conclusion 121

Acknowledgments 121

References 127

I. INTRODUCTION

Lipid polymorphism is one area of membrane research which depends most on freeze-fracture electron microscopy.[1-10] It would not be wrong to say that many lipid structures would not have been known today if the technique of freeze-fracture electron microscopy was not available. While techniques like X-ray diffraction[11] and nuclear magnetic resonance (NMR)[12] have been extremely useful in determining lipid phase behavior, they are not very useful in examining the lipid system at microscopic resolution. Situations where the majority of lipids are in one phase and a small proportion of the lipids is in a different phase, in a separated domain, are especially well suited for freeze-fracture electron microscopy. Some intermediate phases in lamellar to hexagonal transition, like lipidic particles,[2,3,5] are not detectable by X-ray diffraction, and NMR data can at best be ambiguous. Many such phases have been detected by the use of freeze-fracture electron microscopy. In this chapter, I will discuss the various lipid phases as revealed by freeze-fracture electron microscopy.

II. SINGLE LIPID SYSTEMS

All biological lipids can be grouped in one of the two main structural forms: (1) those which form bilayers and (2) those which do not. X-ray diffraction and NMR spectroscopy can be used effectively to distinguish between these two groups,[11,12] however, neither technique can distinguish some structural differences within each group. Let us first consider the bilayer lipids.

A. Bilayer Phases

Figures 1 and 2 are freeze-fracture electron micrographs of aqueous dispersions of 1, 2 distearoylmonogalactosylglycerol (MGDG) and 1, 2 distearoyldigalactosylglycerol (DGDG), respectively. Both MGDG and DGDG form lamellar structures, but there is a difference in the structures formed by the two lipids, a difference which cannot be detected by X-ray diffraction. Bilayers of saturated MGDG form stacks of open bilayers, while bilayers of DGDG form the more common closed vesicles. X-ray diffraction patterns obtained from both lipids are typical of lamellar structures.[13] The electron micrograph of MGDG (Figure 1) shows a large number of cross-fractures and few fracture faces; this is indicative of an interdigitated bilayer.[14] Recent X-ray diffraction studies on a homologous series of 1,2 monoglucosyldiacylglycerol having acyl chains of 12 to 20 carbons also indicate that these lipids form interdigitated bilayers.[15] Figure 3 is an electron micrograph of a freeze-fracture replica of 1,2 dilinolenoyldigalactosylglycerol. Again, the unsaturated DGDG like its saturated counterpart forms bilayer structures. A closer examination of Figures 2 and 3 reveal that the fracture faces of the two lipids are different, those of Figure 3 are smooth and featureless, while those of Figure 2 have some fine structure. This difference in the appearance of the fracture faces can be explained from their X-ray diffraction data. The unsaturated DGDG (Figure 3) is in liquid crystalline phase while the saturated lipid is in a crystalline gel phase.[13] The texture of the fracture faces of bilayers thus reflect the crystallinity of the lipid.

B. Nonbilayer Phases

The common nonbilayer phases formed by lipids in aqueous dispersions are (1) micelles, (2) inverted micelles, (3) inverted hexagonal or hexagonal II (Hex II), and (4) cubic.[11] Micelles are generally too small to be seen by freeze-fracture electron microscopic techniques. Inverted micelles are larger due to the water pocket enclosed by the lipid and can be seen by electron microscopy (see Section III). The inverted hexagonal phase is the most common nonbilayer phase found in biological lipids. Figure 4 is an electron micrograph of a freeze-

FIGURE 1. Freeze-fracture electron micrograph of an aqueous dispersion of 1,2-distearoylmonogalactosylglycerol showing stacked bilayers of the lipid.

fracture replica of 1,2 dilinlenoylmonogalactosylglycerol and is typical of a Hex II phase. The long tube-like structures are clearly evident. MGDG-like phosphatidylethanolamine favor the Hex II phase.[13] The Hex II phase for both lipids appear at lower temperatures with an increase in the unsaturation of the acyl chains and can be easily detected by X-ray diffraction and by ^{31}P NMR, in the case of the phospholipids. Transition from lamellar to Hex II can also be detected by differential scanning calorimetry though the transition enthalpy is small as compared to that of gel to liquid crystalline transition.[6]

The third nonbilayer phase, i.e., inverted cubic, is generally found in mixed lipid system

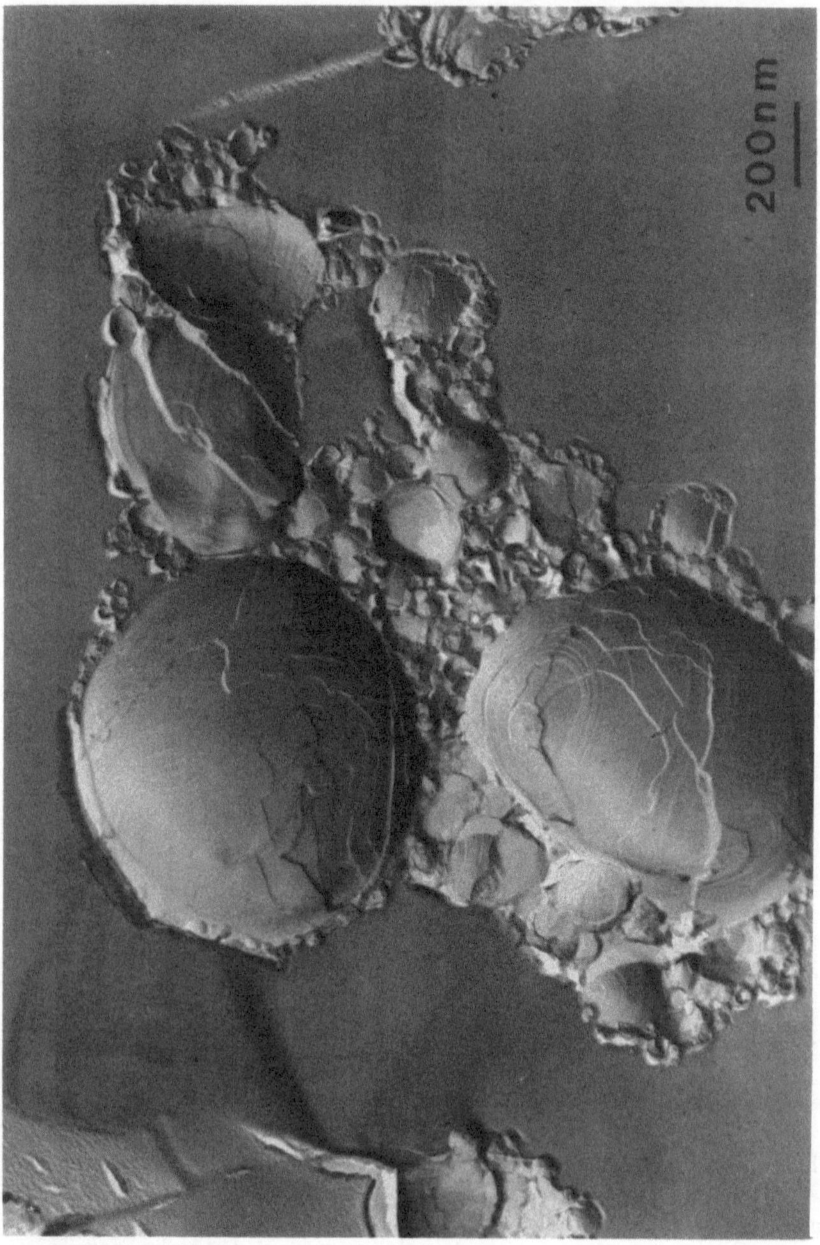

FIGURE 2. Electron micrograph of a freeze-fracture replica of 1,2-distearoyldigalactosylglycerol.

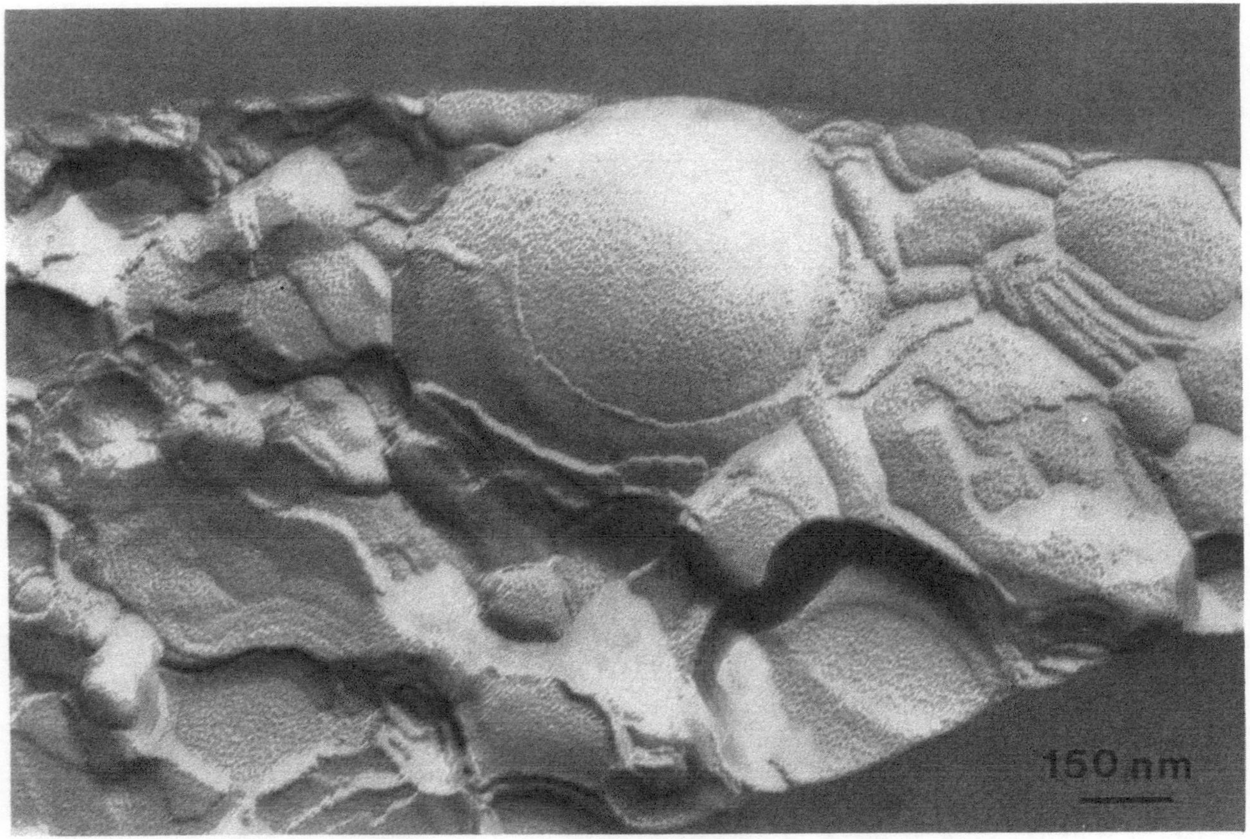

FIGURE 3. Electron micrograph of a freeze-fracture replica of 1,2-dilinolenoyldigalactosylglycerol. The fracture faces are smooth and featureless.

FIGURE 4. Inverted hexagonal (Hex II) structures are seen in this freeze-fracture micrograph of a 1,2-dilinolenoylmonogalactosyldiacylglycerol.

(see Section III) and in some soaps and monoglycerides.[16] In a recent report workers found an inverted cubic phase in dioleoylphosphatidylelthanolamine after repeated cycling between Hex II and lamellar L_α-phase.[17] It appears that during cooling cycle from Hex II to L_α a fraction of the lipid gets trapped in the cubic phase and this trapped cubic phase grows through repeated heating and cooling cycles. The workers found that after about 150 such heating-cooling cycles the majority of the lipid is in cubic phase as determined by X-ray diffraction. Cubic phase of a single pure lipid has not been reported so far using freeze-fracture electron microscopy.

III. MIXED LIPID SYSTEMS

Mixed lipid systems consisting of bilayer and nonbilayer lipids have drawn much attention during the past decade.[18] It was found that when two lipids, one of which is bilayer forming and the other Hex II forming, are mixed, the mixed system have many different structures as revealed by freeze-fracture electron microscopy. The structures range from bilayer to Hex II with intermediate structures like lipidic particles, cylindrical inverted micelles, inverted micelles, and cubic phases. The structures found in any mixed lipid system were found to depend on the lipid composition and environmental factors like temperature, pH, cations, and the presence of cryoprotectants.[1-10] In mixed system of unsaturated MGDG and DGDG, pH and cation do not play a role due to the absence of any ionizable groups in those lipids, but temperature and cryprotectants have very noticeable effect.[10]

A. Bilayer Defect Structures

Lipid bilayers often have defect structures present. The most common of these are confocal domains and point defects, i.e., screw dislocations. These defects are generally observed when a lipid-like DPPC is dispersed in small amounts of water,[19] or after the addition of dehydrating agents like polyethylene glycol.[20] Similar defect structures are also found in mixed lipid systems like MGDG:DGDG,[9] and total lipid extract of calf lung surfactants.[21] Figure 5 shows an electron micrograph of a freeze-fracture replica prepared from a mixed MGDG:DGDG (2:1) system dispersed in excess water. The waves are regions adjacent to confocal domains. Such structures have also been reported in PC:PI mixed systems.[19] The screw dislocation generally found in bilayers under dehydrating conditions can however be found in some fully hydrated systems also, like aqueous dispersions of a calf lung surfactant total lipid extract as shown in Figure 6. In the case of calf lung surfactant extract, the origin of the screw dislocation is not known, but a small hydrophobic peptide present in the extract is thought to cause bilayer destabilization leading to defect formation.[21]

B. Lipidic Particles

A ring of lipidic particles are seen an electron micrograph of a freeze-fracture replica of a 2:1 mixture of MGDG and DGDG dispersed in water (Figure 7). These lipidic particles are 90 to 120 Å in diameter and are similar to the intermembrane attachment sites found in mixtures of soy PE and egg PC (Figure 8).[5] Another type of lipidic particles found in mixed MGDG/DGDG systems are inverted micelles sandwiched between lipid bilayers (Figure 9). The morphology of these particles are different from those of the "attachment sites" shown in Figures 7 and 8. These are more spherical, uniform in size, and generally occur in clusters. These inverted micelles are probably formed through a "pinching off" process from tubular or cylindrical micelles as indicated from the appearance of the particles in Figure 10.[4]

C. Effect of Environmental Factors on Lipid Polymorphism

The addition of a cryoprotectant like ethylene glycol causes a transformation in the structures found in the freeze-fracture replicas of mixed MGDG and DGDG (2:1).[10] Instead

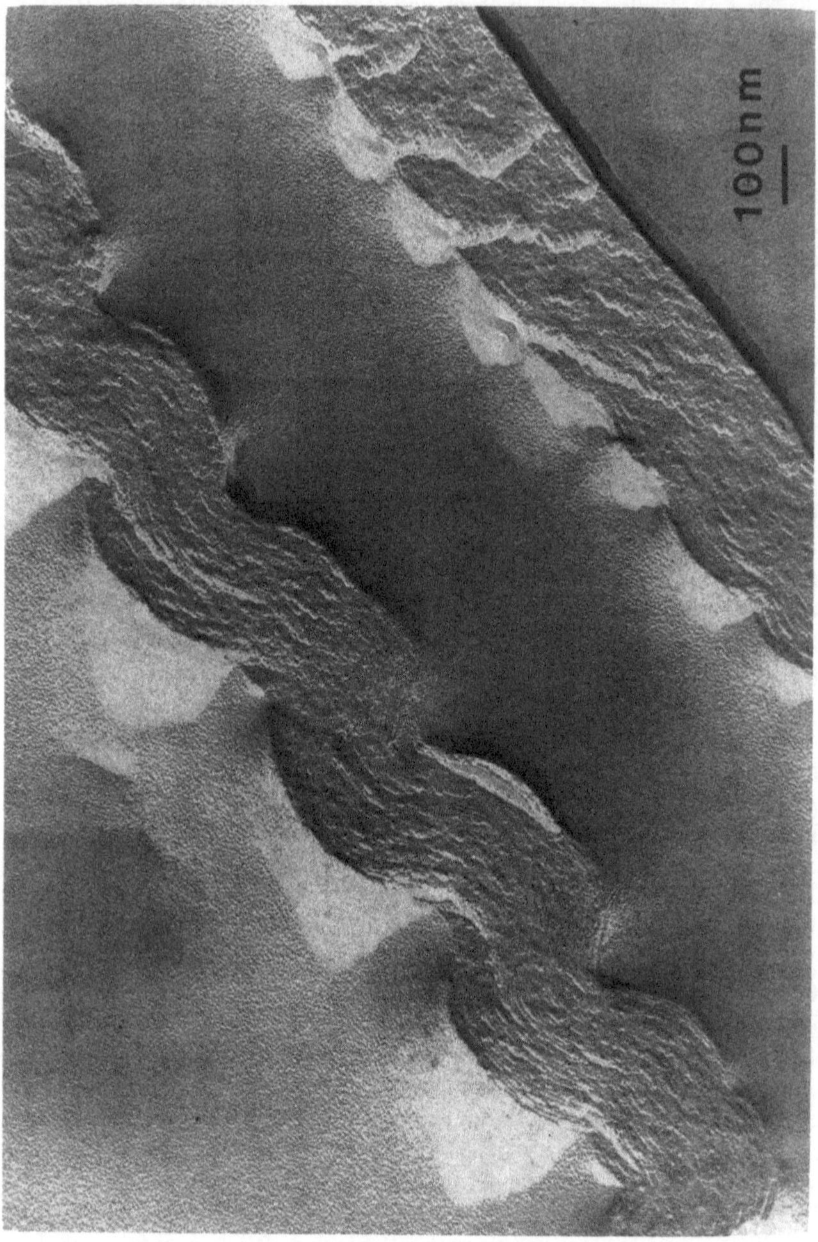

FIGURE 5. A freeze-fracture electron micrograph of a 2:1 mixed MGDG:DGDG dispersion. The wave-like structures are bilayers between confocal domains commonly observed in these dispersions.

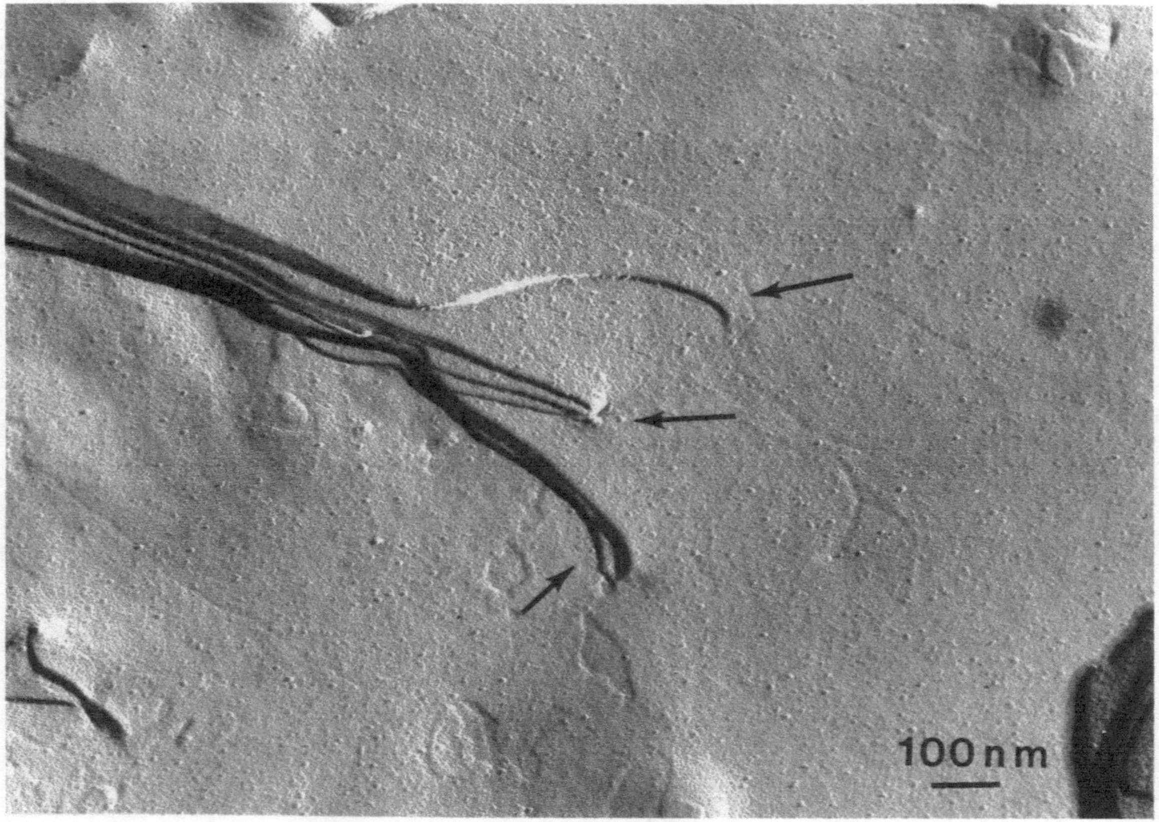

FIGURE 6. Screw dislocations (arrows) are seen in this freeze-fracture electron micrograph of a dispersion of a total calf lung surfactant extract.

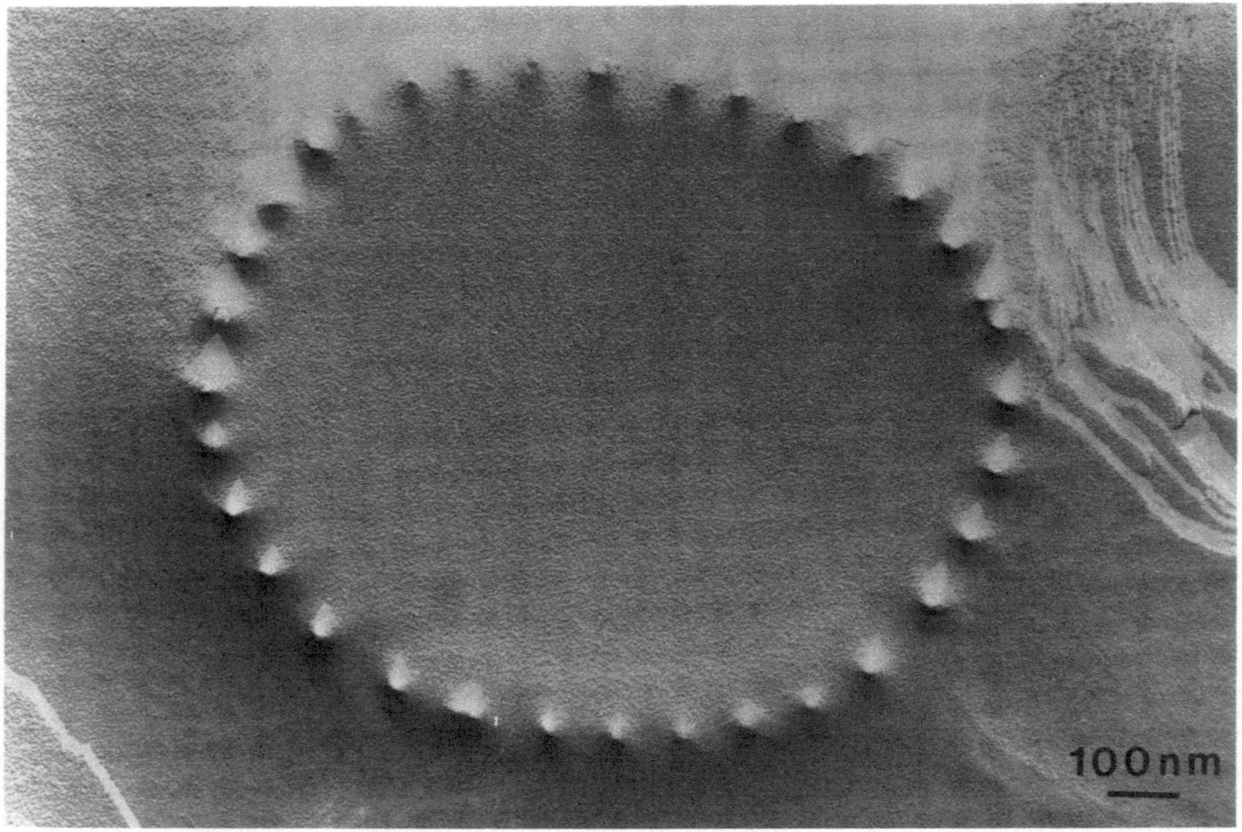

FIGURE 7. Electron micrograph of a MGDG:DGDG (2:1 molar ratio) mixed dispersion showing a ring of lipidic particles. These particles are similar to those found in PE:PC systems and classified as intermembrane attachment sites.

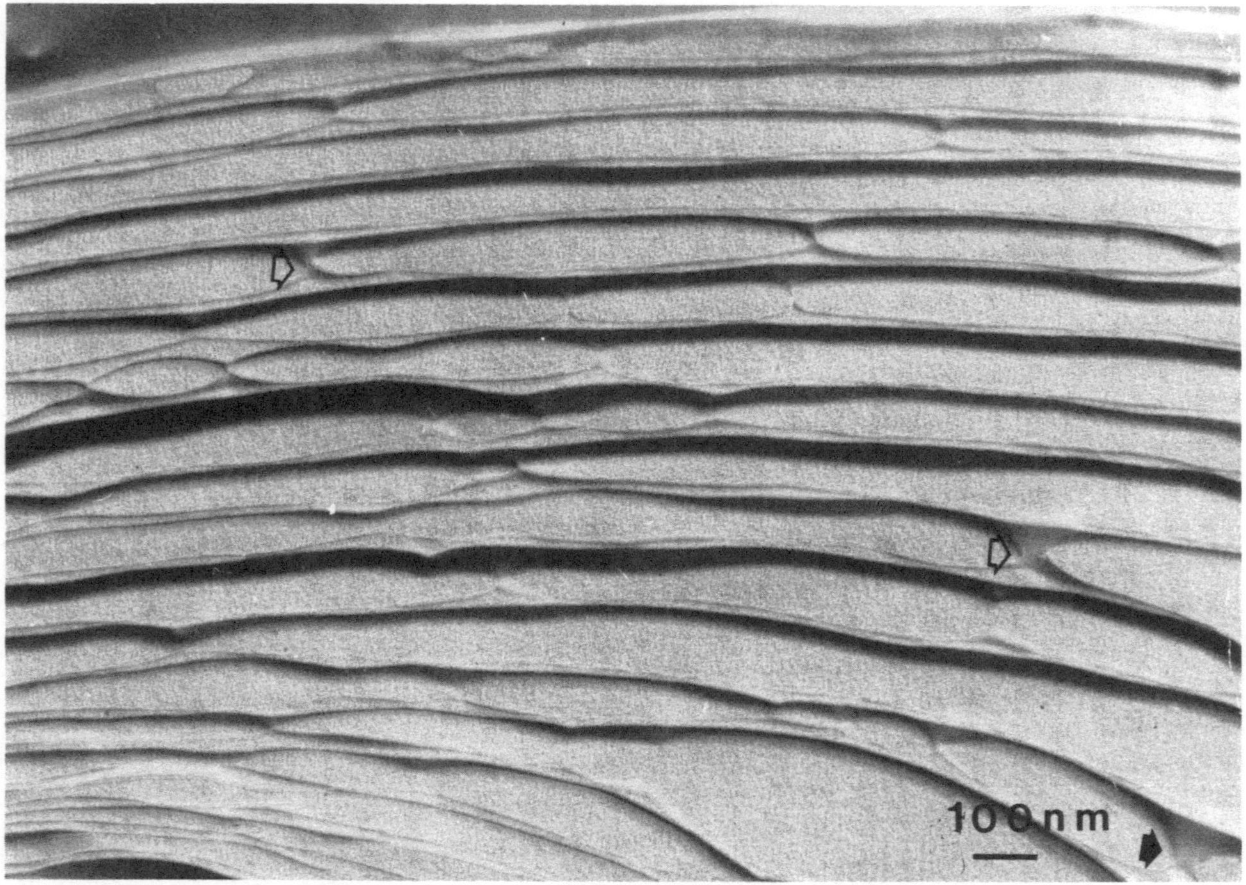

FIGURE 8. A freeze-fracture electron micrograph of a PE:PC dispersion showing the structure of lipidic particles as seen in a cross-fracture. The intermembrane attachments sites are clearly visible (arrows). (From Hui, S. W., Stewart, T. P., Boni, L. T., and Yeagle, P. L., *Science*, 212, 921, 1981. With permission).

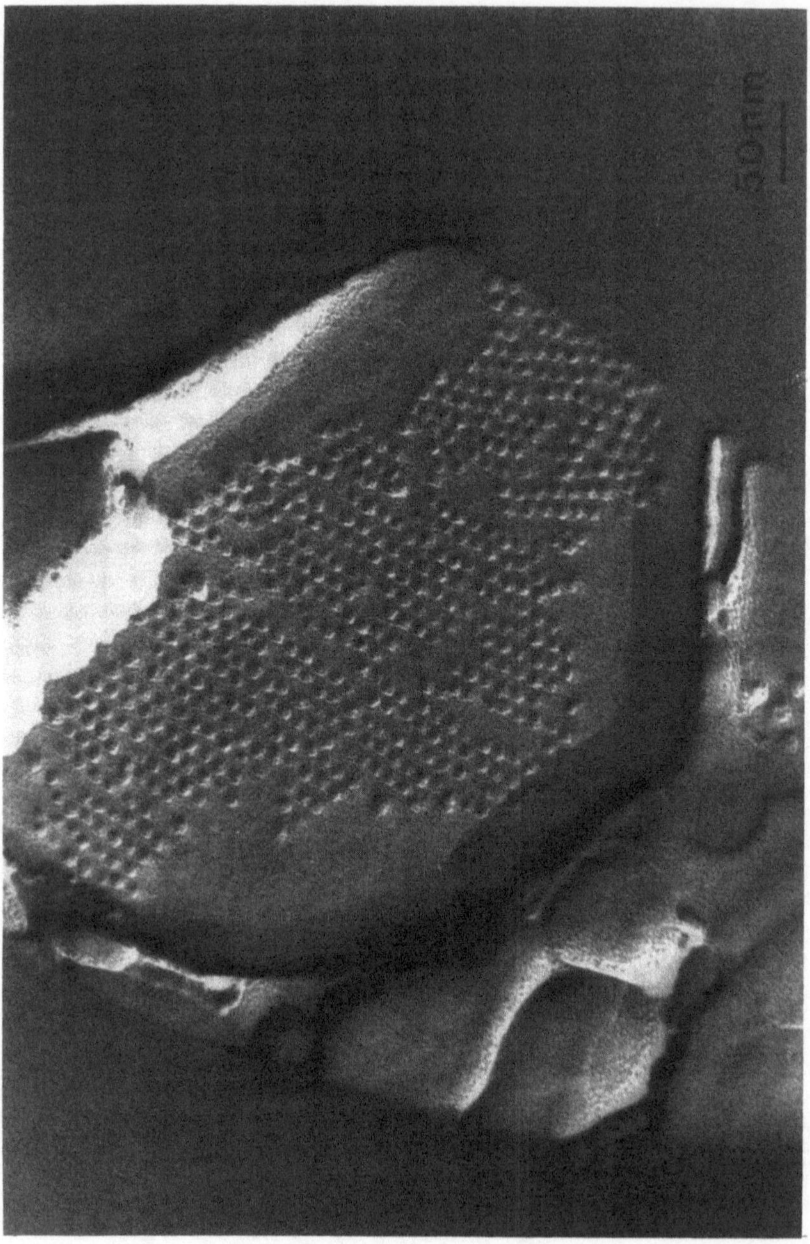

FIGURE 9. Inverted lipid micelles are seen in this electron micrograph of a freeze-fracture replica of a 2:1 mixture of MGDG and DGDG.

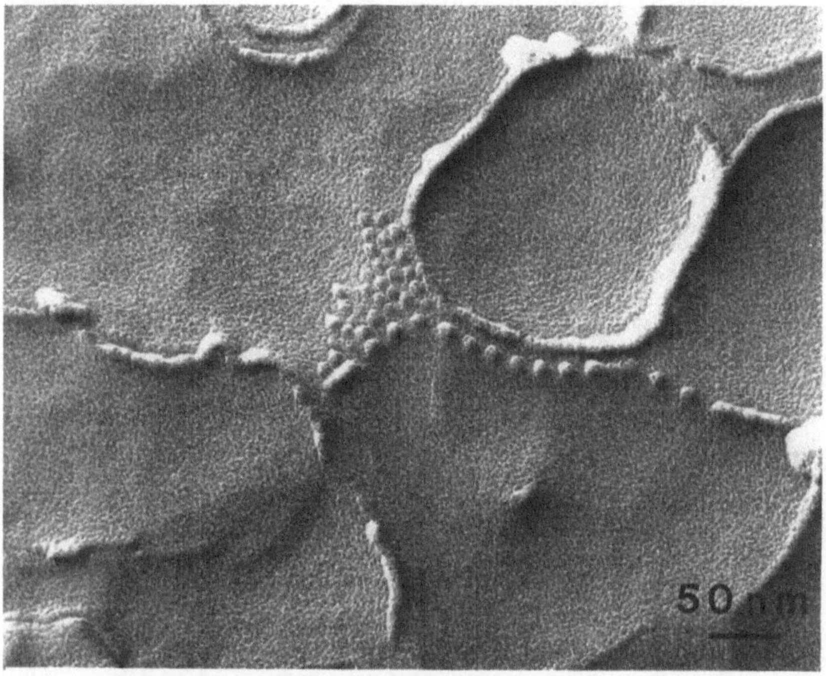

FIGURE 10. Electron micrograph showing the pinching off of inverted lipid micelles from cylindrical inverted micelles in MGDG:DGDG mixed dispersion.

of predominantly bilayer structures as seen in MGDG:DGDG dispersed in water, crystalline aggregates of spheres and vesicles with inverted micelles are found (Figure 11). The crystalline aggregates are very similar to those found in PE:PC systems and which have been characterized as cubic phase from X-ray diffraction studies.[8]

When the mixed MGDG:DGDG dispersions are fast frozen from 50°C, two different crystalline forms are found in freeze-fracture replicas. Figure 12 shows a herringbone-like structure where alternate layers appear to be aligned in different direction. Other regions of the replica do not show this herringbone pattern but instead show a structure very similar to the cubic phase (Figure 13). Sometimes coexisting bilayer and cubic phase can be found in these replicas of dispersions frozen from 50°C. Figure 14 shows one such area where bilayers are seen to merge into a region of cubic phase lipid. An array of inverted micelles can be seen at the edge of one such bilayer surface. These replicas of MGDG:DGDG mixed systems often show regions of Hex II phase lipids coexisting with both bilayer and cubic phase lipid (Figure 15).

Phase separation and coexistence of different phases is observed in other mixed lipid system too. Figure 16 shows an area of a freeze-fracture replica prepared from a soy PE egg PC (80:20 molar ratio) which had been quick frozen from 50°C. The area shows planar bilayers with a phase separated region of Hex II lipid. The circular morphology of the Hex II lipid suggests the formation of that phase from a multilamellar vesicle.

IV. TOTAL MEMBRANE LIPIDS

The interest in studying mixed lipid system containing Hex II lipids is due to a recognition that most biological membranes, especially the highly active membranes like those of mitochondria, chloroplasts, and retinal rod outer segment disks contain very large proportions of Hex II forming lipids. The proportions of these types of lipids can be as high as 50% as

FIGURE 11. Electron micrograph of a 2:1 mixture of MGDG and DGDG dispersed in 50% ethylene glycol showing the lipids in a cubic phase (C) and inverted lipid micelles at the fracture face of a vesicle (V).

FIGURE 12. Freeze-fracture micrograph 2:1 mixture of MGDG and DGDG dispersed in 50% ethylene glycol showing a herringbone-like structure. The sample was quick frozen from 50°C.

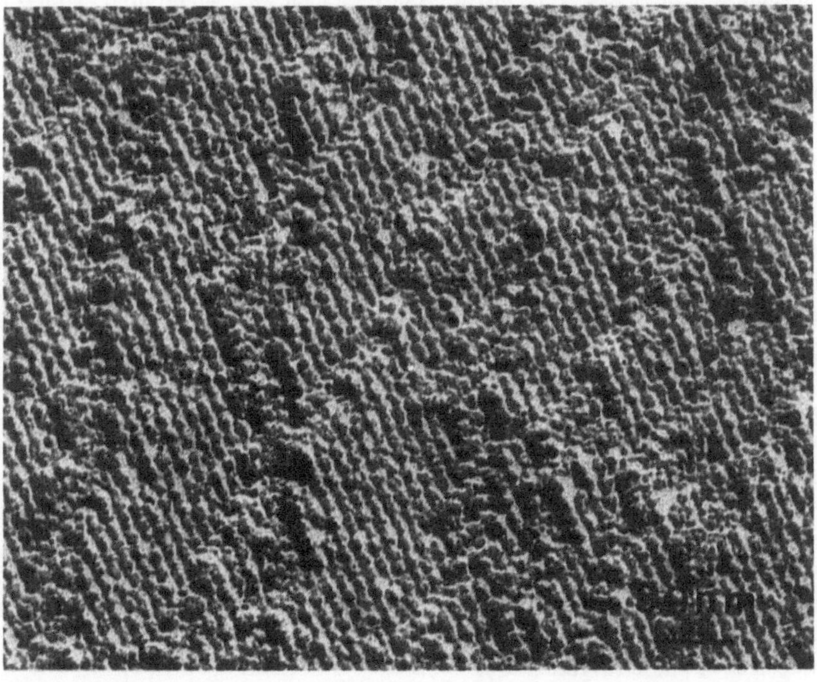

FIGURE 13. Another region of the replica, as in Figure 12, shows cubic-like packing.

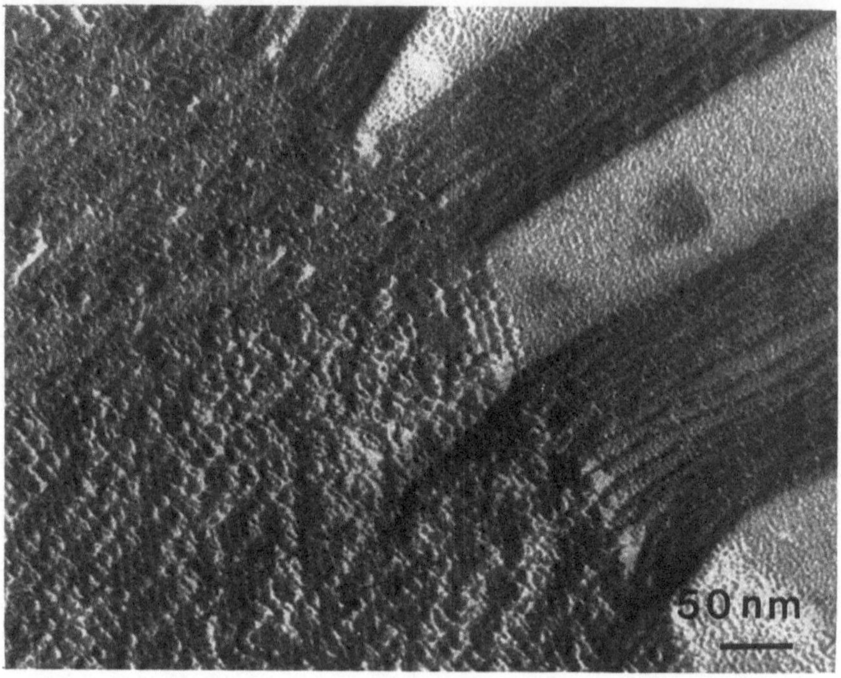

FIGURE 14. Electron micrograph showing the transformation from bilayers to cubic phase in this replica of a MGDG:DGDG (2:1) dispersion in 50% ethylene glycol quenched from 50°C.

in the chloroplast membranes.[22] The obvious question then was, would lipid extracts of such membranes form bilayers or would they form nonbilayer structures?

A. Chloroplast Total Lipid Extract

Total chloroplast lipid, when dispersed in water, form small unilamellar vesicles, SUV, (Figure 17a) which aggregate and fuse together in the presence of divalent cations like Mg^{2+} or Ca^{2+}. The fusion and aggregation process starts at as low as 1 m*M* Mg^{2+} and by 7.5 m*M* large fusion products are seen in the replicas (Figure 17b).[23] These large lipid bodies have different types of lipidic particles and some Hex II lipid as well. Increasing the concentration of Mg^{2+} to 15 and 30 m*M* results in the formation of even larger lipid aggregates. In these aggregates Hex II, tightly packed inverted lipid micelles, and cubic phase lipids can be found (Figure 17c).[23]

B. Retinal Disk Lipid Extract

A similar cation effect is also found in the total lipid extracts of retinal rod outer segment disk membranes.[24] Figure 18A is an electron micrograph of a total disk lipid extract in the absence of added Ca^{2+}. Only unilamellar vesicles are found in the replica. After the addition of Ca^{2+} to a final concentration of 4 m*M* the appearance of the dispersion, as revealed by freeze fracture, is completely altered. Now large lipid bodies are found and they are no longer hollow as in the previous case but are filled with intermediate Hex II like stuctures (Figure 18B). Increasing the concentration of Ca^{2+} further to 8 m*M* results in the complete transition to Hex II phase (Figure 18C).

Total lipid extracts of both chloroplasts and retinal disks are thus found to behave in similar fashion and to exhibit cation dependent polymorphic behaviour. In the absence of divalent cations both lipid system form unilamellar vesicle. This is most probably due to the presence of small (15%) proportions of acidic lipid which are present in these total lipid

FIGURE 15. An electron micrograph showing phase-separated domains of Hex II phase (H) together with region of cubic phase (C) in a replica of a 2:1 mixture of MGDG:DGDG in ethylene glycol (50%) frozen from 50°C.

extracts. The acidic lipids present are phosphatidylglycerol and sulphoquinovosyldiacylglycerol in the case of chloroplasts and phosphatidylserine in the case of retinal disks. It is thus expected that any teriary lipid mixture having acidic lipids and Hex II lipids along with some bilayer lipids will exhibit cation dependent polymorphic behavior.

Freeze-fracture replicas of intact membranes of chloroplast and retinal rod outer segment disks however do not show the presence of any nonbilayer lipid phases. Since cations are generally present in both retinal disks and in chloroplasts it would be safe to suggest that the presence of the membrane proteins must lead to the stabilization of the bilayer structures in these membranes. Is it possible to destabilize these protein stabilized bilayers and create nonbilayer lipid phases in these membrane systems?

V. DESTABILIZATION OF MEMBRANE BILAYERS

A. Effect of Membrane Proteins

One of the first reports of destabilization of membrane lipid bilayer was on chloroplast thylakaloid membranes.[25] These workers treated thylakoid membranes with guanidine thiocyanate to remove some of the membrane peptide and found that after the removal of the peptides, large areas of phase separated lipids were present in freeze-fracture replicas. The phase separated lipids were of lamellar and tubular structures. The tubular structures reported

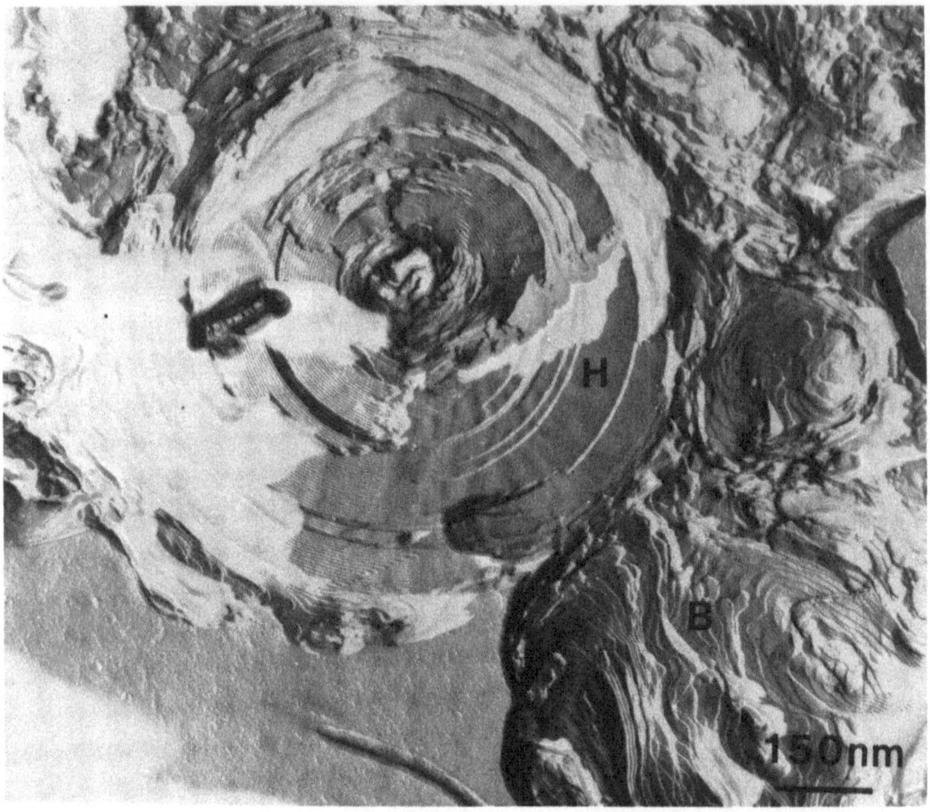

FIGURE 16. Phase separation of Hex II phase (H) from bilayers (B) is found in this replica of a PE:PC (80:20) dispersion frozen from 50°C.

by these workers were indeed Hex II phase lipids. It would thus appear that membrane proteins, at least those of chloroplasts, do stabilize lipid bilayer and on their removal the mixed lipids form bilayers and nonbilayer structures as is predicted from studies on mixed lipids.

B. Destabilization by Environmental Factors

Envrionmental factors like cations, low pH, dehydration, and high temperatures, which are known to help promote bilayer to Hex II phase transition, are also likely to promote destabilization of biological membranes. The structure of chloroplasts after the addition of Mg^{2+} to high concentrations (~30 m*M*) has been examined by freeze-fracture electron microscopy. It was reported that exposure to high cation concentration resulted in breakdown of the thylakoid structure and phase-separated domains of Hex II lipids appeared.[26] The Hex II structures reported by these authors closely resembled those published by Machold et al.[25]

Williams and co-workers[27-29] have shown that the exposure of normal chloroplast thylakoids (Figure 19A) to high temperature resulted in the breakdown of the thylakoids and the appearance of some phase-separated lipid domains (Figure 19B).[27] The phase-separated lipid domains identified by the authors as cylindrical inverted micelles (CIM) bear a striking resemblence to Hex II phase seen in other lipid system. A similar lipid phase separation also occured if chloroplasts were subjected to low pH.[28]

Dehydration, another environmental factor which is favorable for Hex II formation, was found to cause destabilization of retinal rod outer segment disk membranes resulting in a phase separation of Hex II lipids.[30]

These studies on phase separation of membrane lipids leads to an interesting conclusion that biological membranes are not very stable structures and even small perturbations may cause a complete disintegration of the membrane system. Since the more active membranes seem to be more susceptible it is tempting to suggest that this inherent instability of the lipid bilayers in these membranes may be important for their activity.

VI. CONCLUSION

Figure 20 is an electron micrograph of a freeze-fracture replica of a solution of PE in hexane. The fiber-like structures seen in the replica are the extended hydrogen-bonded phase formed by PE when dissolved in a completely nonpolar solvent like a hydrocarbon.[31] This is just to illustrate a possible use of the freeze-fracture technique. Freeze-fracture electron microscopy which has been so valuable to our understanding of lipid phase behavior is by no means limited to aqueous system and it is possible to obtain information about lipids in nonaqueous systems using this technique.

ACKNOWLEDGMENTS

I wish to thank Dr. Steve Hui for his suggestions and encouragement, and also for the many helpful discussions. This work was supported by NIH Grants GM-28120 and GM-30969.

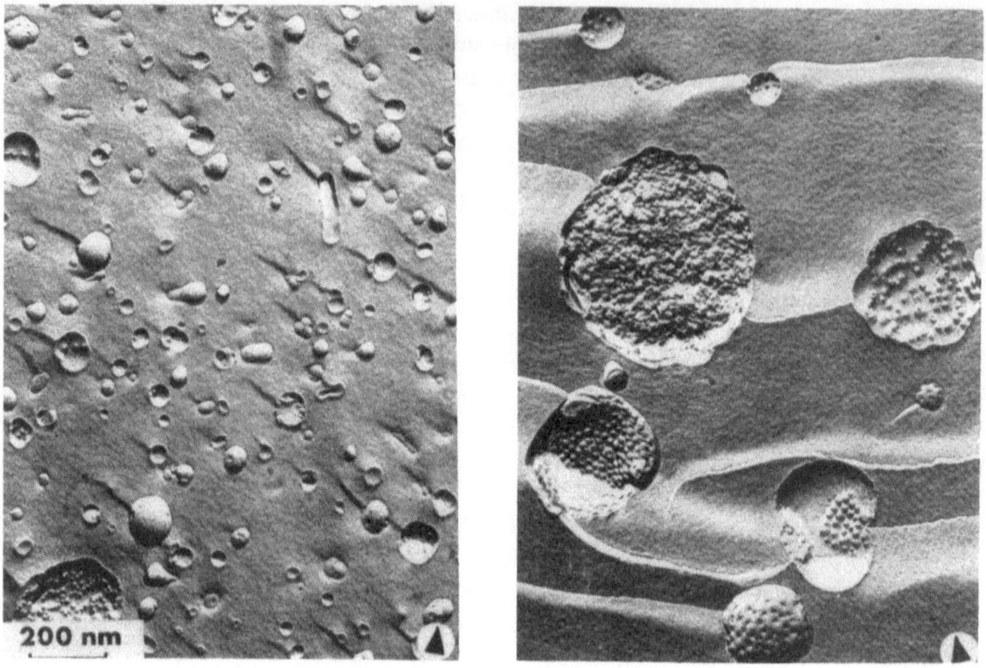

A B

FIGURE 17. (A) Electron micrograph of total chloroplast lipid extract with 1 m*M* Mg^{2+} showing unilamellar vesicles. (B) After addition of Mg^{2+} to 7.5 m*M* to the dispersion large aggregates of lipids are found in the replicas. The lipid aggregates have lipidic particles and some cubic phase lipids. (C) The addition of Mg^{2+} to 30 m*M* to the dispersion results in the formation of very large aggregates. Hex II and cubic phase now fills the lipid aggregates. Arrows indicate shadowing direction. (From Gounaris, K., Sen, A., Brain, A. P. R., Quinn, P. J., and Williams, W. P., *Biochim. Biophys. Acta*, 728, 129, 1983. With permission.)

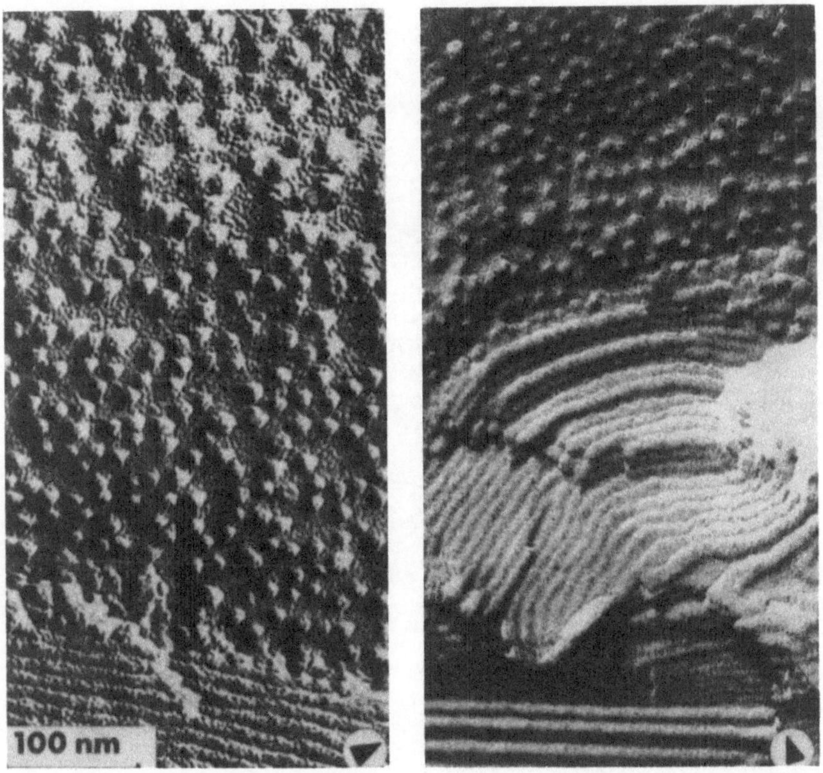

FIGURE 17C.

A

FIGURE 18. (A) Freeze-fracture electron micrograph of a freeze-fracture replica of a total lipid extract of bovine retinal rod disk membranes in the absence of added Ca^{2+}. Only unilamellar vesicles are seen in the replicas. (B) After Ca^{2+} (4 m*M* is added to the disk lipid dispersion, the lipid vesicles are not longer found instead lipid aggregates showing modified bilayer structures are seen. (C) Hex II phase lipid are seen in this replica of total disk lipids after the addition of Ca^{2+} to 8 m*M*. (From Albert, A. D., Sen, A., and Yeagle, P. L., *Biochem. Biophys. Acta*, 771, 28, 1984. With permission.)

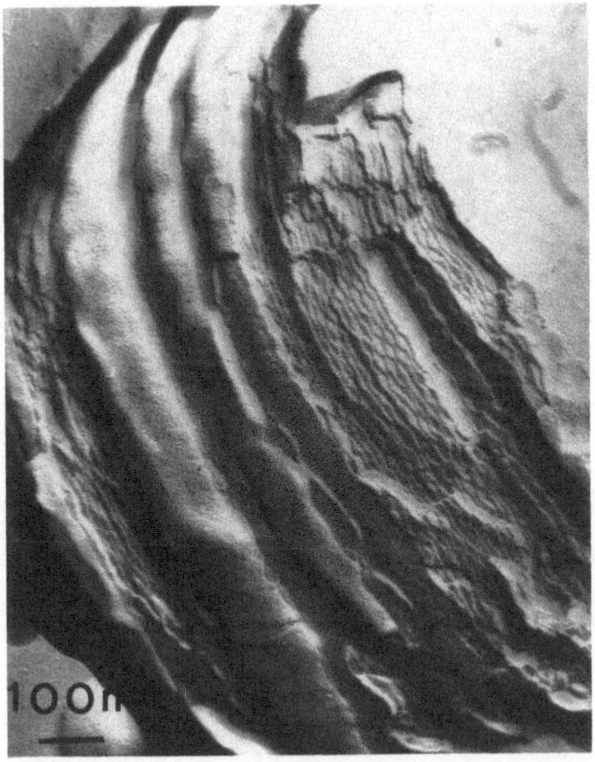

FIGURE 18B.

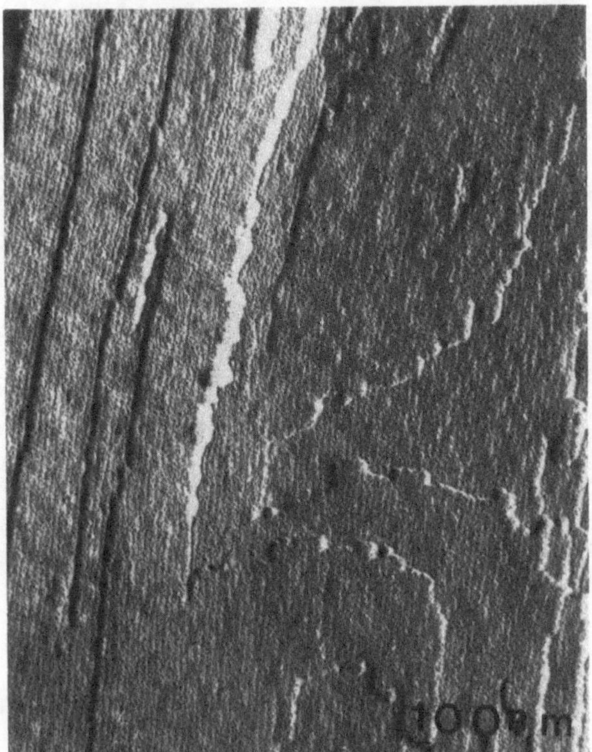

FIGURE 18C.

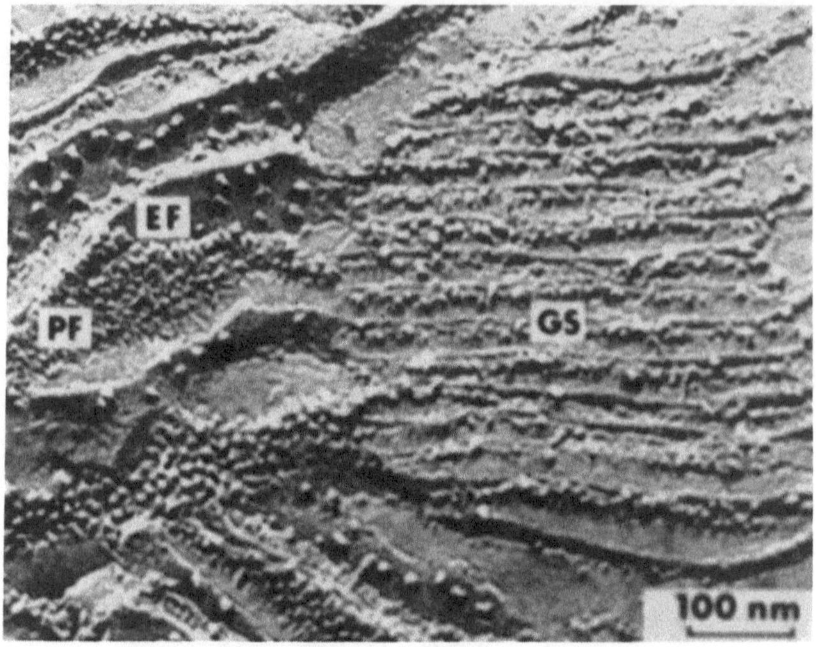

A

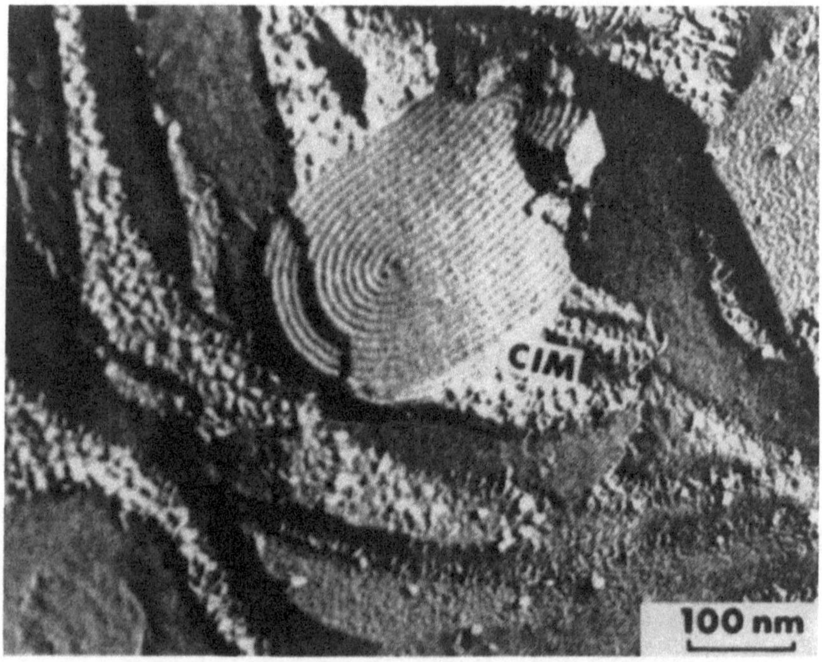

B

FIGURE 19. (A) Electron micrograph of a freeze-fracture replica of pea chloroplast frozen from 20°C. Stacked grana (GS) are found as in normal chloroplasts. (B) Heating the pea chloroplasts to 50°C results in the phase separations of lipids seen here as cylindrical micelles (CIM). The stacked grana regions are no longer found in the replicas. (From Gounaris, K., Brain, A. P. R., Quinn, P. J., and Williams, W. P., *FEBS Lett.*, 153, 47, 1983. With permission.)

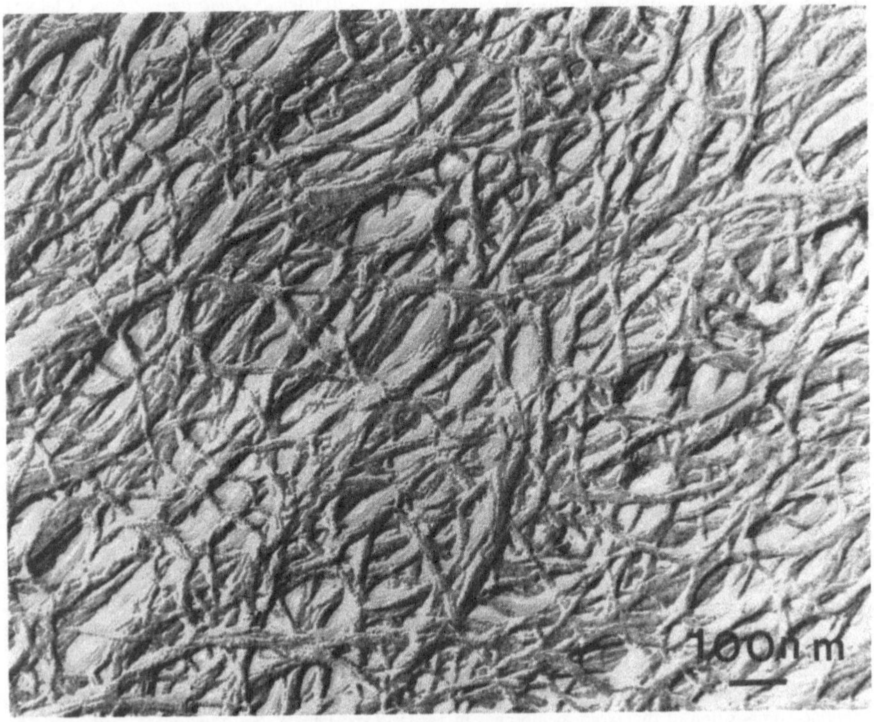

FIGURE 20. An electron micrograph of a freeze-fracture replica of PE dissolved in hexane at a concentration of 10 mg/ml. The long intertwined fiber-like structures are of PE in an extended hydrogen bonded structure.

REFERENCES

1. **Vail, W. J. and Stollery, J. G.,** Phase change in cardiolipin vesicles mediated by divalent cations, *Biochim. Biophys. Acta,* 551, 74, 1979.
2. **Verkleij, A. J., Mombers, C., Leunissen-Bijvelt, J., and Ververgaert, P. H. J. T.,** Lipidic intramembranous particles, *Nature (London),* 279, 162, 1979.
3. **Miller, R. G.,** Do "lipidic particles" represent intramembranous attachment sites?, *Nature (London),* 287, 166, 1980.
4. **Sen, A., Williams, W. P., Brain, A. P. R., Dickens, M. J., and Quinn, P. J.,** Formation of inverted micelles in dispersions of mixed galactolipids, *Nature (London),* 293, 488, 1981.
5. **Hui, S. W., Stewart, T. P., Boni, L. T., and Yeagle, P. L.,** Membrane fusion through point defects in bilayers, *Science,* 212, 921, 1981.
6. **Cullis, P. R. and De Kruijff, B.,** Lipid polymorphism and the functional roles of lipids in biological membranes, *Biochim. Biophys. Acta,* 559, 399, 1979.
7. **Van Venetie, R. and Verkleij, A. J.,** Analysis of hexagonal II phase and its relation to lipidic particles and the lamellar phase: a freeze fracture study, *Biochim. Biophys. Acta,* 645, 262, 1981.
8. **Boni, L. T. and Hui, S. W.,** Polymorphic phase behavior of dilinoleoylphosphatidylethanolamine and palmitoyloleoylphosphatidylcholine mixtures: structural changes between hexagonal, cubic and bilayer phases, *Biochim. Biophys. Acta,* 731, 177, 1983.
9. **Sen, A., Williams, W. P., Brain, A. P. R., and Quinn, P. J.,** Bilayer and non-bilayer transformations in aqueous dispersions of mixed sn-3-galactosyldiacylglycerols isolated from chloroplasts, *Biochim. Biophys. Acta,* 685, 297, 1982.
10. **Sen, A., Brain, A. P. R., Quinn, P. J., and Williams, W. P.,** Formation of inverted lipid micelles in aqueous dispersions of mixed sn-3-galactosyldiacylglycerols induced by heat and ethylene glycol, *Biochim. Biophys. Acta,* 686, 215, 1982.

11. **Luzzati, V.,** X-ray diffraction studies of lipid water systems, in *Biological Membranes*, Vol. 1, Chapman, D., Ed., Academic Press, London, 1968, 371.
12. **Seelig, J.,** ^{31}P nuclear magnetic resonance and the head group structure of phospholipids in membranes, *Biochim. Biophys. Acta*, 515, 105, 1978.
13. **Sen, A., Williams, W. P., and Quinn, P. J.,** The structure and thermotropic properties of pure 1,2-diacylgalactosylglycerols in aqueous systems, *Biochim. Biophys. Acta*, 663, 380, 1981.
14. **Hui, S. W., Mason, J. T., and Huang, C. H.,** Acyl chain interdigitation in saturated mixed-chain phosphatidylcholine bilayer dispersions, *Biochemistry*, 23, 5570, 1984.
15. **Sen, A., Mannock, D. A., Hui, S. W., and McElhaney, R. N.,** unpublished data.
16. **Luzzati, V., Tardieu, A., and Gulik-Krzywicki, T.,** Polymorphism of lipids, *Nature (London)*, 217, 1028, 1968.
17. **Shyamsunder, E., So, P. T. C., Tate, M. W., Turner, D. C., and Grunner, S. M.,** Cubic phase in DOPE, *Biophys. J.*, 53, 211a, 1988.
18. **Verkleij, A. J.,** Lipidic intermembranous particles, *Biochim. Biophys. Acta*, 779, 43, 1984.
19. **Kléman, M., Williams, C. E., Costello, M. J., and Gulik-Krzywicki, T.,** Defect structures in lyotropic smectic phases revealed by freeze fracture electron microscopy, *Philos. Mag.*, 35, 33, 1977.
20. **Boni, L. T., Stewart, T. P., Alderfer, J. L., and Hui, S. W.,** Lipid-polyethylene glycol interactions. II. Formation of defects in bilayers, *J. Membr. Biol.*, 62, 71, 1981.
21. **Sen, A.,** unpublished data.
22. **Nishihara, M., Yokoto, K., and Kito, M.,** Lipid molecular species composition of thylakoid membranes, *Biochim. Biophys. Acta*, 617, 12, 1980.
23. **Gounaris, K., Sen, A., Brain, A. P. R., Quinn, P. J., and Williams, W. P.,** The formation of non-bilayer structures in total polar lipid extracts of chloroplast membranes, *Biochim. Biophys. Acta*, 728, 129, 1983.
24. **Albert, A. D., Sen, A., and Yeagle, P. L.,** The effect of calcium on the bilayer stability of lipids from bovine rod outer segment disk membranes, *Biochim. Biophys. Acta*, 771, 28, 1984.
25. **Machold, O., Simpson, D. J., and Høyer-Hansen, G.,** Correlation between the freeze fracture appearance and peptide composition of thylakoid membranes in barley, *Carlsberg Res. Commun.*, 42, 499, 1977.
26. **Carter, D. P. and Staehelin, L. A.,** Proteolysis of chloroplast thylakoid membranes. II, *Arch. Biochem. Biophys.*, 200, 374, 1980.
27. **Gounaris, K., Brain, A. P. R., Quinn, P. J., and Williams, W. P.,** Structural and functional changes associated with heat-induced phase separation of non-bilayer lipids in chloroplast thylakoid membranes, *FEBS Lett.*, 153, 47, 1983.
28. **Thomas, P. G., Brain, A. P. R., Quinn, P. J., and Williams, W. P.,** Low pH and phospholipase A_2 treatment induce the phase-separation of nonbilayer lipids wthin pea chloroplast membranes, *FEBS Lett.*, 183, 161, 1985.
29. **Thomas, P. G., Dominy, P. J., Vigh, L., Mansourian, A. R., Quinn, P. J., and Williams, W. P.,** Increased thermal stability of pigment-protein complexes of pea thylakoids following catalytic hydrogenation of membrane lipids, *Biochim. Biophys. Acta*, 849, 131, 1986.
30. **Grunner, S. M., Rothschild, K. J., and Clark, N. A.,** X-ray diffraction and electron microscope study of phase separation in rod outer segment photoreceptor membrane multilayers, *Biophys. J.*, 39, 241, 1982.
31. **Sen, A., Yang, P. W., Mantsch, H. H., and Hui, S. W.,** Extended hydrogen bonded structures of phosphatidylethanolamine, *Chem. Phys. Lipids*, 47, 109, 1988.

Chapter 7

THE KINETICS OF ARTIFICIALLY INDUCED MEMBRANE FUSION

Sek Wen Hui

TABLE OF CONTENTS

I. Introduction 130

II. Sequential Stages of Fusion 130
  A. Prefusion Membrane Attachment 130
  B. The Intermediate Stages of Fusion 130
  C. Lumen Formation 131

III. Determination of the Ultrastructural Kinetics of Fusion 132
  A. Calcium-Induced Fusion of Phospholipid Vesicles 132
  B. Fusion Induced by Polyethylene Glycol (PEG) 134
  C. Kinetics of Electrofusion 135

IV. Discussion 143

Acknowledgments 145

References 145

I. INTRODUCTION

Membrane fusion occurs commonly in nature, although we do not yet understand the molecular mechanism. Even at the ultrastructural level, the morphological sequence of fusion between membranes is understood only in general terms. It is believed that adjacent membranes come in close contact before fusion occurs. The attached membranes then become momentarily unstable, and a breakdown of the membrane integrity occurs at the fusion site. The site then develops into a lumen which joins the two initially separated compartments together. The fusion event ends at the enlargement of the lumen. This general picture is deduced partly from logical reasoning and partly from static electron microscopic samplings of fusion events such as endocytosis, secretion, fertilization, and cell division at various stages of action. The sequence of events is then reconstructed based on imagination. The chronicle description and timed measurements of changes of membrane morphology during fusion are possible only recently, using synchronized fusion triggering and rapid cryofixation techniques.

II. SEQUENTIAL STAGES OF FUSION

A. Prefusion Membrane Attachment

The close contact between fusion partners usually precedes fusion. The adjacent membranes are held together by electrostatic or hydration forces, sometimes via molecular bridges. At this stage, neither the contents of two compartments nor the components of two membranes mix. Membrane attachment without fusion is difficult to prove by electron microscopy morphology alone, but is well recognized by fluorescence assays.[1] In some cases, membrane juxtaposition is so tight that it is barely resolvable by electron microscopy. Some molecular exchange between attached membranes is possible.

For membrane vesicles or cells in free suspension, the prefusion membrane attachment leads to aggregation. The rate of aggregation, dissociation, and fusion of vesicle groups x_i and x_j consisting of i and j primary vesicles, into an aggregate x_{ij} and then a fusion product F_{ij} may be quantitatively defined using a mass action model.[1]

$$x_i + x_j \underset{D_{ij}}{\overset{C_{ij}}{\rightleftharpoons}} x_{ij} \xrightarrow{f_{ij}} F_{ij} \qquad (1)$$

where C_{ij}, D_{ij}, and f_{ij} denote the association (aggregation), dissociation, and fusion rate constants, respectively. These constants can be deduced from intermembrane forces, or measured experimentally by fusion assays.[1] The time courses for membrane attachment and fusion between these vesicles or cells can then be calculated. Synchronized cryofixation experiments may then be performed to verify the process morphologically.

B. The Intermediate Stages of Fusion

The actual fusion event starts when the membrane integrity of both partners is locally and momentarily disrupted. The membrane planes of the fusion partners are adjoined momentarily through point or line defects in the bilayers.[2] This stage occurs very rapidly. To capture an ongoing fusion event during the transient membrane destabilizing stage is not a trivial task, especially since that stage is believed to last only a fraction of a second. Therefore, high temporal as well as spatial resolution are required. Video recording of light microscope images at the TV rate ($^1/_{30}$ s) has been applied. It has a distinct advantage of being able to record the same sample continuously over a long period of time, so that the sequence of fusion event may be followed unambiquously. Much valuable information has been obtained by this method.[3] The light microscope is limited in spatial resolution, and the temporal resolution is limited by the rate of video frame, which is in turn limited by the signal-to-noise ratio at a given light level. Traditional electron microscopy requires long specimen fixation time. Only with the advance of rapid-freezing techniques can events at given time points be preserved.

With the rapid-freezing technique, a specimen can be quenched in less than a millisecond, to an accuracy of milliseconds in timing.[4] In combination of the high resolution of the electron microscope, the rapid-freezing technique offers a mean to achieve both high temporal and spatial resolution in imaging. The disadvantage is that, like all electron microscopic studies, one cannot follow the same specimen through a sequence of events. The temporal order of structural events must then be reconstructed using statistical approaches.[4]

To arrest an intermediate stage of membrane fusion poses additional problems. Most observed fusion phenomena occur over a period of time after the triggering. A single fusion event may initiate at any time point during this period, and lasts for only a short interval. The collection of individual events represents the "kinetics" of fusion. The following analysis is applied in order to maximize the probability of arresting fusion events by the rapid-freezing technique.

Let the number of individual fusion events initiating at time t be x(t). This function represents the fusion rate of the macroscopic fusion assays. Assume each individual event to last a duration of τ, within which an intermediate structure is detectable, and the sample is frozen "instanteneously" within a time interval f. The probability to arrest a fusion intermediate in the middle of a fusion event at time t = a may be estimated as:

$$P(a) = \int_{a-\tau}^{a+f} x(t)dt \Big/ \int_{0}^{\infty} x(t)dt \qquad (2)$$

The numerator represents the number of ongoing events which are initiated during the freeze-quenching time a to a + f, and previously occurred events which last until t = a. The freezing time f is limited by the time resolution ($<\tau$), and is further limited by the need to vitrify ice for morphological preservation. In reality, the time f can be regarded as the time taken to bring the temperature down to a point such that all morphological changes cease. Here we assume that x(t) is independent of temperature for the sake of simplicity. In a typical experiment, f is in the order of 1 ms. It now becomes obvious that the best chance to arrest an intermediate stage of fusion is to freeze-quench the sample at a time when x(t) is at a maximum. When the macroscopic fusion kinetics is known, the probability of arresting a fusion event can then be maximized.

Membrane fusion may be triggered using natural or artificial means. Examples of natural triggers include secretion initiators[5] and electric excitation for synaptic vesicle release.[4] In this chapter, I shall discuss our experiments of observing *in vitro* fusion events by freeze-fracture electron microscopy, using artificial triggers such as calcium ion, polyethylene glycol, or electric pulse to trigger membrane or bilayer fusion.

C. Lumen Formation

After the transient membrane destabilization step, the adjacent membranes, or at least the outer leaflets of them, are permanently adjoined. Membrane components such as lipids and membrane proteins are able to diffuse through the junction. Further enlargement of the fusion site leads to the formation of a lumen, through which contents from two compartments are intermixed. The enlargement of the lumen is a relatively slow process, and is therefore relatively easy to be studied by electron microscopy. This step does not follow the previous one automatically. In most cases, a differential osmotic pressure facilitates the formation of the lumen. The osmotic pressure is necessary for polyethylene glycol induced cell fusion.[6] Osmotic pressure is helpful also in virus-induced fusion[7] and electrofusion.[8]

The following examples illustrate how freeze-fracture electron microscopy is used to follow the sequences of morphological changes upon membrane fusion, and their time courses.

III. DETERMINATION OF THE ULTRASTRUCTURAL KINETICS OF FUSION

A. Calcium-Induced Fusion of Phospholipid Vesicles

Large unilamellar vesicles of mixtures of bovine brain phosphatidyl serine and transacylated phosphatidylethanolamine (from egg phosphatidylcholine) at a ratio of 1:3 were made by the reverse phase evaporation method.[9] These vesicles were induced to fuse by the addition of calcium. The kinetics of this system has been measured experimentally.[10] However, we remeasured the fusion kinetics for freeze-fracture experiments because the concentration of the sample required for practical electron microscopic sampling is much higher than that used for previous experiments. The standard terbium/dipocholinic acid (Tb/DPA) fusion assay[10] were used. Two ratios of Tb and DPA containing vesicles were used to estimate the aggregation and fusion constants.[1] The fluorescence curve, after corrected for leakage, represents the cumulative fusion events. The rate of fusion x(t) is the derivative of this curve. For 0.315 m*M* of lipid, the maximal rate was found to be within the first minute of the addition of calcium, and the process was aggregation limited.[11]

For electron microscopic study, the lipid suspension was mixed with calcium on the sample holding sandwich, and freeze-quenched in liquid propane after a given time. For longer time points, calcium was injected and rapidly mixed with lipid vesicles for a given time period, and the sample was placed in a copper sandwich for freeze-quenching. The timed-quenching series of freeze-fracture electron micrographs in Figure 1 shows a variety of time-dependent morphology. Since the sample morphology is not homogeneous, one can only derive the sequence of events indirectly from the frequency of occurence of various structural features. One obvious fact is that the vesicles aggregated immediately after the addition of calcium (Figure 1A). The barrier between aggregated vesicles can be seen breaking into disjoined, bead-like structures in 37 s (Figure 1B) and continued to disintegrate at 1 min after the trigger (Figure 1C). The remaining partitions between fusing vesicles were either vanished or transformed into strings of beady or tubular remnants in 2.5 min (Figure 1D). Since these remnants were not often seen between 0- and 37-s time points when the fusion rate was the highest, they may be side or end products rather than features representing an intermediate stage of membrane fusion. We did not detect any unusual increase of any types of structural features at or near the connecting areas between aggregated vesicles during the peak fusion time. Occasional occurrence of lipidic particle-like features may not represent fusion intermediates, in agreement with the observation by Rand et al.,[3] Miller and Dahl,[12] and Bearer et al.[13]

Using Equation 2 as a guide, we estimated the probability of observing short-lived ($\tau \sim f$) intermediates in this system. Taking the freezing time f to be 1 ms, the fusion rate (taken as the derivative of the fluorescence curve) at the 30-s time point to be 3% per s and the probability of fracturing through a contact area between two vesicles to be 0.1, the total P(t) is about 3×10^{-7}. Assuming each vesicle fuses once, then the chance of seeing intermediate states of fusion while two vesicles are fusing is 1 in 3 million. This is based on the estimate that $\tau = f$, a reasonable assumption. If $\tau > f$, then the chance of seeing an intermediate is slightly better. Therefore, the absence of features pertaining to fusion intermediates in freeze-fracture electron micrographs is inconclusive with regard to the nonexistence of fusion intermediate in this system. At such a low probability level, a search for ultrastructural evidence of fusion intermediates is subject to much artifact.

The rapid-freezing technique does provide information on the aggregation and fusion rates in the microscopic sense. The mass action model, which has successfully described the

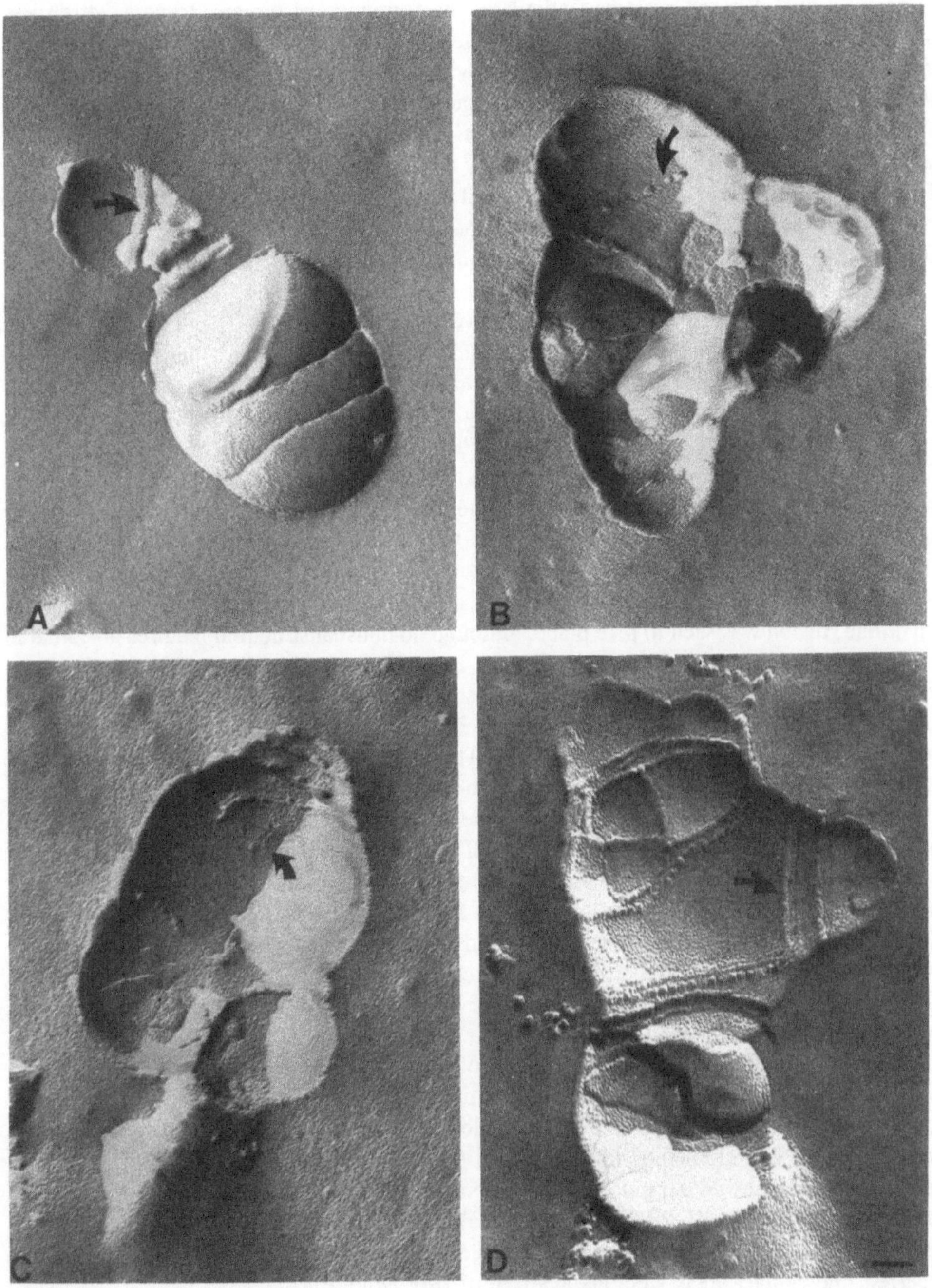

FIGURE 1. Different stages of aggregation and fusion of mixed lipid REVs composed of bovine brain PS and transesterified egg PE (molar ration 1:3), frozen-arrested at (A) 2 s, (b) 37 s, (C) 64 s, and (D) 154 s after the addition of 4 m*M* $CaCl_2$. Arrows indicate possible remnants of partition diaphrams. Bar = 100 nm. (Figures 1A and 1B from Hui, S. W., Stenger, D. A., and Huang, S. K., *Molecular Mechanisms of Membrane Fusion*, Plenum Press, 1988, 306. Figure 1D from Hui, S. W., Nir, S., Stewart, T. P., Boni, L. T., and Huang, S. K., *Biochim. Biophys. Acta*, 941, 137, 1988. With permission.)

fusion reactions of many systems,[1] stipulates the formation of dimers, trimers, tetromers, etc. as the result of aggregation; these aggregates then fuse into larger monomers at the next stage. Freeze-fracture electron microscopy provides an ideal way to test this theory. At each given instant of freeze-quenching, the percentage of vesicles in the dimer, trimer, ... oligomer forms can be measured. The degree of fusion can be estimated from the increase of the mean vesicle size. This is done statistically by measuring 400 electron microscopic images of vesicles.[11] Both the state of aggregation and the size increase with time agree with those predicted by mass action modeling, using the aggregation and fusion constants derived independently from the fluorescence measurement.

B. Fusion Induced by Polyethylene Glycol (PEG)

PEG induces fusion both between lipid vesicles and between cells. Lipid vesicles fuse as soon as a fusogenic concentration of PEG is added to the suspension. This can be monitored by light scattering, by NMR,[14] as well as by fluorescence lifetime measurements.[15] The highest fusion rates occur within 1 to 2 min of PEG treatment depending on the concentration of PEG. The rate is approximately 5%/s at low PEG concentration and may increase an order of magnitude at high concentration. The probability for arresting a fusion event by rapid-freezing is therefore comparable to that for calcium-induced fusion described in the last section.

At 1 s after the mixing of 50% PEG with small unilamellar vesicles of egg phosphatidylcholine, fusion was seen to take place, although no unusual structural features attributable to fusion intermediate was observed (Figure 2A). There are circular marks on the large vesicle, suggesting remnants of fused vesicles before the freeze quench. The increasingly large vesicles became multilamellar (Figure 2B) within 5 min, most likely through the mechanism of defect formation (Figure 2C). These defects are common and are well characterized in dehydrated lipids.[16] Figure 2C shows a confocal domain which connects the inner layers to the outer ones, thus relaxing the constraint of lamination. After dilution, the fusion product remain multilamellar (Figure 2D).

The formation of multilamellar products does not occur in the fusion of biomembranes. After the addition of PEG, the fusion partners aggregate. Cell fusion does not occur during PEG treatment, although in some cases lipid exchange may take place between fusion partners.[17] The opening of lumens and free exchange of cytoplasm between fusion partners are initiated only during the dilution step after the PEG treatment.[18] Therefore, the highest probability to arrest a fusion intermediate is at the beginning of the dilution step.

A simple example is the fusion between reconstituted vesicles containing a single integral membrane protein, glycophorin, and a single lipid component, egg PC. Cytoskeletal effect is absent in this system. In freeze-fracture electron micrographs, multimeric glycophorin were visualized as intramembranous particles (IMP) in freeze-fractured faces. The reconstituted vesicles showed a random distribution of IMP on their fracture faces before the treatment of PEG.[6] At 6% PEG, before the majority of the vesicles began to aggregate, the IMP on their fracture faces aggregated regardless of contact with other vesicles. If contact occurred, the contact areas were always free of IMP (Figure 3A).

A similar response to PEG treatment was also observed for cytoskeleton-free membrane vesicles[6] budded from human erythrocytes (Figure 3B). Intact erythrocytes behave similarly except that the threshold concentration of PEG was higher for similar effects to be observable[19] (Figure 3C). The effects were reversible up to 12% of PEG. It should be emphasized that membrane fusion had not occurred at this stage, either by fluorescence assays or by size criteria. The observed intermembrane attachment over IMP-free areas was strictly a membrane aggregation phenomenon. It is interesting to note that the capability of a fusogen to induce IMP aggregation was found to parallel their fusion efficiency,[19] and that induced IMP aggregation facilities PEG-induced membrane fusion.[20] Thus the creation of IMP-free patches is the single, most important prerequisite in PEG-induced fusion, which is believed to be centered on the lipid bilayer. This is in contrary to protein-induced fusions where a single or a specific group of protein is seen right at the fusion site.

Fusion between cells or between membrane vesicles began immediately after the dilution of PEG, but the kinetics measured by fluorescence was rather slow.[18] The distribution of IMP was reverted to random after PEG dilution (Figure 3D). No unusual morphological features have been found to be unambiguously associated with fusion intermediates.

C. Kinetics of Electrofusion

Electrofusion provides an ideal condition for the detection of fusion intermediates because the fusion events are well synchronized.[21] Consequently, the probability of arresting fusion intermediates is high at the peak of fusion.

The kinetics of electrofusion of erythrocyte membranes was determined by a fluorescence contents mixing assays. A special front-face fluorometer curvette with built-in electrodes was constructed for this purpose.[22] Human erythrocyte ghost membranes were loaded with Tb or DPA as described by Hoestra et al.,[23] and placed between electrodes in the curvette. The loaded cells were prealigned by dielectrophoresis in a 1-MHz electric field of the strength 500 V/cm. A short (15 to 50 μs) square pulse at 2.5 to 5.0 kV/cm was applied to induce fusion between aligned cells. Fusion was measured as the increase of fluorescence upon the mixing of Tb and DPA, whereas leakage was measured as the decrease of fluorescence from vesicles loaded with mixed Tb + DPA and suspended in a EDTA-containing medium. Fluorescence intensity increased immediately after the pulse application. Within 2 s, the fluorescence reached its peak value, then declined slowly over a period of tens of seconds.[22] The maximum rate of fusion occurs within the first second and reaches a percentage rate of 50%/s.

Taking again the freezing time f to be 1 ms, and the fracturing probability of 0.1, we arrive at a $P(t) = 5 \times 10^{-5}$. The probability is still too low to allow detection by freeze-quenching unless $\tau \gg f$. It there are ten intermediate features per fusing cell, and τ for these features is in the order of 0.1 s, then P(t) becomes 0.05 or 5%, which should allow experimental observation.

For freeze-quenching the samples, a rapid-freezing copper sandwich device was modified to double as electrodes. The two opposite copper plates were spaced apart by thin strips of insulating material. Pronase-treated human erythrocytes suspended in isotonic sucrose solution were aligned by dielectrophoresis as described and were induced to fuse by a single 15 μs pulse of 5 kV/cm field strength. For early time points, the electric pulse was transmitted while the sample was on its way dropping to the freezing coolant. The timing between the pulse and the freezing was controlled to within 0.1 s. At least 40 micrographs were taken at each time point and the features on them were analyzed.[24]

Distinct features at each time point are noted. At the 0.1-s time point, the intermembrane aqueous spacing is reduced from 15 nm during dielectrophoresis to less than 2 nm at some points. There are many patches on the cell surface which appear torn from the surrounding membrane (Figure 4A). These discontinuous areas (DA) are seen on about 20% of the fractured cells. Intramembranous particles (IMP) are less dense, and often absent from many DA. At 2 s after the pulse, which corresponds to the maximum of fluorescence, few DA are observed, and are replaced by lipid-particle-like point defects (PD) as the predominant membrane abnormality. These PD and DA seem to be confined to the flattened areas of the membrane which suggest a contact area with an underlying cell (Figure 4B and C). The intercellular distance at this time point returns to the prepulse value of 15 nm, except at the attachment sites.[24] All but a few PD disappear at 10 s after the pulse; the remaining ones develop into intermembrane attachment sites. Continuity between membranes of fusion partners can be seen through these attachment sites (Figure 4D and E). Because of a peculiar

FIGURE 2. Different stages of aggregation and fusion of egg PC induced by 50% (w/v) of PEG (mw 8000). The samples were frozen-arrested at (A) 1 s after the addition of PEG, (B) in PEG for 1 min; and (d) in 100-fold dilution after treated in PEG for 5 min. Defect structures including a cross-fracture of confocal domain (arrowed) from egg PC MLV in 50% PEG is shown in (C). Bar = 100 nm. (2A from Boni, L. T., Stewart, T. P., Alderfer, J. L., and Hui, S. W., *J. Membr. Biol.*, 62, 69, 1981. With permission.)

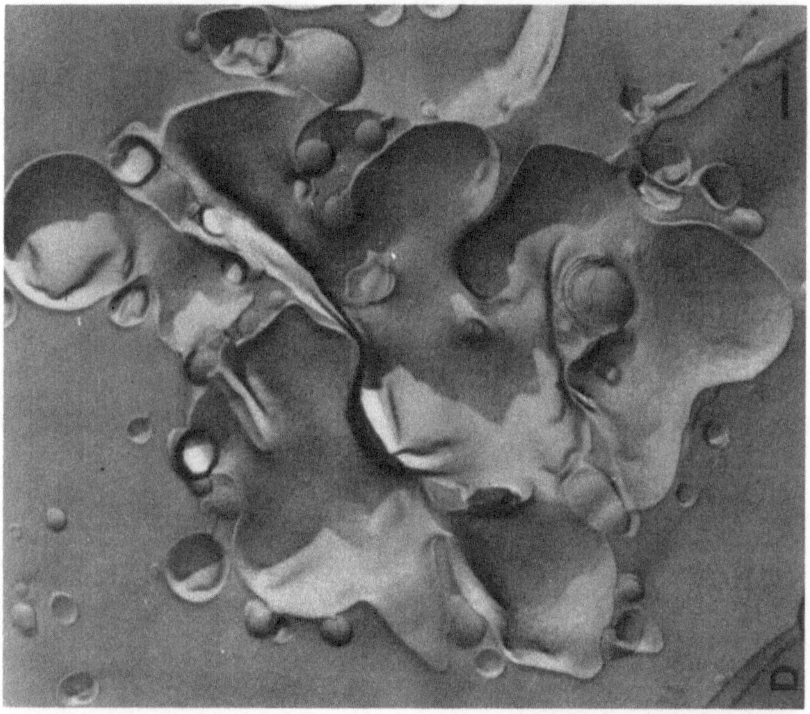

FIGURE 2D.

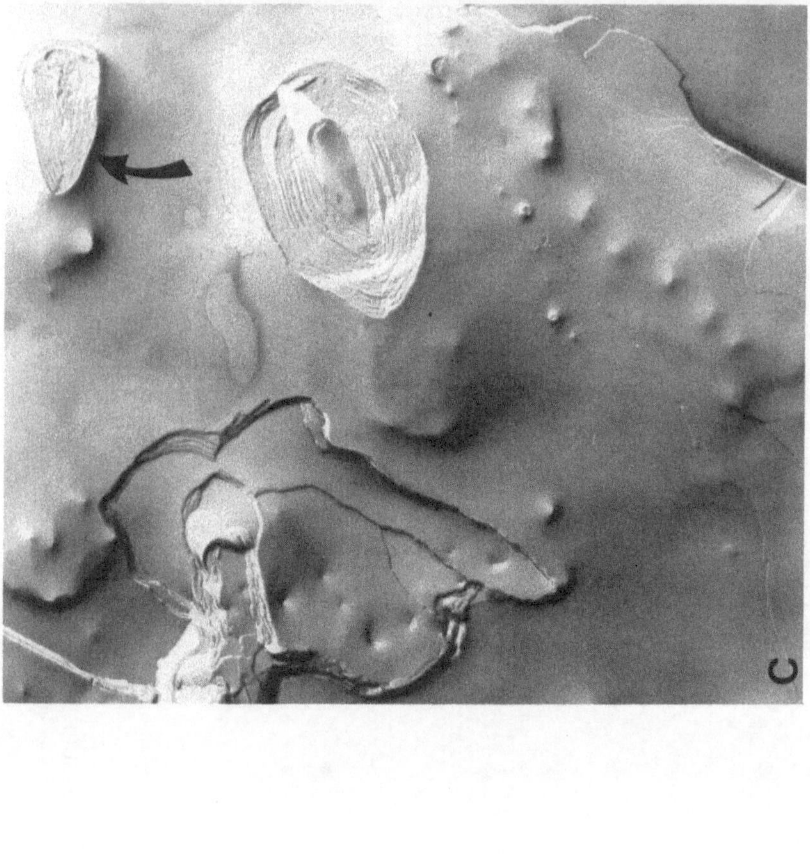

FIGURE 2C.

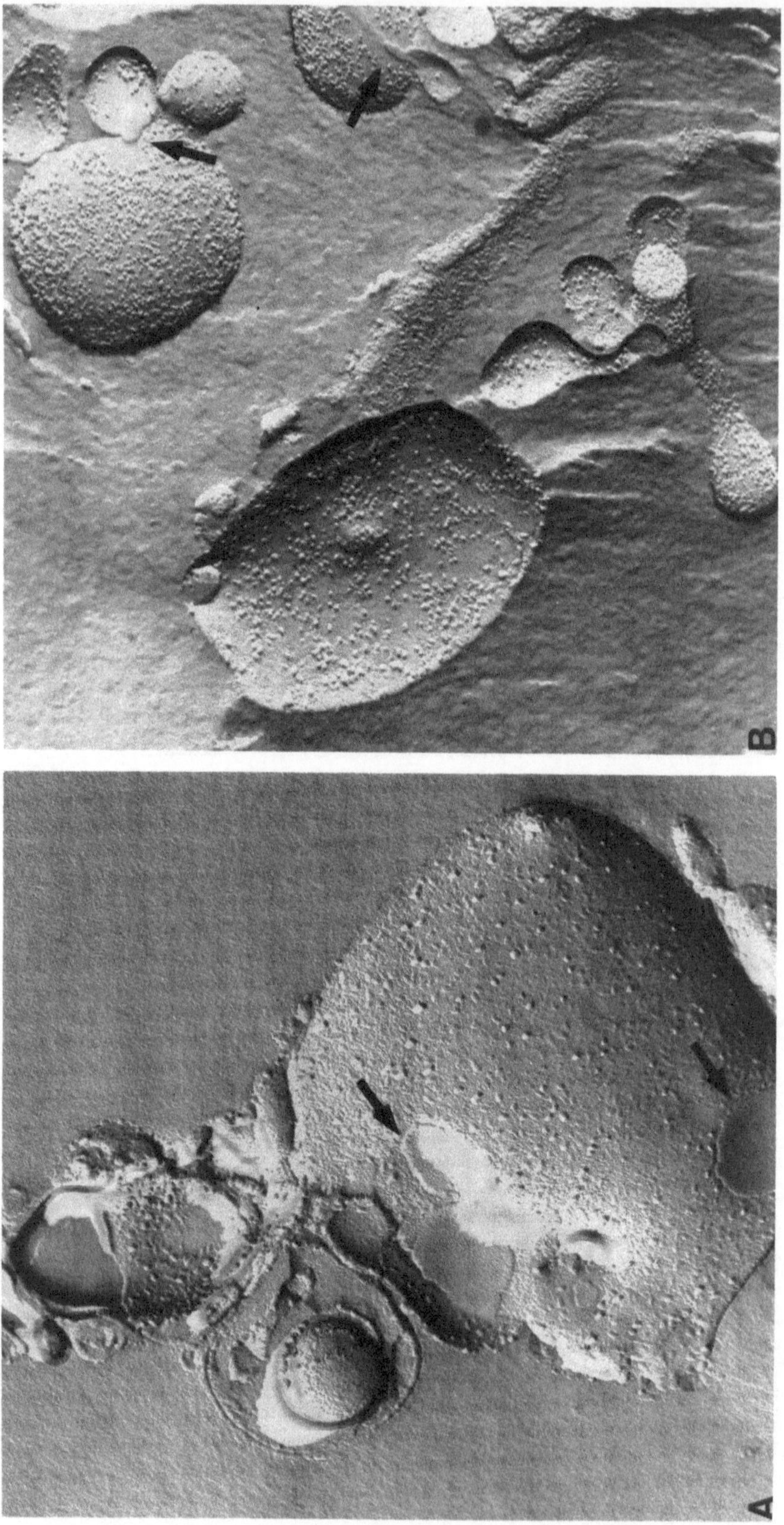

FIGURE 3. Aggregation of IMP as a result of PEG (mw 8000) treatments. (A) Reconstituted vesicles containing glycophorin and egg PC in 12% PEG; (B) cytoskeleton-free erythrocyte vesicles in 6% PEG; (C) erythrocyte in 12% PEG; and (D) cytoskeleton-free erythrocyte vesicles after dilution from 12 to 2% PEG. Note IMP-free contact areas (arrowed). Bar = 100 nm.

FIGURE 3C.

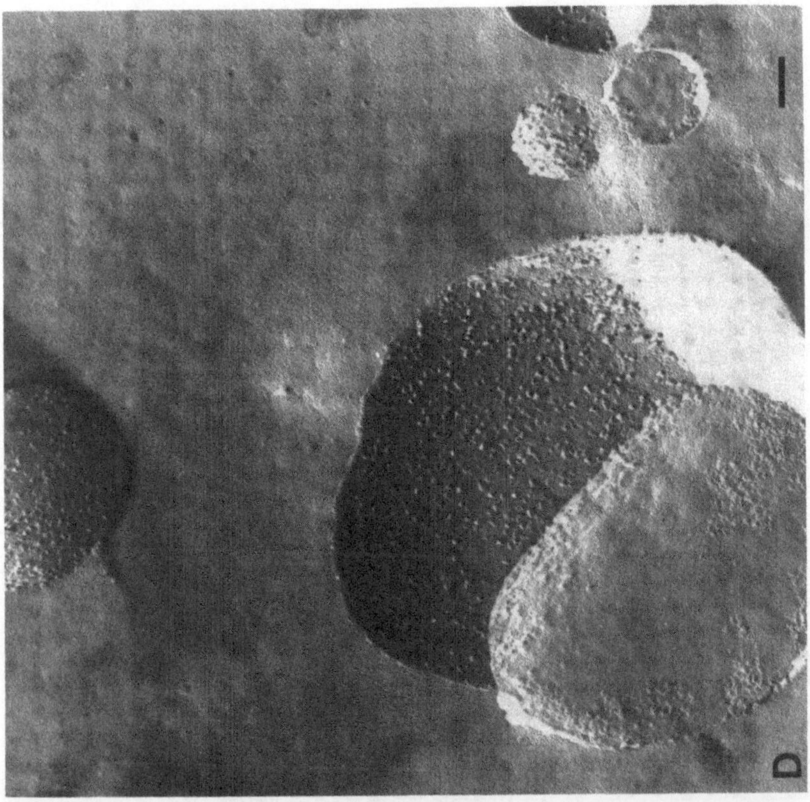

FIGURE 3D.

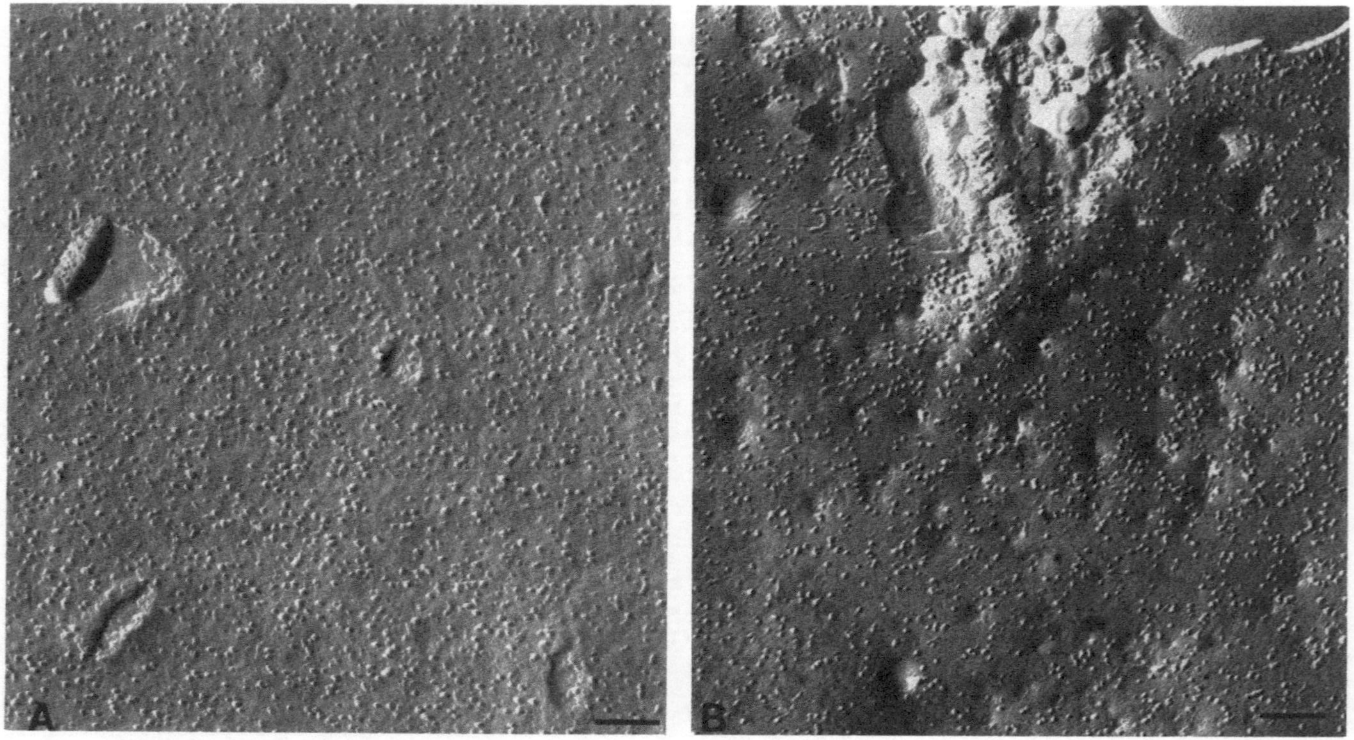

FIGURE 4. Various stages of electrofusion of human erythrocytes. (A) At 0.1 s after a single 5 kV/cm 15 μs electric pulse, showing discontinous areas (DA); (B) at 2 s after the pulse, showing point defects and remnants of DA; (C) low magnification image at 0.1 s after the pulse showing DA are restricted within a presumed area of close contact with an underlying cell along the pearl chain; (D) continuation of the P-face of one cell to the E-face of another cell through a line defect as a result of electrofusion, 60 s after the pulse; (E) stereo pair of images along the membrane fracture planes showing the connection of E-face of one cell to the P-face of another through a point defect, and (F) stereo pair of images showing a cross-fracture through a cytoplasmic bridge which connects the contents (hemoglobin) of two adjacent cells, at 60 s after the pulse. Bar = 100 nm. (4A from Hui, S. W., Stenger, D. A., and Huang, S. W., *Molecular Mechanisms of Membrane Fusion*, Plenum Press, 1988, 310. 4D from Stenger, D. A., and Hui, S. W., *J. Membr. Biol.*, 93, 48, 1986. With permission.)

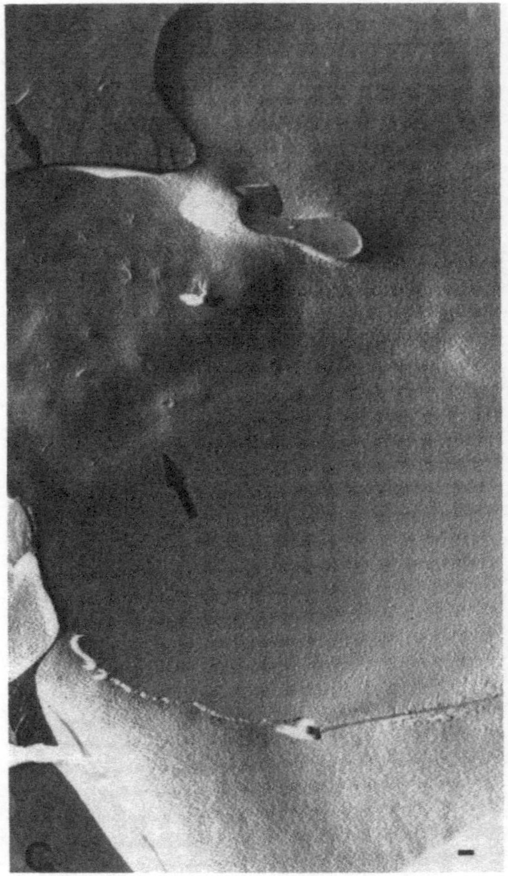

FIGURE 4C.

FIGURE 4D.

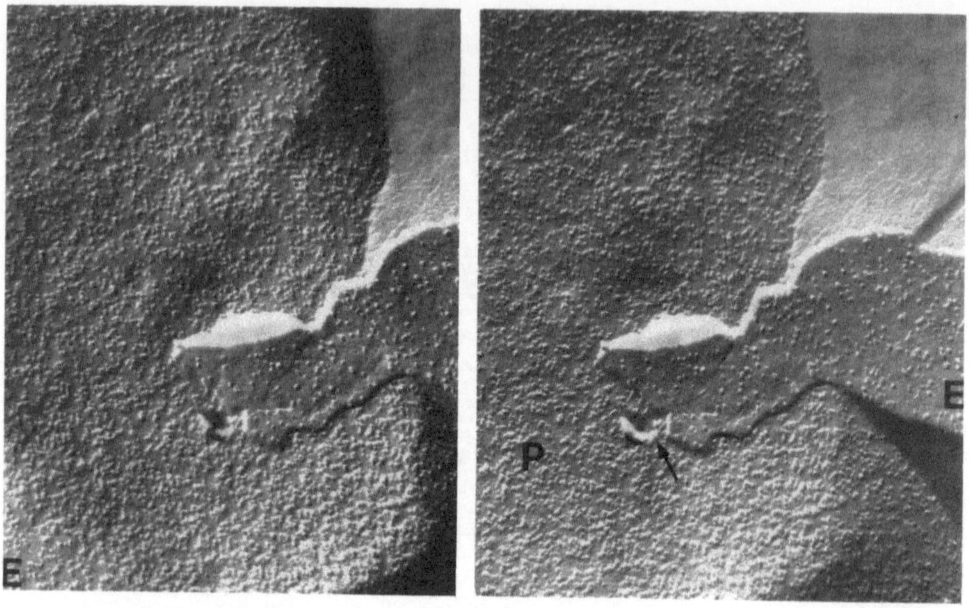

FIGURE 4E.

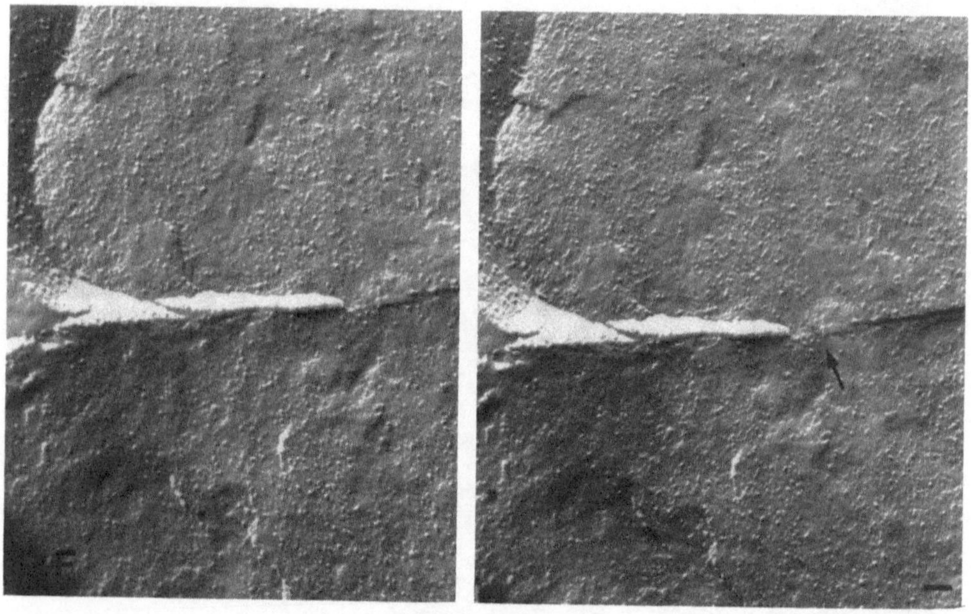

FIGURE 4F.

arrangement of line defects, the fracture plane may pass from the central plane of the membrane of one cell to that of another cell through a common point or line defect in the partition. This topography is illustrated in Figures 5B and C. Only through these types of connections can the E-face of one cell be seen continuous to the P-face of another cell (Figure 4D and E). The stereoscopic pair images in Figure 4E show the continuation through a point defect instead of a line defect. Some of these will develop into cytoplasmic connections

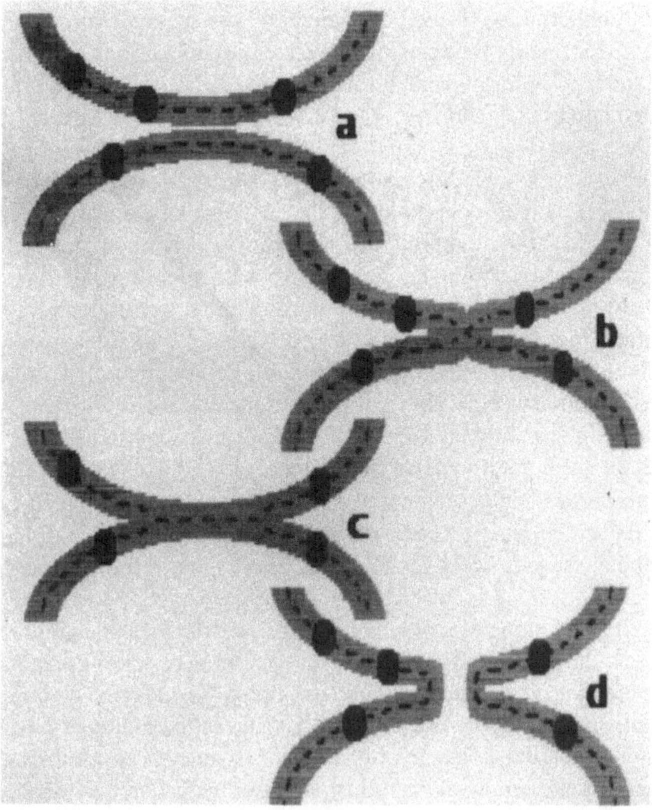

FIGURE 5. A schematic model describing the features seen in a fusion sequence. (a) Close membrane contact is facilitated by the exclusion of IMP from the contact area; (b) a point defect is formed through which the fracture plane (dotted lines) may pass from one cell to another, thereby joining the E-face of one cell to the P-face of another; (c) a line defect depicting similar phenomena; and (d) a lumen is formed through which the contents of the fusion partner intermix.

or lumens (Figure 4F). The spaces occupied by cytoplasmic content (hemoglobin) in either cell are seen to be connected through this bridge in this stereoscopic pair. These cytoplasmic connections or lumens become well established between pairs of fusion partners in 60 s.[24] Apart from these connections, the rest of the cells are indistinquishable from the control population.

The dominant features which appear at the maximum fusion time points are DA and PD. A statistical presentation of the appearance of these features at given time points has been presented elsewhere.[24] The frequency of appearence in 20% of cells indicates that their lifetime, τ, is much longer than f. Most of these DA and PD disappear at a latter time. Some of these DA could be precursors for fusion, leading to a subsequent cytoplasmic bridging. The widening of the lumen, which determines the success of cell fusion, is subject to the influence of osmotic swelling,[25] in agreement with a previous finding.[8]

IV. DISCUSSION

All fusion processes we considered here seem to follow a common sequence of events, although the timing of each step may vary. These steps are (1) close membrane apposition (aggregation), (2) transient membrane destabilization, and (3) lumen formation. Step 1 usually lasts long enough to be observed by electron microscopy (as in the case of calcium-induce liposome fusion), and even remains stable indefinitely (as in PEG-induced cell fusion). Direct contacts between lipid bilayers at this stage are facilitated by the removal of integral proteins from the contact area (Figure 5A). This step is reversible, and in general no fusion is detected by the content-mixing or the membrane-mixing criterion. The transition from step 1 to step 2 is a crucial kinetic process, which is governed by the nature of the destabilizing trigger. In our liposome experiments, it obeys the chemical kinetics of calcium-lipid or PEG-lipid interaction. In cell fusion experiments, osmotic pressure and other factors also play an important role in triggering.[8,18,25] The local membrane instabilities of fusion intermediates take the form of transient interbilayer connections such as a point defect (Figure 5B), or a line and area defect (Figure 5C). Through these bilayer defects, a freeze-fracture plane may pass from one fracture face (say the P-face) from one cell continuously to the other fracture face (say E-face) of another cell.[2] This is illustrated in Figure 4E. Step 2 is very short-lived, and is macroscopically inseparable from step 3, which may take minutes to hours to complete. The enlargement of the lumen (Figure 5D) from a point defect in the bilayer depends on the mechanical properties of the membrane, and therefore varies from cell type to cell type. The cartoon model in Figure 5 depicts the general features of this fusion sequence, with marked fractured planes from which features may be recognized from freeze-fracture electron micrographs.

The rapid freezing technique contributes to clarify the functions of a number of prerequisite factors associated with fusion. The importance of prior vesicle aggregation and protein clearence from potential fusion sites are good examples. The results from rapid cryofixation experiments supplied microscopic evidence for the mass action model of fusion, and provided a structural rationale for the potencies of various fusogens. The importance of dielectrophoretic alignment in electrofusion was illustrated by microscopic evidence.

The most challenging problem is the documentation, by the electron microscope, of fusion intermediates during a kinetic event. Although the advance of the rapid freezing technique enables us to arrest instanteneously morphological features in the kinetic process of membrane fusion, the probability of freezing and revealing such features depends on a number of other factors. The most important factor is the rate of fusion, which determines the frequency of individual fusion events occuring per unit time. Since the freezing time is relatively short, the fraction of fusion events initiated during that time is small, unless the remains of previously initiated fusion intermediates are also counted to increase the probability. Thus the second most important factor is the lifetime of fusion intermediates. The longer the lifetime, the easier it is to visualize the structure. In some cases, the lifetime is infinitely long (the intermediate is stable); then the detection becomes a certainty. Paradoxically, these structures cannot be defined as fusion intermediates as such. There are various reports asserting that those stable structures are equivalent to fusion intermediates, but the association is at best circumstantial. The true intermediate structure must vanish after the fusion is completed, and an incomplete fusion is not strictly a fusion.

Even for the most synchronized fusion system like the electrofusion, the highest percentage fusion rate is only 0.5/s as measured by fluorescence methods. A freezing rate of 1 ms reduces the probability to 5×10^{-4} unless the lifetime of fusion is much longer than a millisecond. The DA feature that we observed is such a long-lived intermediate or byproduct. For slower fusion rates such as in the calcium or PEG-induced fusions, the probability is even lower for detecting such long-lived features. Thus the rarity of observed features does not imply that fusion intermediates do not exist, nor that these features are indistinguishable from the normal membrane structure.

Our previous understanding of membrane fusion depended heavily on electron microscopy. With the advance of triggering and rapid cryofixation technology, our appreciation of the fusion kinetics and the molecular mechanism of membrane fusion will continue to benefit from electron microscopy and freeze-fracturing in particular.

ACKNOWLEDGMENTS

This work is supported by grants GM28120 and GM20969 from the NIH. Parts of the results presented here are from collaborations with Dr. S. Nir. The unpublished work from this laboratory represents the contributions from L. T. Boni, T. P. Stewart, T. Isac, D. A. Stenger, and S. K. Huang.

REFERENCES

1. **Nir, S., Bentz, J., Wilschut, J., and Duzgunes, N.,** Aggregation of phospholipid vesicles, *Prog. Surf. Sci.,* 13, 1, 1983.
2. **Hui, S. W., Stewart, T. P., Boni, L. T., and Yeagle, P. L.,** Membrane fusion through point defects in bilayers, *Science,* 212, 921, 1981.
3. **Rand, R. P., Kachar, B., and Reese, T. S.,** Dynamic morphology of interacting phosphatidylserine vesicles, *Biophys. J.,* 47, 483, 1985.
4. **Heuser, J. E., Reese, T. S., Dennis, M. J., Jan, Y., Jan, L., and Evans, L.,** Synaptic vesicle exocytosis captured by quick-freezing and correlated with quantal transmitter release, *J. Cell Biol.,* 81, 275, 1979.
5. **Chandler, D. E. and Heuser, J.,** Membrane fusion during secretion: cortical granule exocytosis in sea urchin eggs as studied by quick-freezing and freeze fracture, *J. Cell Biol.,* 83, 91, 1979.
6. **Hui, S. W., Isac, T., Boni, L. T., and Sen, A.,** Action of polyethylene glycol on the fusion of human erythrocyte membranes, *J. Membr. Biol.,* 84, 137, 1985.
7. **Knutton, S. and Bachi, T.,** The role of cell swelling and hemolysis in Sandai virus induced cell fusion and in the diffusion of incorporated viral antigens, *J. Cell Sci.,* 42, 153, 1980.
8. **Ahkong, Q. F. and Lucy, J.,** Osmotic forces in artificially induced fusion, *Biochim. Biophys. Acta,* 858, 206, 1986.
9. **Szoka, F., Olson, F., Heath, T., Vail, W., Mayhew, E., and Paphadjopoulos, D.,** Preparation of unilamellar liposomes of intermediate size (0.1-0.2 μm) by a combination of reverse phase evaporation and extrusion through polycarbonate membranes, *Biochim. Biophys. Acta,* 601, 559, 1980.
10. **Duzgunes, N., Wilschut, J., Fraley, R., and Paphadjopoulos, D.,** Studies on the mechanism of membrane fusion: role of head-group composition on calcium- and magnesium-induced fusion of mixed phospholipid vesicles, *Biochim. Biophys. Acta,* 642, 182, 1981.
11. **Hui, S. W., Nir, S., Stewart, T. P., Boni, L. T., and Huang, S. K.,** Kinetic measurement of fusion of phosphatidylserine-containing vesicles by electron microscopy and fluorometry, *Biochim. Biophys. Acta,* 941, 130, 1988.
12. **Miller, D. C. and Dahl, G. P.,** Early events in calcium induced liposome fusion, *Biochim. Biophys. Acta,* 689, 165, 1982.
13. **Bearer, E. L., Duzgunes, N., Friend, D. S., and Papahadjopoulos, D.,** Fusion of phospholipid vesicles arrested by quick-freezing, *Biochim. Biophys. Acta,* 693, 93, 1982.
14. **Boni, L. T., Hah, J. S., Hui, S. W., Mukherjee, P., Ho, J. T., and Jung, C. Y.,** Aggregation and fusion of unilamellar vesicles by poly-(ethylene glycol). *Biochim. Biophys. Acta,* 775, 409, 1984.
15. **Parente, R. A. and Lentz, B. R.,** Rate and extent of poly-(ethylene glycol)-induced large vesicle fusion monitored by bilayer and internal contents mixing, *Biochemistry,* 25, 6678, 1986.
16. **Boni, L. T., Stewart, T. P. and Hui, S. W.,** Alterations in phospholipid polymorphism by polyethylene glycol, *J. Membr. Biol.,* 80, 91, 1984.
17. **Wojcieszyn, J. W., Schlegel, R. A., Lumley-Sapanski, K., and Jacobson, K. A.,** Studies on the mechanism of polyethylene glycol-mediated cell fusion using fluorescent membrane and cytoplasmic probes, *J. Cell Biol.,* 96, 151, 1983.
18. **Huang, S. K. and Hui, S. W.,** Kinetics of polyethylene glycol-induced fusion of erythrocytes by fluorometry and video-enhanced fluorescence microscopy, unpublished results, 1989.
19. **Roos, D. S. and Choppin, P. W.,** Biochemical studies on cell fusion, *J. Cell Biol.,* 101, 1578, 1985.
20. **Huang, S. K. and Hui, S. W.,** Chemical co-treatments and intramembrane particle patching in the poly(ethylene glycol)-induced fusion of turkey and human erythrocytes, *Biochim. Biophys. Acta,* 860, 539, 1986.
21. **Zimmermann, U.,** Electric field-mediated fusion and related electrical phenomena, *Biochim. Biophys. Acta,* 694, 227, 1982.
22. **Stenger, D. A. and Hui, S. W.,** Human erythrocyte electrofusion kinetics monitored by aqueous contents mixing, *Biophys. J.,* 53, 833, 1988.

23. **Hoestra, D., Wilshut, J., and Schorphof, G.,** Fusion of erythrocyte ghosts induced by calcium phosphate, *Eur. J. Biochem.*, 146, 131, 1985.
24. **Stenger, D. A. and Hui, S. W.,** Kinetics of ultrastructural changes during electrically induced fusion of human erythrocytes, *J. Membr. Biol.*, 93, 43, 1986.
25. **Stenger, D. A. Kubiniec, R. T., Purucker, W. J., Liang, H., and Hui, S. W.,** Optimization of electrofusion parameters for efficient production of murine hybridoma, 7, 505, 1988.

Chapter 8

FREEZE-FRACTURE (-ETCHING) ANALYSIS OF EXO-ENDOCYTOSIS

Helmut Plattner

TABLE OF CONTENTS

I. Introduction 148

II. General Methodical and Functional Aspects 148

III. The *Paramecium* Cell 150

IV. Chromaffin Cells from the Adrenal Medulla 155

V. Comparison of Exo-Endocytotic Systems with or without IMP Aggregates 163

VI. Transbilayer IMP Rearrangement 167

VII. Fusion Model 168

References 170

I. INTRODUCTION

Exo- and endocytosis are frequently coupled events, when a cell releases its secretory contents by reversible fusion of a secretory granule membrane with the cell membrane. Since the freeze-fracture and freeze-etching techniques expose preferably membranes, these techniques have frequently been used to study exo-endocytosis. Of particular interest was the analysis of the restructuring of the membranes involved in fusion, since freeze-fracturing offers the opportunity to analyze the rearrangement of intramembranous particles (IMP). IMP are considered mainly as ultrastructural equivalents of membrane integrated proteins,[1] since their reconstitution with pure lipid membranes resulted in the formation of IMP; IMP may correspond either to one large protein or to a homo- or hetero-oligomer.[1] The situation has become more complicated since focally disordered lipids[2-4] or lipids in a micellar arrangement[5,6] were also shown by freeze-fracturing to form IMP-like structures. This kind of IMP is now frequently discussed to be possibly engaged in membrane fusion.

II. GENERAL METHODICAL AND FUNCTIONAL ASPECTS

Although the process of membrane fusion during exo- and endocytosis has been analyzed for a long time by biochemical and ultrastructural techniques, the actual fusion process remained obscure. Also, for a long time the attention was focused merely on the lipids, since it is evident that it is finally the lipids that fuse. In addition, insufficient freezing techniques have led to a substantial body of freeze-fracture work with a broad range of secretory cells that seemed to support the exclusion of IMP from fusion zones. Furthermore, standard fixation and ultrathin section work also seemed to support the exclusion of membrane-associated proteins from fusogenic sites. Although this again appeared feasible, since lipid bilayers have to come into close contact, this assumption was nevertheless also wrong. Since these method-based ultrastructural aspects had been reviewed already some time ago,[7] I do not intend in this article to revitalize old controversies, but rather to summarize ultrastructural aspects based on adequate fast-freezing and freeze-fracture techniques.

The precise understanding of ultrastructural aspects was important with regard to a possible regulatory role of proteins (membrane integrated and/or associated), which they could effectuate only when they would stay close to the site of actual membrane fusion. This review is largely based on two extreme situations, where fusion sites either display regular IMP aggregates (as in *Paramecium* cells) or a quite irregular membrane structure (as in chromaffin cells).

Fast-freezing turned out to be the only feasible approach, but the kind of fast-freezing technique used is not essential. These techniques have been reviewed by Plattner and Bachmann,[8] by Menco,[9] and Gilkey and Staehelin,[10] and most recently by Sitte et al.[11] The earliest attempts to analyze exo-endocytosis with fast-freezing have been performed by Heuser and coworkers.[12,13] They slammed nerve terminal preparations onto a polished copper block cooled by liquid helium; electric contacts allowed them to freeze the tissue at defined time periods after electrical triggering. Other cell types were triggered and frozen in suspension, like adrenal medullary chromaffin cells[14] and *Paramecium tetraurelia* cells[15,16] (for a summary see Reference 17). The freezing method used was mostly sandwich-propane-jet freezing. If suspended cells can be attached on the surface of a carrier platelet, they may also be frozen by the metal-mirror method; this was performed with sea urchin eggs[18] and with mast cells,[19] while they were stimulated for secretory activity. Since the cooling rate achieved (which is a measure for the quality of cryofixation and for the time resolution to be expected for dynamic processes[8,20] is essentially the same with either one of these methods,[11,21] one might use either one for the analysis of exo-endocytotic processes. Sitte and co-workers[11] also correctly point out that there is no need to use expensive liquid helium

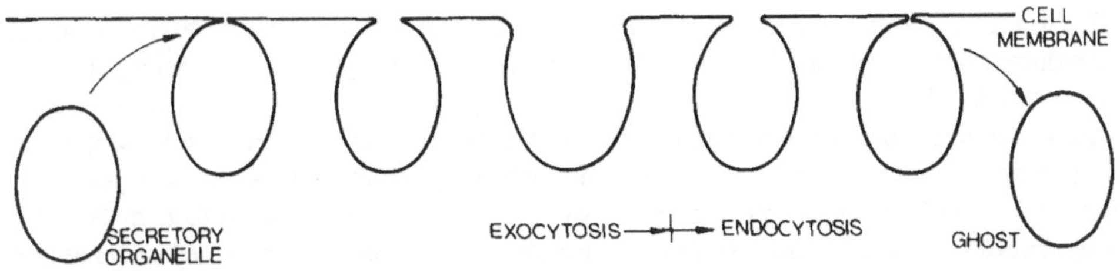

FIGURE 1. Scheme of exo-endocytosis coupling. Note that exocytosis represents a PS-PS fusion (see text), whereas endocytosis is a ES-ES type fusion process. The scheme also illustrates, that it would be difficult to differentiate between these two types of membrane fusion on freeze-fracture replicas.

	nerve terminals	mast cells	non-polar gland cells	polar gland cells	*Paramecium*
response time	≪1 sec	30 sec	>10 min	>10 min	≤1 sec
% released per trigger cycle	small	total release	small	small	up to 95%

FIGURE 2. Comparison of different exocytotic systems, indicating the different response time as well as the percentage of secretory organelles, the contents of which can be released in response to a single trigger (thick lines = cell membrane). Secretory granules are drawn with non-interrupted or interrupted outlines, depending on whether they are available for immediate release or not.

to obtain high cooling rates. In addition, slamming methods might destroy very delicate structures which could be of potential interest.[22]

Another important methodical detail is the possibility to visualize and identify an open connection between the secretory organelle lumen and the extracellular space by deep-etching. The absence of a lipidic diaphragm (long postulated) as a fusion intermediate) can thus be ascertained (see also Reference 7 for this controversy). Further support comes from freeze-substitution studies which should allow one to see any diaphragm formation, if it would occur (which is not the case).[7,23,24]

It occurs only rarely in nature that the membrane of a secretory granule remains permanently integrated in the cell membrane (e.g., with cortical granule membranes during "cortical reaction" in the course of fertilization of an egg cell). This holds at least for the type of exocytosis to be covered here, i.e., "regulated exocytosis" in the sense of Kelly's[25] classification. ("Constitutive" exocytosis, as required for the permanent cell membrane turnover, would be rather difficult to identify). In most cases exocytosis is coupled with endocytosis, which allows the cell to maintain the area size and the compositional specificity of the cell membrane. Again only with a few systems (e.g., with paramecia) one can discern exo- from endocytosis. For this pragmatic reason I shall frequently use the term exo-endocytosis (e.g., for chromaffin cells) without discriminating these two processes. In such systems both aspects can be discerned only on a statistical basis, when ultrastructural data (e.g., size distributions of exo-endocytotic openings) are compared with functional data (e.g., catecholamine output) obtained at the same time.[14]

One has to bear in mind, however, that these two types of membrane fusion are of a totally different kind (Figure 1). According to the freeze-fracture terminology generally accepted,[26] exocytosis is a PS-PS fusion, while endocytosis is a ES-ES fusion[27] (PS = plasmatic surface, ES = exoplasmic surface).

Figure 2 compares some of the cell types frequently used for the analysis of exo-endocytosis by freeze-fracturing and freeze-etching. The scheme takes into account on the one hand the time required for immediate response and for the release of the majority of secretory contents on the other hand.

"Fast secretors" are nerve terminals and paramecia, but only paramecia release most of their secretory contents at a time in a synchronous way. Both these systems appear very attractive for studying the sequence of events; they should allow one to discern between phenomena during exocytosis as well as during endocytosis. Both systems have been analyzed by fast-freezing and freeze-fracturing (see below). Recently, it has been recognized, however, that one has to know with some precision the level of the nerve terminal tissue actually analyzed;[28] otherwise false data may be obtained on the course of exocytosis-coupled endocytosis. In the case of paramecia, the distinct morphology of the cell membrane and of the secretory organelle ("trichocyst") membrane involved in membrane fusion,[15,17,29,30] as well as the characteristic appearance of these membrane regions during and after completing endocytosis ("resealing"), allows one to identify very clearly all stages of exo- and endocytosis, i.e., to recognize their directionality from the actual appearance.

Figure 2 also takes into account the occurrence of "compound exocytosis" in mast cells which allows them to release their secretions by granule-granule fusion, once some granules have fused with the cell membrane. Therefore, total degranulation can be completed within less than 1 min. The "cortical reaction" occurring in egg cells during fertilization is also rather fast (though also not comparable with nerve terminals or paramecia).

Finally, Figure 2 reminds us of the sluggish reaction of the "average gland cells", which might either have a polar or a nonpolar organization. Examples are pancreatic acinar cells or chromaffin cells of the adrenal medulla. To my knowledge, acinar cells have not been subjected to fast-freezing, but important data have been obtained just by applying "regular freezing" techniques (plunging a piece of tissue into a liquid cryogen) to nonchemically pretreated pieces of tissues.[31] Chromaffin cells, which we have analyzed by fast-freezing and freeze-etching,[14] will serve us as an example that exo- and endocytotic processes can be discerned with some degree of reliability only on a statistical basis.

It is important to note that early stages of membrane fusion may be so ephemeral that they might be "caught" only with the fastest freezing methods available. For example, they might last for only 1 ms or less.[16] Fast-freezing methods may allow for a "time resolution" just adequate for a "physical fixation" of these intermediates. These aspects, including the relative value of cooling rates, are discussed in more detail elsewhere.[20] I should underscore that methods to be applied must yield high cooling rates in the uppermost temperature regions and not just below 0°C.[32] "Cold metal block" freezing was also sufficient to seize events in the millisecond range,[12] but early fusion stages have been reported only exceptionally.[23] This has so far been achieved best with chromaffin cells[14] processed by the sandwich-propane jet method (critically evaluated in Reference 32).

III. THE *PARAMECIUM* CELL

Ultrastructural aspects pertinent to this system have recently been summarized.[17] This includes the following details:

(1) 95% of all trichocysts are docked to the cell membrane.
(2) Therefore, most of them can be rapidly released (see below) and the ultrastructural changes occurring during this synchronous exocytosis (as described below) can easily be analyzed.
(3) Preformed release sites are arranged in regular distances at precisely defined sites, i.e., on "cross-ridges" of the regular egg box-like pattern of the cell surface (Figure 3).

FIGURE 3. Freeze-fracture replica (exoplasmic fracture face = EF[26]) of the cell membrane of a fast-frozen *Paramecium tetraurelia* cell, whose cell surface displays a characteristic surface relief. Only "perpendicular ridges" show IMP aggregates ("rosettes") indicating the presence of a dischargeable trichocyst.[29] ["Ring" IMP are only faintly visible (as pits) and form a diameter of 300 nm[29]]. RI = rings, RO = rosettes (arrowheads). Unpublished micrograph. As in all subsequent micrographs, the encircled arrow indicates the shadowing direction. Bar = 1 μm.

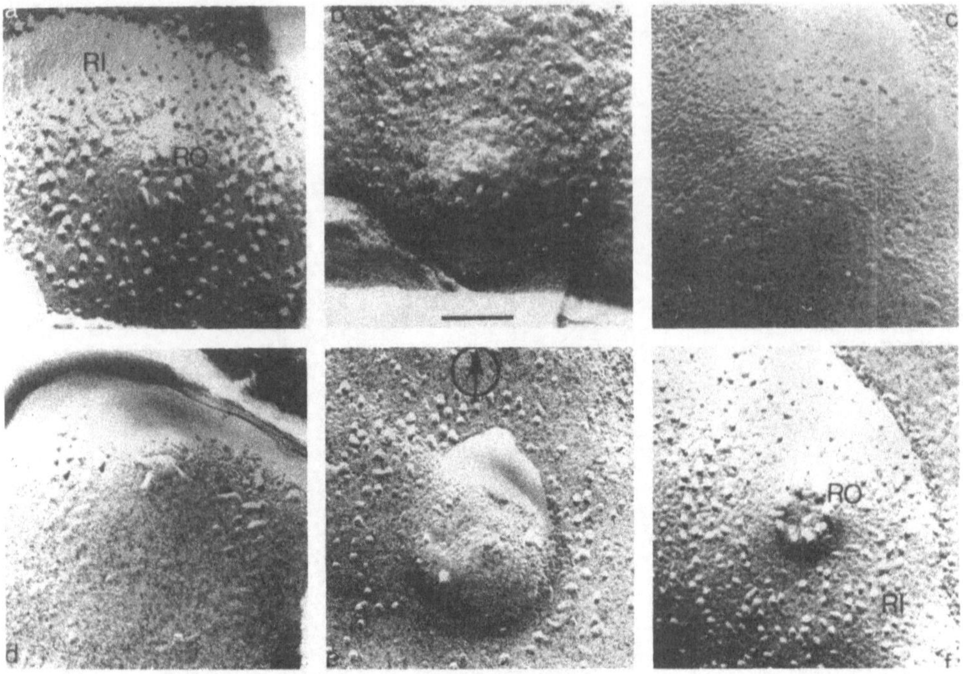

FIGURE 4. Docking sites of trichocysts, as recognized on the PF-face of the cell membrane of slightly chemically fixed and freeze-fractured *P. tetraurelia* cells. Even weak aldehyde fixation causes that ring (RI) as well as rosette (RO) IMP remain associated with the PF-face during freeze-fracturing [(a) control, (b) slight pronase digestion, (c) extensive pronase digestion, (d) extensive trypsin digestion, (e) digestion with Sepharose-bound pronase, and (f) esterase treatment]. Only proteases digest rosette and some ring IMP (c to e), whereas controls [buffer only, (a)] or esterase (f) remain without any visible effect. Pronase bound to Sepharose beads (larger than a cell) can digest only IMP beads because of steric accessibility (e). Bar = 0.1 μm. (From Vilmart, J. and Plattner, H., *J. Histochem. Cytochem.*, 31, 626, 1983. With permission.)

(4) These areas display characteristic IMP assemblies, including a 300-nm double "ring" of IMP and, in its center, a "rosette" of about 9 IMP.
(5) It is known that the presence of a rosette indicates the presence of a trichocyst underneath that can be discharged by exocytosis.[33] The ring appears as a "parenthesis" type structure, when a docking site is not occupied by a trichocyst.[30,33,34]
(6) These data are supplemented by ultrastructural studies with mutants defective in membrane fusion. For instance, *nd* ("nondischarge") mutants have perfect ring structures, but they do not at all or not completely assemble rosettes above docked trichocysts;[34,35] therefore, the capacity of membrane fusion is paralleled by the presence of a rosette.
(7) Always associated with the occurrence or absence of rosettes is the presence or absence of "connecting material" between the trichocyst and the cell membrane.[35,36] These two ultrastructural features, therefore, determine the fusion capacity of these membranes. (For possible functional equivalents see Reference 37.)

Also important is the fact that both, rosette (and ring) IMP (Figure 4) as well as "connecting materials", are sensitive to proteolytic enzymes.[7,38,39] Rosette IMP can also be digested with carrier-bound proteases[39] because they are located on top of the perpendicular elevations of the cell surface (Figure 4). These structures were, therefore, assumed to represent proteins.

The aspect of synchronous exocytosis in this system has recently been reviewed in more detail.[17] From electrophysiological recordings of surface capacitance changes we know that the membrane of a trichocyst remains fused with the cell membrane for only 50 ms; within

1 s all dischargeable trichocysts can be discharged when a cell is triggered by adding an appropriate polyamino-compound (like aminoethyldextran = AED).[17]

To analyze ultrastructural changes during synchronous exocytosis cells have been pipetted onto one of the two platelets used to make up a sandwich sample for fast-freezing; AED was added onto the other platelet (made of copper). Both can be rapidly assembled and, within 1 s or less, a jet of melting propane (−150°C) is shot onto the copper platelet. Since the other platelet is a thermal insulator, this allows for very high cooling rates, particularly in the uppermost temperature regions,[32] which are the most relevant for membrane fusion.[20]

This method allowed us to obtain snapshots of the earliest presumable fusion stages (Figure 5). The situation is plotted schematically in Figure 6.

In the experiments presented in Figure 5 the formation of larger exocytotic openings was, in addition, inhibited by adding antibodies against cell surface antigens which, thus, become partly immobilized.[16] Earliest fusion stages are formed in the center of rosettes. They are represented by 10-nm large irregularities, but it is difficult to decide whether crater-, particle-, or pit-like structures are formed first. (It appears useless to me to look for further details in this dimension on freeze-fracture replicas since plastic deformations, decoration, and condensation phenomena can distort the true geometry (cf. Reference 40); 10-nm intermediates are sometimes located on a nipple-like depression of the cell membrane (Figures 5b and c) directed toward a similar protrusion of the trichocyst membrane (documented in Figure 3c in Reference 17). A larger exocytotic opening (achieved under normal trigger conditions, i.e., without simultaneous inhibition) is schematically shown in Figure 7; it shows rosette IMP still closely opposed to the fusion membranes.

The formation of a diaphragm was never observed.

Rosette IMP are presumably still present on the fusion site, as the two membranes fuse (Figures 5a, 6, and 7). They must then spread rapidly (as in Figures 5 b and c which, however, have been obtained by simultaneous maximal trigger and inhibition, as outlined above). We found that rosette IMP spread beyond rings and then might be scattered all over the membrane of the cell body.[15,30]

In later stages of exocytosis one can see contents protrude.[15] Later stages are presented in Figure 8. One sees the empty trichocyst membrane which can easily be identified by its characteristic structural features (Figure 8a). Later on, the opening becomes smaller again (Figure 8b) and resealing (endocytosis) starts. This ends with the formation of an "oval ring" (Figure 8c) which later collapses to a "parenthesis"-type structure (Figure 8d). The collapse may be important to get trichocyst membranes again together so they can be fused and detached (see also below). These stages are devoid of a rosette.

This sequence of events can be ascertained by morphometric evaluation of freeze-fracture replicas (Figure 9). These experiments were hampered by the time required to assemble sandwich samples, though freezing was then very fast. We now aim at applying fast-freezing at very precise time points of synchronous exocytosis performance, which altogether takes only 1 s (see above). These restrictions, however, do not invalidate the sequence of events outlined here.

Another important aspect we found with this system is the high specificity of the process of membrane resealing.[30] Only in 0.3% of all exocytotic events analyzed could we find a partial incorporation of elements of trichocyst membranes (clearly recognizable by their typical "texture") into a resealed cell membrane (Figure 10). Moreover, such exceptions were restricted to one cell. This small failure rate would, thus, roughly indicate the error probability the system works with. No such estimates would be possible with biochemical methods since one could not discriminate false resealing sites from sites of ongoing secretory activity and since contamination problems are still too serious with cell fractionation. The behavior of IMP arrays can, of course, yield no information on any possible exchange of lipid materials. Interestingly, with aberrant resealing rings do not collapse; this could be the reason why the trichocyst membrane cannot be fully internalized as normally (see above).

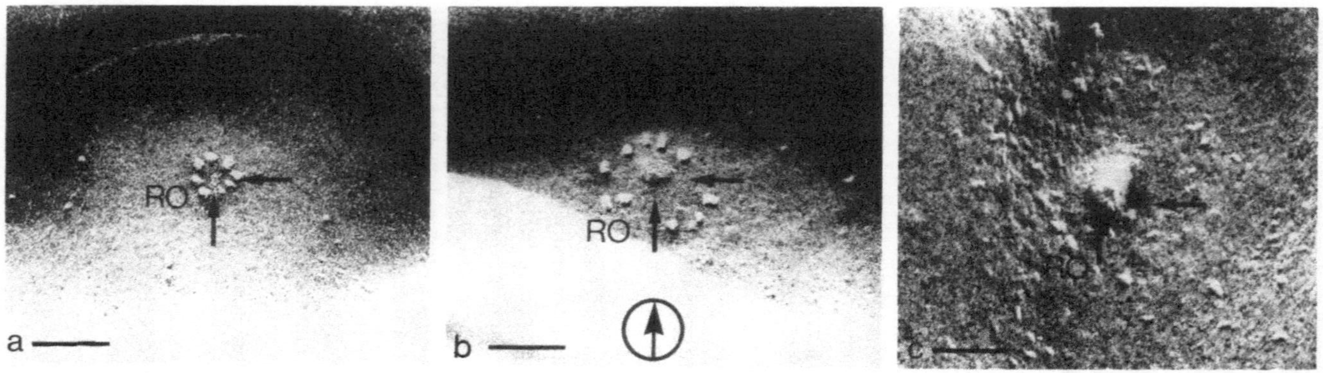

FIGURE 5. EF-face[26] of a *P. tetraurelia* cell, quick-frozen in the native state during simultaneous exocytosis triggering and exocytosis inhibition by antibodies directed against cell surface antigens (see Reference 16). RO = ''rosette'' IMP. The crossing points of the arrows indicate irregularities in the membrane, 10 nm in size, which might represent early fusion stages (for more details, see text). Also note the formation of a nipple-like elevation in (b) and (c). In (c) a relatively large pit is found on top of the ''nipple'', possibly indicating an expanding early exocytotic opening. Bars = 0.1 μm. (From Momayezi, M., Girwert, A., Wolf, C., and Plattner, H., *Eur. J. Cell Biol.*, 44, 247, 1987. With permission.)

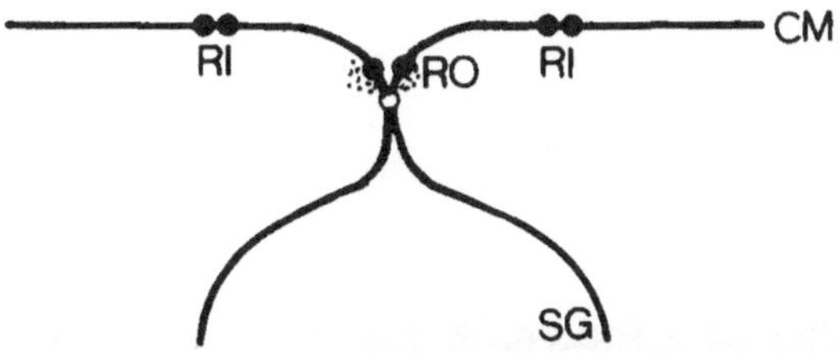

FIGURE 6. Summary of early fusion events during trichocyst exocytosis in *P. tetraurelia* cells; to be compared with Figure 5, CM = cell membrane; RI = ring IMP; RO = rosette IMP (closed circles); and SG = secretory granule (trichocyst). Trichocysts have a nipple-like elevation already in the resting state,[17] and a similar structure is formed within the cell membrane as fusion goes on (see Figure 5); the profile is exaggerated here. Stippled area indicates membrane-associated proteins. The open circle indicates the focal fusion spot.

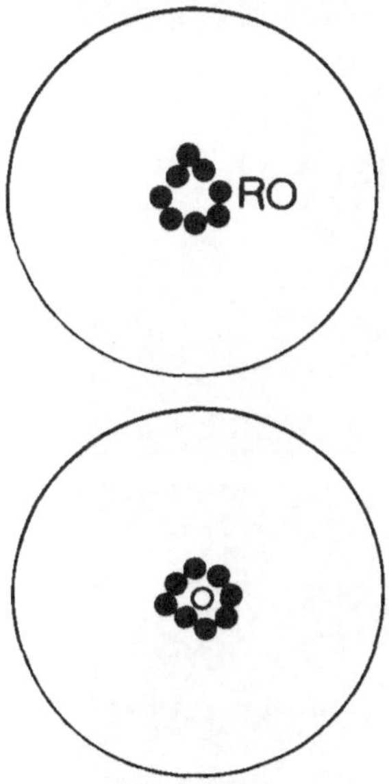

FIGURE 7. Organization of the membrane fusion in *P. tetraurelia* cells at the site of trichocyst release, (a) before fusion and (b) at the earliest fusion event (open circle). Solid discs represent rosette IMP. The maximal size of the exocytotic opening (corresponding to the size of a ring) is, for comparison, indicated by the circles (beyond which rosette IMP would spread).

The sequence of events as it normally occurs in a *Paramecium* cell is schematically summarized in Figure 11. Figure 11 (A) represents the resting state and (B) shows the occurrence of a focal fusion event amidst a rosette IMP aggregate. As rosette IMP spread, the exocytotic opening enlarges (C and D). Rosette IMP are spread beyond rings (E through G), that only seem to delineate the extrusion site, but not to contribute to fusogenicity. Figures 11 (F) and (G) are early and late resealing stages, respectively.

IV. CHROMAFFIN CELLS FROM THE ADRENAL MEDULLA

The chromaffin cells used for this study were isolated from the bovine adrenal medulla.[14]

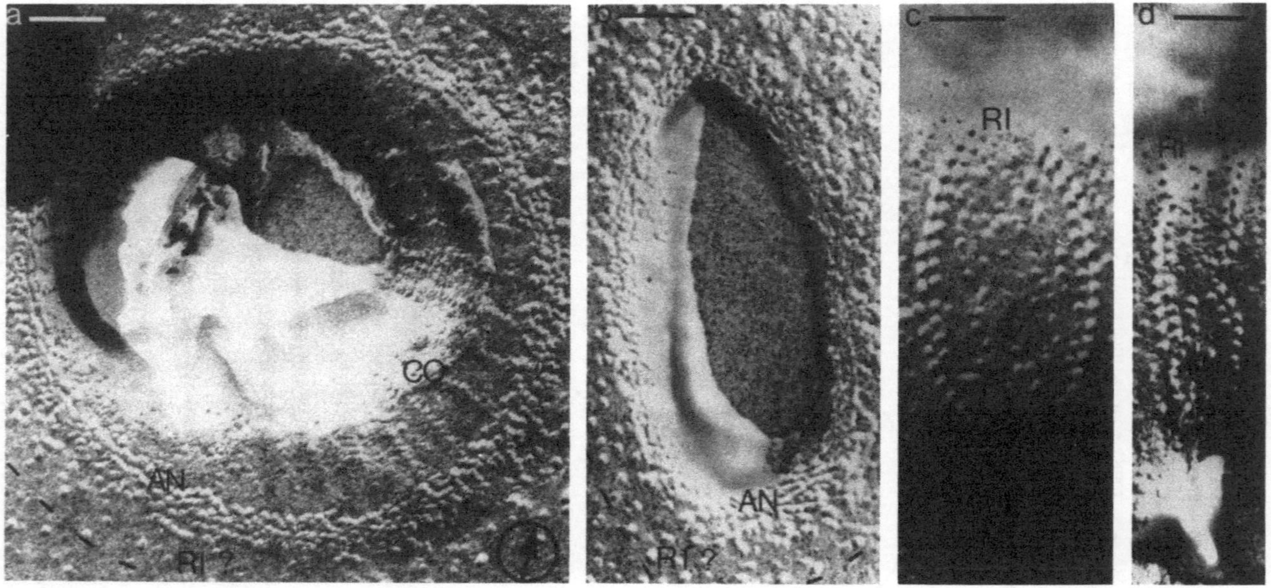

FIGURE 8. Freeze-fracture morphology of exo-endocytosis coupling in *P. tetraurelia* cells. (a) Maximal size of an exocytotic opening, showing a presumable dilated ring (RI) as well as elements of the uppermost region of the trichocyst membrane (AN = "annulus" IMP, CO = "collar" striations). (b) Collapse of the trichocyst release site, which leads to early (c) and late (d) resealing stages (c = "oval ring", d = "parenthesis"). Note the absence of rosette IMP or of trichocyst membrane components in the resealed cell membrane. Bars = 0.1 μm. (From Plattner, H., Pape, R., Haacke, B., Olbricht, K., Westphal, C., and Kersken, M., *J. Cell Sci.*, 77, 1, 1985. With permission).

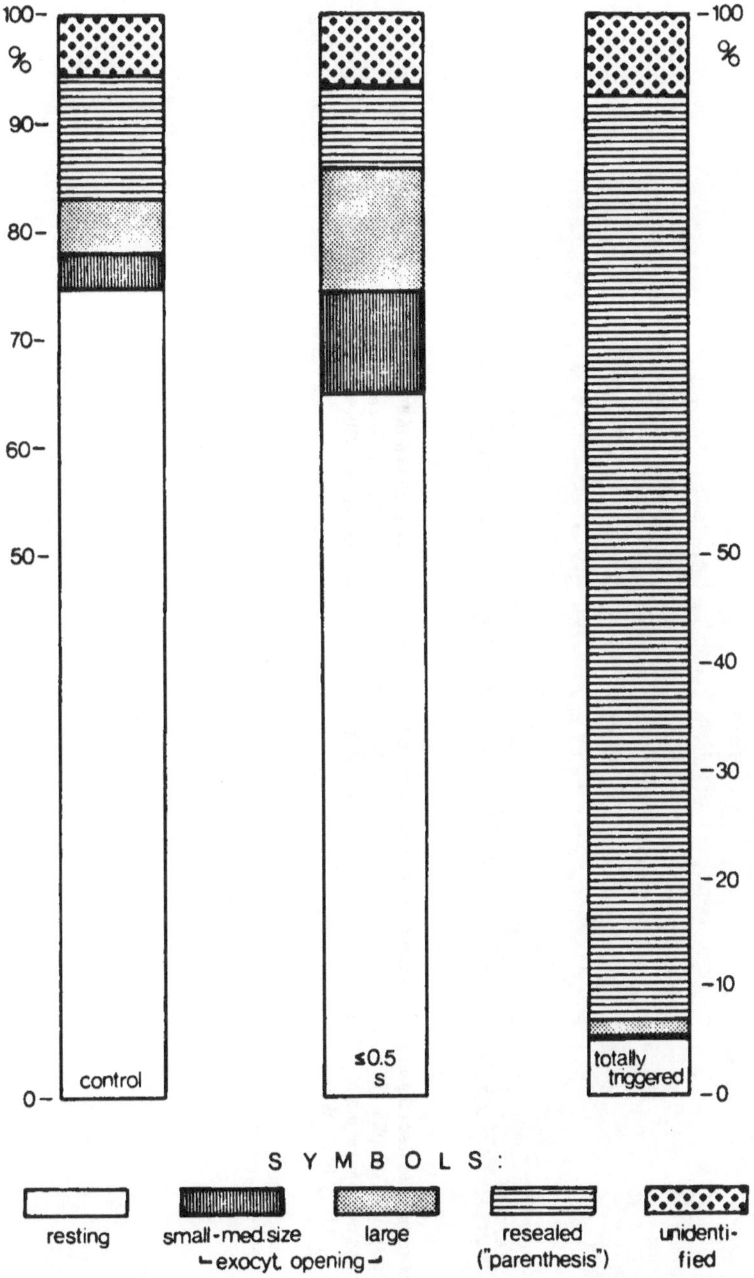

FIGURE 9. Quantitative evaluation of exocytosis and resealing stages observed in the course of synchronous exocytosis in *P. tetraurelia* cells. The earliest trigger stages ($\leq$0.5 s) we could get hold of already allowed us to observe an increase in the number of exocytotic openings. Later on resealed stages are abundant. (From Olbricht, K., Plattner, H., and Matt, H., *Exp. Cell Res.*, 151, 14, 1984. With permission.)

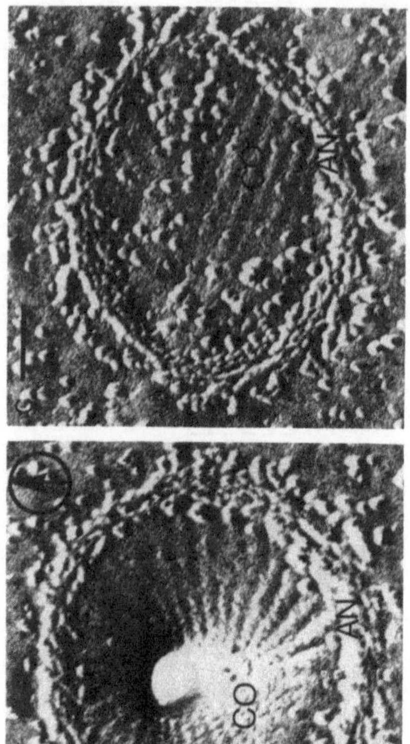

FIGURE 10. Aberrant membrane resealing in a *P. tetraurelia* cell; to be compared with normal resealing as shown in Figure 8. Aberrant resealing is rare; it involves the integration of collar (CO) and annulus (AN) elements into the cell membrane. (a,b,c) Could represent a sequence of events, but all these stages were found only in one out of 178 cells. Bars = 0.1 μm. (From Plattner, H., Pape, R., Haacke, B., Olbricht, K., Westphal, C., and Kersken, H., *J. Cell Sci.*, 77, 1, 1985. With permission.)

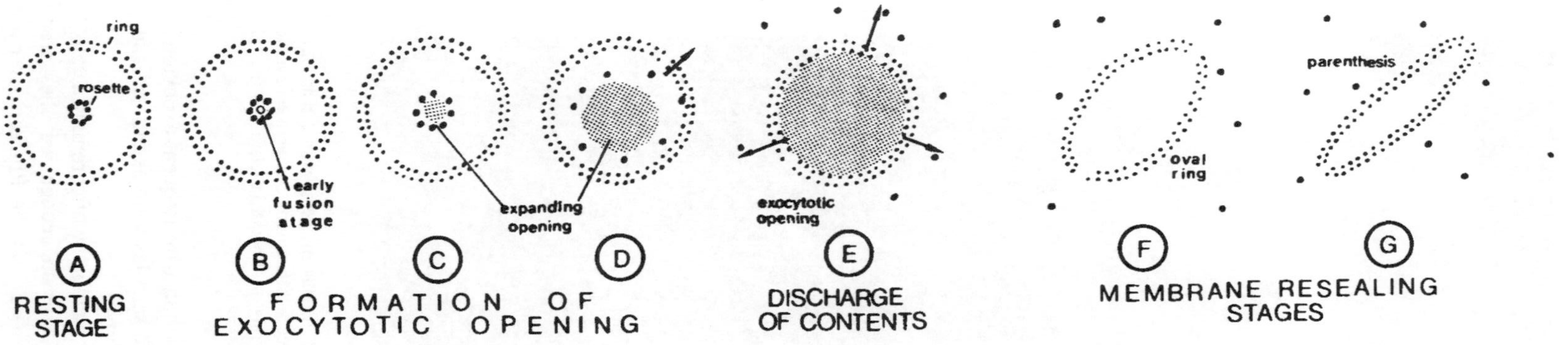

FIGURE 11. Scheme of freeze-fracture appearance of exo-endocytosis in *P. tetraurelia*. Note the occurrence of an early focal fusion event amidst rosette IMP, which subsequently spread as the exocytotic opening widens. Rosette IMP spread beyond rings that delineate an exocytosis site. During resealing first an oval ring, then a parenthesis stage is formed. Stages (A through G) may occur even within one cell (stage G from earlier exocytotic activity). (From Momayezi, M., Girwert, A., Wolf, C., and Plattner, H., *Eur. J. Cell Biol.*, 44, 247, 1987. With permission.)

FIGURE 12. PF-face of chromaffin cell isolated from the bovine adrenal medulla, triggered with carbachol (2 min, 38°C) and subjected to propane-jet-sandwich freezing. Note the occurrence of different size classes of exocytotic openings (as explained in Figures 15 and 16) and the absence of IMP clustering or dispersal in relation to sites of ongoing exocytotic activity. Bar = 1 μm.

They were triggered by carbachol for 2 min (38°C), a time period with frequent exocytotic but less frequent endocytotic events.[41] Still at 38°C, sandwich samples were assembled and subjected to fast-freezing by a jet of melting propane (see above).

Figures 12 and 13 show the PF- and the EF-face of the cell membrane of chromaffin cells thus prepared. Exocytotic openings of very different size classes are recognized; they are represented by depressions, when viewed from the outside (i.e., on the PF-face) or by

FIGURE 13. Same as in Figure 12, but showing an EF-face. Small nipple-like elevations with focal fusion stages on top are recognized several times. Bar = 1 μm.

concomitant elevations, when viewed from the inside (i.e., on the EF-face). (This also allows one to discriminate microvilli from sites of ongoing secretory activity).

The correlation of this cell membrane morphology with ongoing exocytotic activity is supported by the occurrence of some occasional cross-fractures through cells which show some typical Ω-shaped exocytotic profiles (Figure 14).

Figure 15 gives a collection of all sizes of exo-endocytotic events (with the chance to have more exocytotic than endocytotic processes, as outlined above). Although a real time sequence of exo-endocytosis stages cannot be obtained with a nonsynchronous system, one can subdivide all stages actually observed into subgroups according to size. These stages

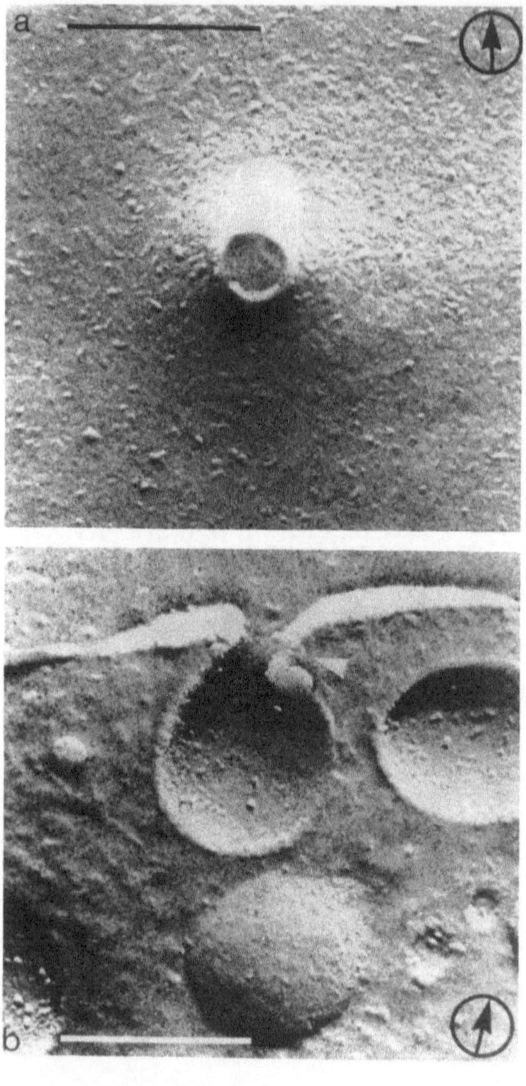

FIGURE 14. Same preparation as in Figure 12, showing a correlation between a cell membrane fracture (a) and a cross-fracture through the cytoplasm with an Ω-shaped exocytotic profile (b). Note the occurrence of some IMP close to the exocytotic opening in (a). Bars = 0.25 μm. (From Schmidt, W., Patzak, A., Lingg, G., Winkler, H., and Plattner, H., *Eur. J. Cell Biol.*, 32, 31, 1983. With permission.)

can then be pooled into different categories of size range. These, finally, can be compared with the actual output of catecholamines (secretory materials, that can easily be determined). Figure 16 was thus obtained. It clearly documents that all size classes of exocytotic profiles occur with increased frequency when the cells are triggered, as compared to untriggered aliquots. The data obtained with stimulated cells have to be compared with basal release rates. This cannot be neglected since cells freshly isolated from the tissue by treatment with enzymes and omission of calcium are "leaky" to some extent, though their vitality is high. Only this quantitative evaluation justifies the arrangement of the details, as shown in Figure 15, from the left to the right, in the sense of "morphogenetic series".

In Figure 15, the series on top represents the EF-face, and that on the bottom represents the PF-face. The series in the middle corresponds to that on top, with the exception that these aliquots were deep-etched for 1.5 min at −100°C. We thus could ascertain the absence of a lipidic diaphragm.

Most interesting is the appearance of *bona fide* early fusion stages ("fusion intermediates"). They are represented by 10-nm large pits which are mostly located on top of a funnel-like depression of the cell membrane (or an elevation, when viewed on EF-faces) (Figures 15a, g, and m). Beyond the "early fusion intermediates" one can find all sorts of size transitions (Figures 15b, h, and n or c, i, and o, respectively). These openings are always etchable (Figures 15 h and i). From these data we concluded that the smallest fusion stages simply enlarge to yield larger ones.[14] This is compatible only with a lipidic nature of the earliest fusion stages according to the "focal fusion scheme" presented before[7] and at the end of this chapter.

Further expansion of the opening is shown in Figures 15 d through f, j through l, and p through r. The final size may correspond to the size of a chromaffin granule (0.2 to 0.3 μm; 0.22 μm on the average).

None of these stages was ever accompanied by a dispersal or clustering of IMP around the fusion zone.[14]

As mentioned, Figure 16 compares the frequency of these profile sizes with the actual secretory activity measured over a 2-min test period. Again, it must be emphasized that for no individual fusion image can we unequivocally ascertain that it is on the way out (exocytosis) or on the way in (endocytosis). However, since during the time of analysis (2 min) exocytosis is more frequent than endocytosis (which operates with a somewhat longer half-life[41]), we are confident that the stages presented in Figure 15 are representative mainly for ongoing exocytosis.

The data obtained for *bona fide* early fusion events in chromaffin cells are summarized in Figure 17. This includes the formation of a fusion intermediate on top of a funnel-like depression. As mentioned IMP may stay close to the actual fusion site.[14]

V. COMPARISON OF EXO-ENDOCYTOTIC SYSTEMS WITH OR WITHOUT IMP AGGREGATES

Clearly the *Paramecium* cell and the chromaffin cell represent two extreme situations, since the first operates with and the second without clear-cut IMP aggregates on the fusogenic sites. Already in a previous review on this topic[7] I had pointed out that the difference is only marginal for the following reason. With all cell types, which do not display IMP aggregates (when frozen rapidly for freeze-fracturing), one can recognize the presence of some IMP — irregularly scattered though — close to the fusion zone. If one accepts the functional interpretation of these IMP mainly as membrane-integrated proteins,[1] this implies the presence of such proteins at fusion sites. This knowledge, however, is only of restricted value, as long as we do not know in any case the molecular identity of these proteins. IMP along the active zones of nerve terminals have been speculated to represent Ca^{2+} channels.[42] Such a function could not be substantiated for rosettes of paramecia[43] so far.

More important, therefore, is that the presence of membrane-associated proteins on the fusion site has been documented for *Paramecium* cells,[30,36,38] chromaffin cells,[44] the "active zones" of nerve terminals,[45] and with egg cells.[46] For other examples, see References 7 and 38. It might well be the previous failure to detect such materials on ultrathin sections has been due to loss during preparation. Only quite recently could we obtain some evidence for the possible composition of these "connecting materials" located between the two membranes to be fused during exocytosis.[37]

The difference between the "*Paramecium*-type" and the "chromaffin cell-type" ultra-

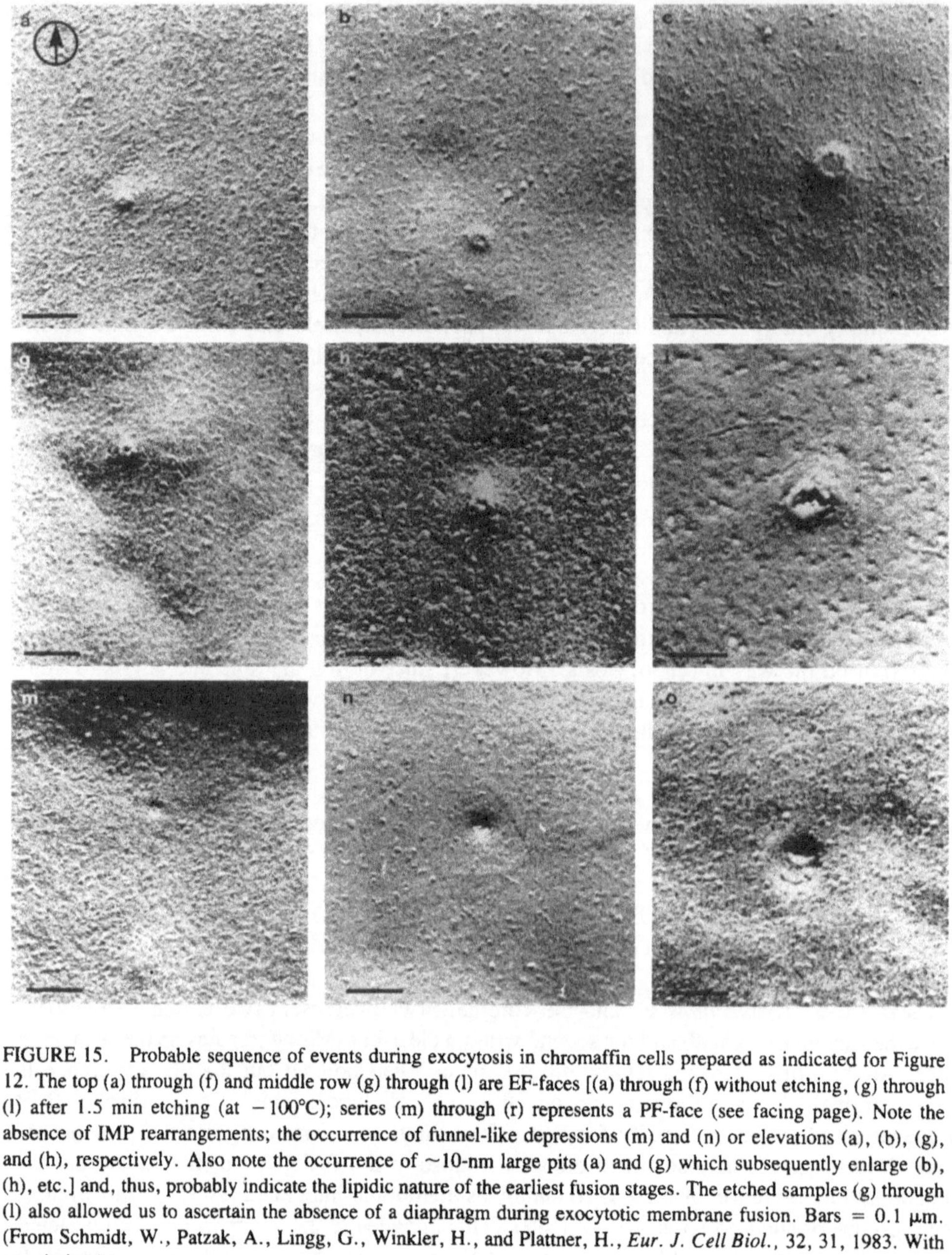

FIGURE 15. Probable sequence of events during exocytosis in chromaffin cells prepared as indicated for Figure 12. The top (a) through (f) and middle row (g) through (l) are EF-faces [(a) through (f) without etching, (g) through (l) after 1.5 min etching (at $-100°C$); series (m) through (r) represents a PF-face (see facing page). Note the absence of IMP rearrangements; the occurrence of funnel-like depressions (m) and (n) or elevations (a), (b), (g), and (h), respectively. Also note the occurrence of ~10-nm large pits (a) and (g) which subsequently enlarge (b), (h), etc.] and, thus, probably indicate the lipidic nature of the earliest fusion stages. The etched samples (g) through (l) also allowed us to ascertain the absence of a diaphragm during exocytotic membrane fusion. Bars = 0.1 μm. (From Schmidt, W., Patzak, A., Lingg, G., Winkler, H., and Plattner, H., *Eur. J. Cell Biol.*, 32, 31, 1983. With permission.)

structure of exocytosis sites would then possibly be only a gradual one. Table 1 summarizes the systems involving regular IMP arrays.

The presence of regular IMP arrays in motor end plates and, less evidently though, in some spinal cord and brain cells (Table 1) might suggest that this precise ultrastructural setup could account for a rapid reaction to a stimulus. For example, Figure 18, from the work of Heuser et al.[12] clearly shows that, in motor end plates of frogs, the exocytotic release of transmitter takes place in close association with perl-strings of IMP. As mentioned, ultrathin sections of this region had shown the occurrence of "connecting materials"[45] of so far unidentified nature (see Section VII). In paramecia the release of trichocysts is also very rapid (though we do not yet know why). Why IMP arrays are needed in other systems

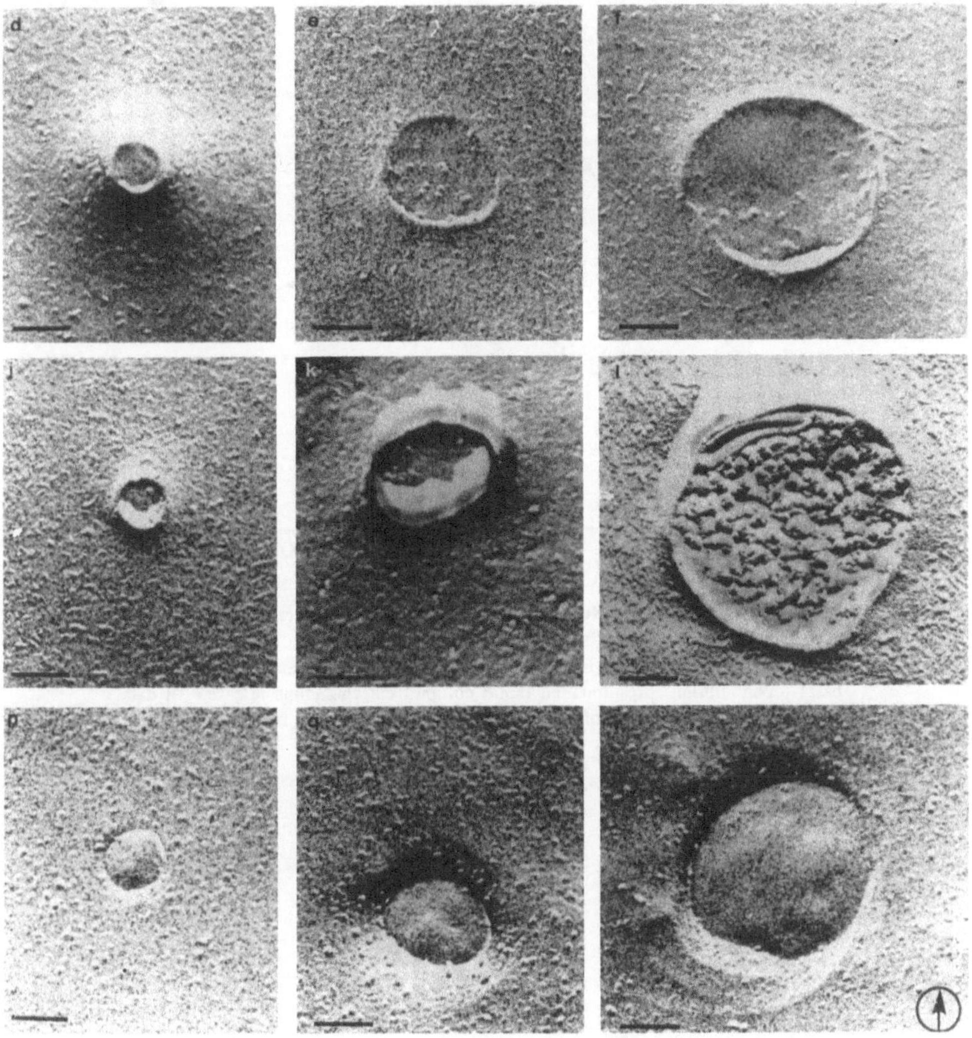

FIGURE 15 (continued).

listed in Table 1 is also poorly understood since not all these cells are known to have to release materials by exocytosis very rapidly. The implication of "rhoptries" (their fusion sites being endowed with rosettes) in host cell invasion also has become questioned.[66-68] But clearly in heterothermic organisms, particularly in protozoa, IMP arrays are much more frequent than in higher organisms (Table 1). Unfortunately, no heterothermic organism has so far been thoroughly analyzed by freeze-fracturing with regard to IMP arrays in different secretory cells. So far we can only stress the possible importance (1) of the occurrence of some IMP and (2) of some kind of "connecting material" as general features relevant for exo-endocytotic membrane fusion. These may occur in a less concentrated form or as regular arrays at the exocytotically active regions of most organisms except for those listed in Table 1. Since regular IMP arrays are more evident in heterothermic organisms (Table 1), the question arises whether they might help there to provide a specific lipid microdomain which would be sufficiently susceptible to fusion triggering even at unfavorable temperatures. We do not know the answer to this question, however.

Older references on the formation of IMP-free zones before/during exocytosis and of IMP enrichment during endocytosis do not appear reliable since they have been shown to be plagued by artifacts caused by the chemical pretreatments used before freezing (for review, see Reference 7). The general assumption was that exocytosis entails IMP dispersal before

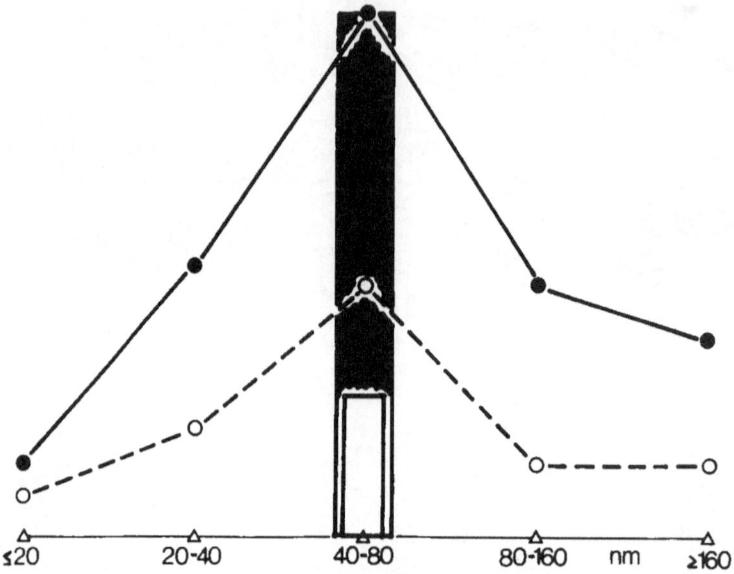

FIGURE 16. Spontaneous (open symbols) and carbachol-triggered (solid symbols) exocytosis in chromaffin cells (leaky after enzymatic isolation). *Abszissa*: size distribution of exo-endocytotic openings. *Ordinate*: relative frequency of profiles and basal (open column) or stimualted (solid column) catecholamine release. All data were obtained within 2 min of stimulation (see text). According to data contained in Reference 14.

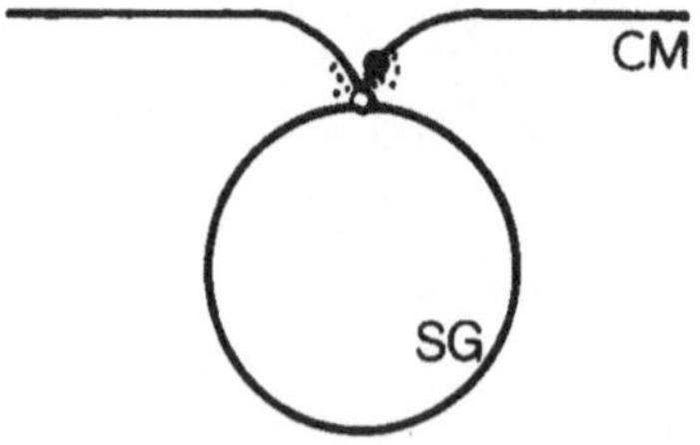

FIGURE 17. Scheme of exocytotic membrane fusion in adrenal medullary gland cells (cf. Figure 15). Note the formation of a funnel-like depression within the cell membrane (CM), where it starts to form an early fusion zone (open circle); also IMP (solid disc) and membrane-associated proteins (stippled area) remain closely associated with this zone of contact between cell membrane and secretory granule (SG), when they fuse.

fusion can occur, while endocytosis was believed to be paralleled by IMP aggregation.[69] The consistency of these findings led Chandler[70] to the explanation that they would represent reproducible artifacts which might reflect increased membrane fluidity before membrane fusion can occur.

However, when fast-freezing techniques were used, exocytosis has never been observed to entail any long-distance IMP dispersal. This holds for nerve terminals in motor end plates of frogs,[12,13] for the cortical reaction in sea urchin eggs,[18] for mast cells,[19] paramecia,[15,17] chromaffin cells,[14] amoebocytes from *Limulus*[23] or for neutrophilic polymorphonuclear granulocytes.[71] Interestingly, for some of these cell types previous publications had shown either IMP dispersal or IMP aggregation to occur before/during exocytosis.[7] This underscores once more the importance of fast-freezing techniques for a reliable analysis of ultrastructural events during exo-endocytosis. However, such techniques are not yet available, e.g., for

Table 1
EXO-ENDOCYTOTIC SYSTEMS INVOLVING REGULAR IMP ARRAYS[a]

Species	Cell type or type of organism	Type of secretory granule or material	Ref.
(A) ***Protozoa***			
Tetrahymena	Ciliate	Mucocysts	47, 48, 49
Paramecium	Ciliate	Trichocysts	29, 34, 35
Didinium	Ciliate	Toxicocysts, mucocysts	50
Spirostomum, Litonotus, Blepharisma, Folliculina, Halteria	Ciliates	Mucocysts	50
Cyclidium	Ciliate	Mucocysts	51
Euplotes	Ciliate	Ampules	52
Acanthocystis, Raphidiophrys, Heterophrys, Ciliophrys	Actinopods	Extrusomes[b]	50, 53
Chilomonas	Flagellate	Small extrusomes	54
Eimeria	Pathogenic Sporozoa	Rhoptries (?)[c]	55
Toxoplasma, Sarcocystis	Pathogenic Sporozoa	Rhoptries (?)[c]	56
Aggregata	Pathogenic Sporozoa	Rhoptries (?)[c]	57
Plasmodium	Pathogenic Sporozoa	Rhoptries (?)[c]	58
(B) ***Metazoa***			
Ascaris (nematode)	Sperm		59[d]
Lithobius (chilopodes, arthrophods)	Neuromuscular junction	Neurotransmitter	61
Frog	Neuromuscular junction	Neurotransmitter (acetylcholine)	12, 13, 62
Petromyzon (lamprey)	Spinal cord	Neurotransmitter	63
Rat	Spinal cord	Neurotransmitter	64
Rat	Hippocampus (brain)	Neurotransmitter (glutamate)	65

[a] Only some of the data collected in this table were obtained by fast-freezing techniques. Because of an analogy with closely related systems, for which regular IMP arrays have been ascertained to be involved in exo-endocytosis, these data are included here with some reservation.

[b] "Extrusome" is the general term for extrusive organelles in protozoa, i.e., for secretory organelles whose release can be followed in the light microscope. Trichocysts, mucocysts, and toxicocysts are special types of extrusomes.

[c] Recent work by Enzeroth et al.[66] suggests that "dense granule" rather than "rhoptry" contents are released during host cell invasion. Work cited here assumes or infers mostly the opposite. (See also Reference 67 and review by Brever.[68])

[d] Secretory cells of the vas deferens of the same organism showed no rosette-like structures.[60]

tissues. In this case even moderate cooling rates (as achieved by dipping small tissue pieces into Freon) can yield reliable data, if one is prepared to accept less aesthetic micrographs (due to ice crystal damage). Making this compromise some authors also observed that no large scale IMP rearrangements occur, e.g., with exocrine pancreatic acinar cells during exocytosis triggering.[31]

This does not invalidate the value of fast-freezing techniques since they are the only ones to get hold of the earliest fusion stages (see Figures 5 and 15). It would also be interesting to reanalyze exocytotic events in zoospores from the phythopathogenic fungus *Phytophthora palmivora*, which has been reported to involve the formation of a diaphragm starting from a central micelle-like knob.[72]

VI. TRANSBILAYER IMP REARRANGEMENT

When fast-freezing techniques have been applied to presynaptic vesicles (derived from

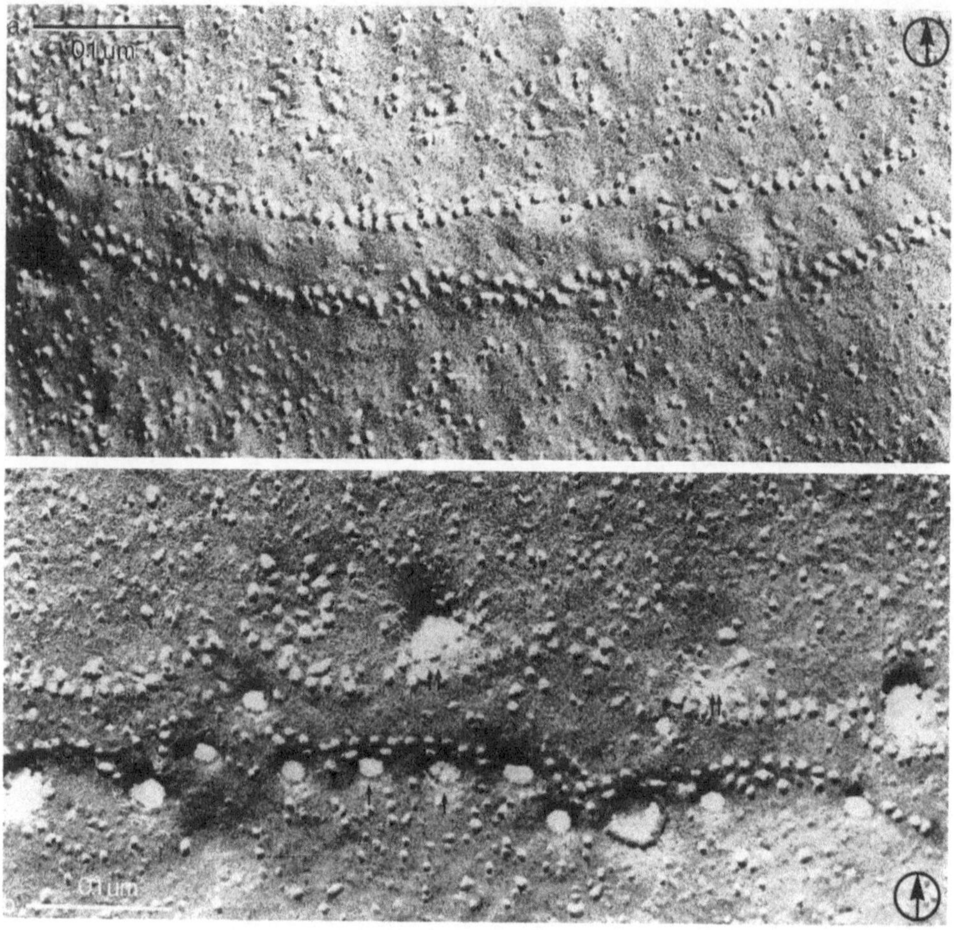

FIGURE 18. PF-face of an "active zone" contained within the presynaptic membrane of a neuromuscular junction in a frog muscle [(a) resting state and (b) during synchronous transmitter release]. An active zone displays IMP rows (with underlying membrane-associated proteins); along their border transmitter vesicles undergo exocytosis (arrow). Double arrows indicate sites where alledgedly vesicle membranes are flattened out and integrated into the cell membrane. (However, more recent work suggests this not to be the case in the most quickly frozen outermost sample layers.[28]) Bars = 0.1 µm. (From Heuser, J. E., Reese, T. S., Dennis, M. J., Jan, V., Jan, L., and Evans, L., *J. Cell Biol.*, 81, 275, 1979. With permission.)

Torpedo electroplaques), a change in the density of IMP was found by several groups to occur under conditions that trigger transmitter release.[73-75] It did not come out unambiguously whether oligomeric particles partly decompose or whether some may be reformed; IMP may also rearrange perpendicular to the plane of the membrane;[73,75] a change of the overall number of IMP on both fracture faces has also been postulated. More precise details would be interesting with regard to a most recent hypothesis on membrane fusion during exocytosis, which postulates the involvement of proteins (see below). Of course, one would also have to show that the above-mentioned IMP shifts would occur precisely at the sites involved in fusion. But this has not yet been ascertained.

VII. FUSION MODEL

Our data, as well as any other data obtained by fast-freezing techniques, are compatible with the model presented in Figure 19C. It is important to note that IMP need not necessarily

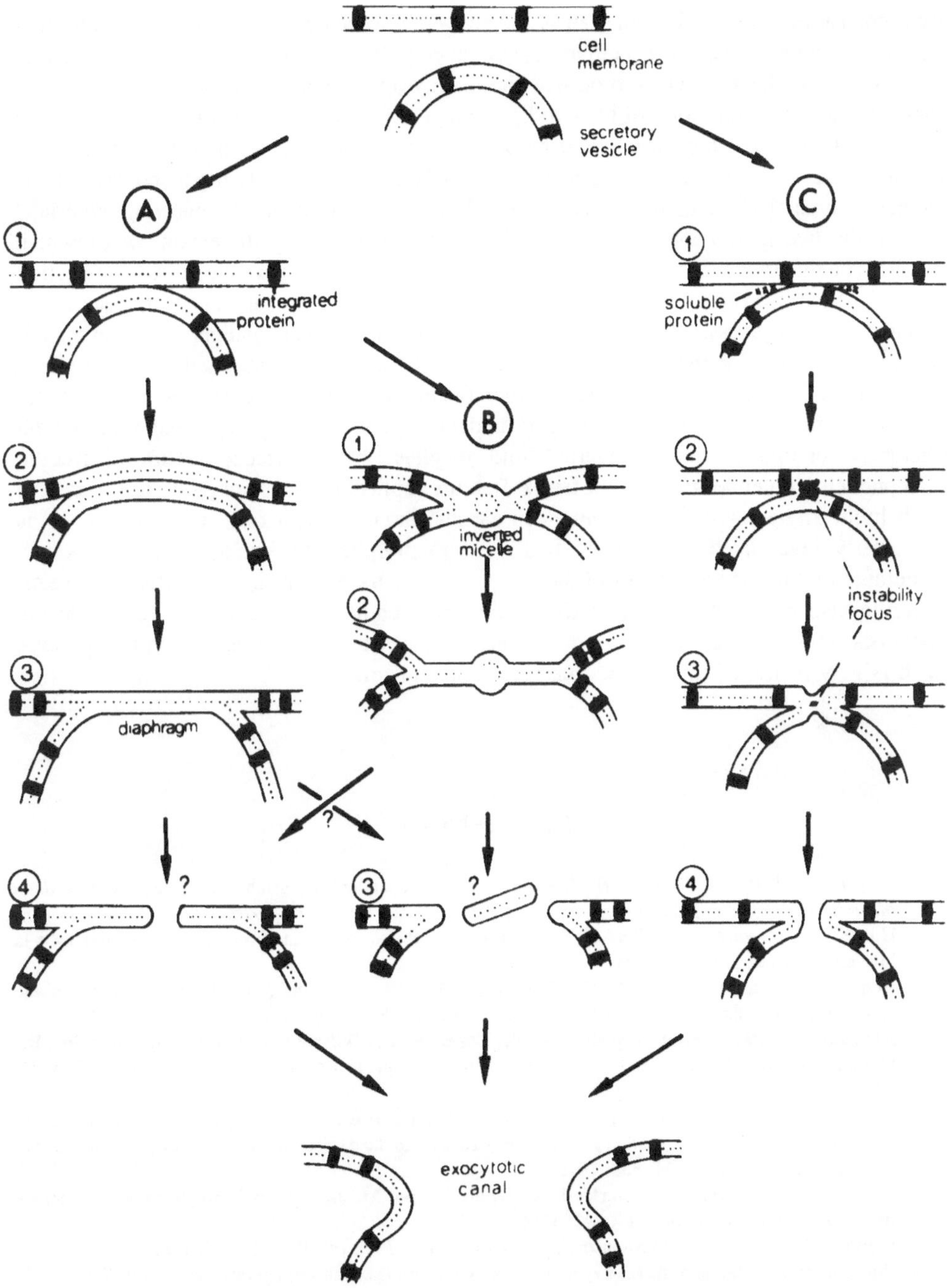

FIGURE 19. Three schemes proposed in the literature for the performance of membrane fusion during exocytosis. (A) is derived from Reference 69, (B) from Reference 72, and (C) from the author of this chapter. The splitting level of membranes is indicated as a dotted line; dark dots within membranes represent IMP *alias* membrane-integrated proteins; little crosses symbolize soluble proteins ("connecting materials") associated with sites of membrane fusion. (A) and (B) assume a large-scale clearing of IMP and of membrane-associated proteins as well as the formation of a "diaphragm" [eventually after formation of a broad pentalaminar zone of close membrane apposition in (A) or with the formation of a fusion nucleus in the form of a long-living lipid micelle (B)]. In the studies performed in my laboratory[14-16] we have found evidence only for the "focal fusion scheme" outlined in (C), as explained in more detail in the text. The nature of the focal fusion event, i.e., the precise arrangement of lipids, is not known. Clearly IMP and "connecting material" remain closely associated with the site of ongoing fusion, which they, thus, might help to regulate. (From Plattner, H., *Cell Biol. Int. Rep.*, 5, 435, 1981. With permission.)

be enriched at fusion sites. But one can assume now that some IMP (believed to be membrane-integrated proteins[1]) and some membrane-associated proteins have to be present on potential fusion sites. We have recently proposed that exocytotic membrane fusion is triggered by cross-linking and perturbation of the two lipid bilayers via hydrophobic binding of fusogenic proteins.[37] In *Paramecium*, this could be achieved by dephosphorylation in response to a stimulus which activates a Ca^{2+} calmodulin stimulated phosphoprotein phosphatase of the calcineurin type.[16] These components are thought to be present in the "connecting material" between the two membranes to be fused.[17] The role of rosettes still remains unknown; it still could be that they would represent sites of selective calcium entry (see, however, Section V). Hydrophobic protein binding has previously been shown so far only to trigger virus-mediated fusion processes,[76-78] and fusion of artificial membrane systems.[79-81] I have summarized above the arguments why I consider the 10-nm fusion intermediate observed in chromaffin cells as a lipid aggregate. Similar fusion structures are discussed now also for artificial membranes, although it remains undecided whether they represent focal defect structures[2-4] or highly oriented inverted lipid micelles.[5,6,82] Our freeze-fracture data could be reconciled with either model, but freeze-fracture replication does not allow one to decide which lipid arrangement really occurs. An earlier statement by Koike and colleagues[83] is now clearly invalidated; they stated that "... no morphological images unambiguously attributable to intermediate stages of the fusion process have ever been identified". Freeze-fracturing also has allowed us to estimate the time required for membrane fusion during exocytosis. Only the fastest cooling techniques allowed us to put a handle on this process which is very restricted in time ($\leq$1 ms) and space (10 nm).

REFERENCES

1. **Zingsheim, H. P. and Plattner, H.,** Electron microscopic methods in membrane biology, in *Methods in Membrane Biology,* Vol. 7, Korn, E. D., Ed., Plenum Press, New York, 1976, 1.
2. **Hui, S. W., Stewart, T. P., Boni, L. T., and Yeagle, P. L.,** Membrane fusion through point defects in bilayers, *Science,* 212, 921, 1981.
3. **Boni, L. T., Hah, J. S., Hui, S. W., Mukherjee, P., Ho, J. T., and Jung, C. Y.,** Aggregation and fusion of unilamellar vesicles by poly(ethylene glycol), *Biochim. Biophys. Acta,* 775, 409, 1984.
4. **Düzgünes, N., Paiement, J., Freeman, K. B., Lopez, N. G., Wilschut, J., and Papahadjopoulos, D.,** Modulation of membrane fusion by ionotropic and thermotropic phase transitions, *Biochemistry,* 23, 3486, 1984.
5. **Verkleij, A. J., Mombers, C., Gerritsen, W. J., Leunissen-Bijvelt, J., and Cullis, P. R.,** Fusion of phospholipid vesicles in association with the appearance of lipidic particles as visualized by freeze fracturing, *Biochim. Biophys. Acta,* 555, 358, 1979.
6. **Cullis, P. R., DeKruijff, B., Verkleij, A. J., and Hope, M. J.,** Lipid polymorphism and membrane fusion, *Biochem. Soc. Trans.,* 14, 242, 1986.
7. **Plattner, H.,** Membrane behaviour during exocytosis, *Cell Biol. Int. Rep.,* 5, 435, 1981.
8. **Plattner, H. and Bachmann, L.,** Cryofixation: a tool in biological ultrastructural research, *Int. Rev. Cytol.,* 79, 237, 1982.
9. **Menco, B. Ph. M.,** A survey of ultra-rapid cryofixation methods with particular emphasis on applications to freeze-fracturing, freeze-etching, and freeze-substitution, *J. Electron Microsc. Tech.* 4, 177, 1986.
10. **Gilkey, J. C. and Staehelin, L. A.,** Advances in ultrarapid freezing for the preservation of cellular ultrastructure, *J. Electron Microsc. Tech.,* 3, 177, 1986.
11. **Sitte, H., Edelmann, L., and Neumann, K.,** Cryofixation without pretreatment at ambient pressure, in *Cryotechniques in Biological Electron Microscopy,* Steinbrecht, R. A. and Zierold, K., Eds., Springer-Verlag, Berlin, 1987, 87.
12. **Heuser, J. E., Reese, T. S., Dennis, M. J., Jan, Y., Jan, L., and Evans, L.,** Synaptic vesicle exocytosis captured by quick freezing and correlated with quantal transmitter release, *J. Cell Biol.,* 81, 275, 1979.
13. **Heuser, J. E. and Reese, T. S.,** Structural changes after transmitter release at the frog neuromuscular junction, *J. Cell Biol.,* 88, 564, 1981.

14. **Schmidt, W., Patzak, A., Lingg, G., Winkler, H., and Plattner, H.,** Membrane events in adrenal chromaffin cells during exocytosis: a freeze-etching analysis after rapid cryofixation, *Eur. J. Cell Biol.*, 32, 31, 1983.
15. **Olbricht, K., Plattner, H., and Matt, H.,** Synchronous exocytosis in *Paramecium* cells. II. Intramembranous changes analyzed by freeze-fracturing, *Exp. Cell Res.*, 151, 14, 1984.
16. **Momayezi, M., Girwert, A., Wolf, C., and Plattner, H.,** Inhibition of exocytosis in Paramecium cells by antibody-mediated cross-linking of cell membrane components, *Eur. J. Cell Biol.*, 44, 247, 1987.
17. **Plattner, H.,** Synchronous exocytosis in *Paramecium* cells, in *Cell Fusion*, Sowers, A. E., Ed., Plenum Press, New York, 1987, 69.
18. **Chandler, D. E. and Heuser, J.,** Membrane fusion during secretion. Cortical granule exocytosis in sea urchin eggs as studied by quick-freezing and freeze-fracture, *J. Cell Biol.*, 83, 91, 1979.
19. **Chandler, D. E. and Heuser, J. E.,** Arrest of membrane fusion events in mast cells by quick-freezing, *J. Cell Biol.*, 86, 666, 1980.
20. **Knoll, G., Verkleij, A. J., and Plattner, H.,** Cryofixation of dynamic processes in cells and organelles, in *Cryotechniques in Biological Electron Microscopy*, Steinbrecht, R. A. and Zierold, K., Eds., Springer-Verlag, Berlin, 1987, 258.
21. **Kopstad, G. and Elgsaeter, A.,** Theoretical analysis of specimen cooling rate during impact freezing and liquid-jet freezing of freeze-etch specimens, *Biophys. J.*, 40, 163, 1982.
22. **Pinto da Silva, P. and Kachar, B.,** Quick freezing vs. chemical fixation: capture and identification of membrane fusion intermediates, *Cell Biol. Int. Rep.*, 4, 625, 1980.
23. **Ornberg, R. L. and Reese, T. S.,** Beginning of exocytosis captured by rapid-freezing of *Limulus* amebocytes, *J. Cell Biol.*, 90, 40, 1981.
24. **Morgenstern, E., Neumann, K., and Patscheke, H.,** The exocytosis of human blood platelets. A fast freezing and freeze-substitution analysis, *Eur. J. Cell Biol.*, 43, 273, 1987.
25. **Kelly, R. B.,** Pathways of protein secretion in eukaryotes, *Science*, 230, 25, 1985.
26. **Branton, D., Bullivant, S., Gilula, N. B., Karnowsky, M. J., Moor, H., Mühlethaler, K., Northcote, D. H., Packer, L., Satir, B., Satir, P., Speth, V., Staehelin, L. A., Steere, R., and Weinstein, R. S.,** Freeze-etching nomenclature, *Science*, 190, 54, 1975.
27. **Stoessel, T. P., Bretscher, M. S., Cecarelli, B., Dales, S., Helenius, A., Heuser, J. E., Hubbard, A. L., Kartenbeck, J., Kinne, R., Papahadjopoulos, D., Pearse, B., Plattner, H., Pollard, T. D., Reuter, W., Satir, B. H., Schliwa, M., Schneider, Y. J., Silverstein, S. C., and Weber, K.,** Membrane dynamics, in *Transport of Macromolecules in Cellular Systems*, Dahlem Konferenzen, Berlin: *Life Sci. Res. Rep.*, 11, 503, 1978.
28. **Torri-Tarelli, F., Grohovaz, F., Fesce, R., and Ceccarelli, B.,** Temporal coincidence between synaptic vesicle fusion and quantal secretion of acetylcholine, *J. Cell Biol.*, 101, 1386, 1985.
29. **Plattner, H., Miller, F., and Bachmann, L.,** Membrane specializations in the form of regular membrane-to-membrane attachment sites in *Paramecium*. A correlated freeze-etching and ultrathin-sectioning analysis, *J. Cell Sci.*, 13, 687, 1973.
30. **Plattner, H., Pape, R., Haacke, B., Olbricht, K., Westphal, C., and Kersken, H.,** Synchronous exocytosis in *Paramecium* cells. VI. Ultrastructural analysis of membrane resealing and retrieval, *J. Cell Sci.*, 77, 1, 1985.
31. **Tanaka, Y., DeCamilli, P., and Meldolesi, J.,** Membrane interactions between secretion granules and plasmalemma in three exocrine glands, *J. Cell Biol.*, 84, 438, 1980.
32. **Plattner, H. and Knoll, G.,** Cryofixation of biological materials for electron microscopy by the methods of spray-, sandwich-, cryogen-jet- and sandwich-cryogen-jet-freezing; a comparison of techniques, in *The Science of Biological Specimen Preparation for Microscopy and Microanalysis*, Revel, J. P., Barnard, T., and Haggis, G. H., Eds., Scanning Electron Microscopy Inc., AMF O'Hare, IL., 1984, 139.
33. **Pape, R. and Plattner, H.,** Synchronous exocytosis in Paramecium cells. V. Ultrastructural adaptation phenomena during re-insertion of secretory organelles, *Eur. J. Cell Biol.*, 36, 38, 1985.
34. **Beisson, J., Lefort-Tran, M., Pouphile, M., Rossignol, M., and Satir, B.,** Genetic analysis of membrane differentiation in *Paramecium*. Freeze-fracture study of the trichocyst cycle in wild-type and mutant strains, *J. Cell Biol.*, 69, 126, 1976.
35. **Pouphile, M., Lefort-Tran, M., Plattner, H., Rossignol, M., and Beisson, J.,** Genetic dissection of the morphogenesis of exocytosis sites in *Paramecium, Biol. Cell*, 56, 151, 1986.
36. **Plattner, H., Reichel, K., Matt, H., Beisson, J., Lefort-Tran, M., and Pouphile, M.,** Genetic dissection of the final exocytosis steps in *Paramecium tetraurelia* cells: cytochemical localization of Ca^{2+}-ATPase activity over preformed exocytosis sites, *J. Cell Sci.*, 46, 17, 1980.
37. **Plattner, H., Lumpert, C. J., Gras, U., Vilmart-Seuwen, J., Stecher, B., Höhne, B., Momayezi, M., Pape, R., and Kersken, H.,** Enzymatic regulation of membrane fusion during synchronous exocytosis in Paramecium cells, in *Molecular Mechanisms of Membrane Fusion*, Ohki, S., Doyle, D., Flanagan, T. D., Hui, S. W., and Meyhew, E., Eds., Plenum Press, New York, 1988, 477.

38. **Westphal, C. and Plattner, H.,** Ultrastructural analysis of the cell membrane-secretory organelle interaction zone in *Paramecium tetraurelia* cells. I. *In situ* characterization by electron "staining" and enzymatic digestion, *Bio. Cell.*, 42, 125, 1981.
39. **Vilmart, J. and Plattner, H.,** Membrane integrated proteins at preformed exocytosis sites, *J. Histochem. Cytochem.*, 31, 626, 1983.
40. **Robards, A. W. and Sleytr, U. B.,** *Low Temperature Methods in Biological Electron Microscopy*, Elsevier, Amsterdam, 1985.
41. **Grafenstein, H. R., Roberts, C. S., and Baker, P. F.,** Kinetic analysis of the triggered exocytosis/endocytosis secretory cycle in cultured bovine adrenal medullary cells, *J. Cell Biol.*, 103, 2343, 1986.
42. **Pumplin, D. W., Reese, T. S., and Llinás, R.,** Are the presynaptic membrane particles the calcium channels?, *Proc. Natl. Acad. Sci. U.S.A.*, 78, 7210, 1981.
43. **Gilligan, D. M. and Satir, B. H.,** Stimulation and inhibition of secretion in *Paramecium:* role of divalent cations, *J. Cell Biol.*, 97, 224, 1983.
44. **Aunis, D., Hesketh, J. E., and Devilliers, G.,** Freeze-fracture study of the chromaffin cell during exocytosis: evidence for connections between the plasma membrane and secretory granules and for movements of plasma membrane-associated particles, *Cell Tissue Res.*, 197, 433, 1979.
45. **Couteaux, R. and Pécot-Dechavassine, M.,** Vésicules synaptiques et poches au niveau des "zones actives" de la junction neuromusculaire, *C.R. Acad. Sci. Ser. D*, 271, 2346, 1970.
46. **Guraya, S. S.,** Recent progress in the structure, origin, composition, and function of cortical granules in animal egg, *Int. Rev. Cytol.*, 78, 257, 1982.
47. **Wunderlich, F. and Speth, V.,** Membranes in *Tetrahymena*. I. The cortical pattern, *J. Ultrastruct. Res.*, 41, 258, 1972.
48. **Satir, B., Schooley, C., and Satir, P.,** Membrane reorganization during secretion in *Tetrahymena, Nature (London)*, 235, 53, 1972.
49. **Satir, B., Schooley, C., and Satir, P.,** Membrane fusion in a model system. Mucocyst secretion in *Tetrahymena, J. Cell Biol.*, 56, 153, 1973.
50. **Bardele, C. F.,** Membrandifferenzierungen an Anheftungsorten von Extrusomen bei Protozoen, in *Verh. Dtsch. Zool. Ges.*, Vol. 70, Rathmayer, W., Ed., Georg Thieme Verlag, Stuttgart, 1977, 264.
51. **Bardele, C. F.,** Mapping of highly ordered membrane domains in the plasma membrane of the ciliate *Cyclidium glaucoma, J. Cell Sci.*, 61, 1, 1983.
52. **Dallai, R. and Luporini, P.,** Membrane specializations in the ciliate Euplotes crassus at the sites of interaction of the ampules with the plasma membrane, *Eur. J. Cell Biol.*, 23, 280, 1981.
53. **Davidson, L. A.,** Ultrastructure of the membrane attachment sites of the extrusomes of *Ciliophrys marina* and *Heterophrys marina* (Actinopoda), *Cell Tissue Res.*, 170, 353, 1976.
54. **Grim, J. N. and Staehelin, L. A.,** The ejectisomes of the flagellate *Chilomonas paramecium:* visualization by freeze-fracture and isolation techniques, *J. Protozool.*, 31, 259, 1984.
55. **Dubremetz, J. F. and Torpier, G.,** Freeze fracture study of the pellicle of an eimerian sporozoite (protozoa, coccidia), *J. Ultrastruct. Res.*, 62, 94, 1978.
56. **Porchet, E. and Torpier, G.,** Etude du germe infectieux de *Sarcocystis tenella* et *Toxoplasma gondii* par la technique du cryodécapage, *Z. Parasitenkd.*, 54, 101, 1977.
57. **Porchet, E., Torpier, G., and Richard, A.,** Etude après cryofracture du mérozoite de la coccidie *Aggregata eberthi, Z. Parasitenkd.*, 66, 257, 1982.
58. **Dubremetz, J. F., Torpier, G., Maurois, P., Prensier, G., and Sinden, R.,** Structure de la pellicule du sporozoite de *Plasmodium yoeli:* étude par cryofracture, *C. R. Acad. Sci.*, 288, 623, 1979.
59. **Burghardt, R. C. and Foor, W. E.,** Membrane fusion during spermiogenesis in *Ascaris, J. Ultrastruct. Res.*, 62, 190, 1978.
60. **Burghardt, R. C.,** Intercellular junctions and exocytosis in the vas deferens of *Ascaris, J. Ulstrastruct. Res.*, 71, 162, 1980.
61. **Reger, J. F. and Fitzgerald, M. E.,** A correlated thin section freeze-fracture study on the neuromuscular junction of fast-acting intersegmental muscle of the chilopodan, *Lithobius forficatus* L, *J. Submicrosc. Cytol.*, 13, 1, 1981.
62. **Dreyer, F., Peper, K., Akert, K., Sandri, C., and Moor, H.,** Ultrastructure of the "active zone" in the frog neuromuscular junction, *Brain Res.*, 62, 373, 1973.
63. **Pfenninger, K. H. and Rovainen, C. M.,** Stimulation- and calcium-dependence of vesicle attachment sites in the presynaptic membrane; a freeze-cleave study on the lamprey spinal cord, *Brain Res.*, 72, 1, 1974.
64. **Venzin, M., Sandri, C., Akert, K., and Wyss, U. R.,** Membrane associated particles of the presynaptic active zone in rat spinal cord. A morphometric analysis, *Brain Res.*, 130, 393, 1977.
65. **Harris, K. M. and Landis, D. M. D.,** Membrane structure at synaptic junctions in area CA1 of the rat hippocampus, *Neuroscience*, 19, 857, 1986.
66. **Entzeroth, R., Dubremetz, J. F., Hodick, D., and Ferreira, E.,** Immunoelectron microscopic demonstration of the exocytosis of dense granule contents into the secondary parasitophorous vacuole of Sarcocystis muris (Protozoa, Apicomplexa), *Eur. J. Cell Biol.*, 41, 182, 1986.

67. **Sibley, L. D., Krahenbuhl, J. L., Adams, G. M. W., and Weidner, E.,** *Toxoplasma* modifies macrophage phagosomes by secretion of a vesicular network rich in surface proteins, *J. Cell Biol.*, 103, 867, 1986.
68. **Brever, W. V.,** How the malaria parasite invades its host cell, the erythrocyte, *Int. Rev. Cytol.*, 96, 191, 1985.
69. **Orci, L. and Perrelet, A.,** Ultrastructural aspects of exocytotic membrane fusion, in *Membrane Fusion*, Poste, G. and Nicolson, G. L., Eds., Elsevier/North-Holland, Amsterdam, 1978, 629.
70. **Chandler, D. E.,** Comparison of quick-freeze and chemically fixed sea-urchin eggs: structural evidence that cortical granule exocytosis is preceded by a local increase in membrane mobility, *J. Cell Sci.*, 72, 23, 1984.
71. **Chandler, D. E., Bennett, J. P., and Gomperts, B.,** Freeze-fracture studies of chemotactic peptide-induced exocytosis in neutrophils: evidence for two patterns of secretory granule fusions, *J. Ultrastruct. Res.*, 82, 221, 1983.
72. **Pinto da Silva, P. and Nogueira, M. L.,** Membrane fusion during secretion. A hypothesis based on electron microscope observation of *Phytophthora palmivora* zoospores during encystment, *J. Cell Biol.*, 73, 161, 1977.
73. **Israël, M., Manaranche, R., Morel, N., Dedieu, J. C., Gulik-Krzywicki, T., and Lesbats, B.,** Redistribution of intramembrane particles related to acetylcholine release by cholinergic synaptosomes, *J. Ultrastruct. Res.*, 75, 162, 1981.
74. **Garcia-Segura, L. M., Muller, D., and Dunant, Y.,** Increase in the number of presynaptic large intramembrane particles during synaptic transmission at the *Torpedo* nerve-electroplaque junction, *Neuroscience*, 19, 63, 1986.
75. **Egea, G., Esquerda, J. E., Calvet, R., Solsona, C., and Marsal, J.,** Structural changes at pure cholinergic synaptosomes during the transmitter release induced by A-23187 in *Torpedo marmorata*. A freeze-fracture study, *Cell Tissue Res.*, 248, 207, 1987.
76. **White, J., Kielian, M., and Helenius, A.,** Membrane fusion proteins of enveloped animal viruses, *Q. Rev. Biophys.*, 16, 151, 1983.
77. **Doms, R. W., Helenius, A., and White, J.,** Membrane fusion activity of the influenza virus hemagglutinin. The low pH-induced conformational change, *J. Biol. Chem.*, 260, 2973, 1985.
78. **Roth, M. G., Doyle, C., Sambrook, J., and Gething, M. J.,** Heterologous transmembrane and cytoplasmic domains direct functional chimeric influenza virus hemagglutinins into the endocytic pathway, *J. Cell Biol.*, 102, 1271, 1986.
79. **Young, T. M., and Young, J. D.,** Protein-mediated intermembrane contact facilitates fusion of lipid vesicles with planar bilayers, *Biochim. Biophys. Acta*, 775, 441, 1984.
80. **Batenburg, A. M., Bougis, P. E., Rochat, H., Verkleij, A. J., and DeKruijff, B.,** The penetration of a cardiotoxin into cardiolipin model membranes and its implications on lipid organization, *Biochemistry*, 24, 7101, 1985.
81. **DeKruijff, B., Cullis, P. R., Verkleij, A. J., Hope, M. J., Van Echteld, C. J. A., and Taraschi, T. F.,** Lipid polymorphism and membrane function, in *The Enzymes of Biological Membranes*, Vol. 1, Martonosi, A. N., Ed., Plenum Press, New York, 1985, 131.
82. **Siegel, D. P.,** Inverted micellar intermediates and the transitions between lamellar, cubic, and inverted hexagonal lipid phases. II. Implications for membrane-membrane interactions and membrane fusion, *Biophys. J.*, 49, 1171, 1986.
83. **Koike, H., Tanaka, Y., Rez, G., and Meldolesi, J.,** Membrane interactions in pancreatic acinar cells: exocytosis, recycling and autophagocytosis, in *Biology of Normal and Cancerous Exocrine Pancreatic Cells*, Ribet, A., Pradayrol, L., and Susini, C., Eds., Elsevier/North-Holland, Amsterdam, 1980, 215.

Chapter 9

THE VITELLINE LAYER AND FERTILIZATION ENVELOPE OF ECHINODERM AND AMPHIBIAN EGGS: VISUALIZATION OF A CELL SURFACE-ANCHORED EXTRACELLULAR MATRIX

Carolyn A. Larabell and Douglas E. Chandler

TABLE OF CONTENTS

I. Introduction 176

II. The Vitelline Layer-Fertilization Envelope Transformation in the Sea Urchin Egg 176

III. The Vitelline Envelope of the Amphibian Egg 183

IV. Conclusion 187

Acknowledgments 189

References 197

I. INTRODUCTION

The extracellular matrix has moved into the limelight as a highly organized structure that is produced by and interacts with cells, often in a highly specific manner.[1] The ultrastructure of the extracellular matrix is now being visualized at high resolution by two techniques: the quick-freeze, deep-etch, rotary-shadow method for tissues and isolated cells, and rotary-shadowing of mica-adsorbed macromolecules for visualization of individual purified proteins.[2]

In the first method, tissue is quick-frozen on contact with a liquid helium-cooled copper block as described by Heuser et al.,[3] freeze-fractured, and deeply etched for 5 min or longer (at -80 to $-100°C$ in a vacuum of 10^{-6} mbar), then rotary-shadowed with platinum/carbon at a low angle (15 to 25°). This method has produced elegant views of cell interiors, of the cytoskeleton in particular, and now is being used to visualize the cell surface and the extracellular matrix. This technique has already provided a detailed model of cell wall structure in *Chlamydomonas*,[4,5] of microdomains in sperm plasma membrane,[6] of fenestrae in endothelial cells of capillaries,[7] and of extracellular coats on sea urchin and mammalian eggs.[8-10]

The second method consists of spraying purified macromolecules onto freshly cleaved mica. The proteins may be sprayed either in a physiological buffer, and the mica plates quick-frozen, freeze-dried, and rotary-shadowed as described by Heuser,[11] or they may be sprayed in 40% glycerol, dried *in vacuo* at room temperature, and rotary-shadowed.[12-14] These methods have been used to visualize single molecules found in the extracellular matrix such as collagen,[15,16] procollagen,[17] laminin,[18] and fibronectin,[19] as well as interactions between these proteins.[20]

II. THE VITELLINE LAYER-FERTILIZATION ENVELOPE TRANSFORMATION IN THE SEA URCHIN EGG

The vitelline layer of the sea urchin represents one of the best studied examples of an extracellular matrix. The vitelline layer is secreted by the immature oocyte during its maturation in the ovary, and in the mature egg consists of a thin coating anchored to the cell surface. This matrix is extremely important physiologically. Incorporated into its structure are numerous glycoprotein complexes that serve as sperm receptors. These receptors bind acrosome-reacted sperm in a species-specific manner,[21-23] by interacting with bindin, a protein found at the tip of the sperm acrosomal process.[24-26] One out of many sperm bound to the vitelline layer fuses with the egg, becomes engulfed, and subsequently contributes its chromosomes to formation of the first zygote nucleus.

Sperm entry initiates cortical granule exocytosis which sweeps as a wave over the entire cell surface.[27-30] The enzymes and structural proteins released by exocytosis modify the vitelline layer (VL) to form the fertilization envelope (FE) through a defined sequence of biochemical reactions.[31] First, a protease, "vitelline delaminase", severs proteins that link the vitelline layer to the plasma membrane while another serine protease destroys sperm receptors in the VE.[32,33] During the same period, hyalin, also released from the cortical granules, polymerizes to form a mesh-like network of filaments that swells and forces the vitelline envelope (VE) away from the egg surface. Next, the VE is chemically cross-linked by ovoperoxidase which is released by exocytosis and becomes incorporated into the VE structure.[34,35] Formation of dityrosine bonds by this enzyme is thought to account for the high chemical stability of the FE ultimately formed.[36-39] Before hardening, the VE structure is quite sensitive to proteases and disulfide reducing agents whereas after hardening the envelope is very resistant to degradation.[31] Finally, the crosslinked VE is coated both inside and out by a self-aggregating protein referred to as "paracrystalline" protein.[40,41] Taken together, all of these changes serve to form a tough, protective coat around the embryo during early development, and to cast off supernumary sperm that might otherwise cause polyspermy and abnormal development.

Remodeling of the vitelline layer to form the fertilization envelope is, as in other systems for extracellular matrix assembly, based on self assembly. Hyalin, secreted as a monomer, undergoes self association to form the intricate polymer that hydrates and fills the perivitelline space.[42] Ovoperoxidase binds to another secreted protein, proteoliasin, which in turn binds to the elevated vitelline envelope.[35] Paracrystalline protein aggregates to form sheets covering both the inner and outer surfaces of the fertilization envelope. All three of these assembly steps are calcium dependent and are triggered the moment these proteins are secreted into sea water.[35,41,42] The vitelline layer serves as the solid state support for organizing this self assembly.[31,43]

Ultrastructural details of this self assembly are an ongoing interest in our laboratory. In this communication we provide high resolution, stereo views of vitelline layer structure in both sea urchin and *Xenopus laevis* eggs, and document the structural changes that occur in the sea urchin VL as it is transformed into a FE.

A three-dimensional view of the sea urchin egg cortex (Figure 1) shows that the VL consists of a network of filaments that adheres closely to the plasma membrane. Just under this microvilli-studded membrane lies a cortical granule (CG) that has been cleaved open to reveal a highly organized interior. Each sea urchin egg contains about 18,000 such granules arranged in a single layer and it is these granules that undergo exocytosis just after fertilization. In this particular species, *Strongylocentrotus purpuratus*, the contents of the cortical granules are arranged in concentric layers like an onion. At high magnification (Figure 2), these layers are seen to consist of shelves of tightly packed particles, each shelf connected by a regular array of filaments to the shelf above and below it. The organization of these layers contrasts with the meshwork of filaments found in other regions of the granule (arrow, Figure 2). The organized nature of the granule interior reflects the fact that enzymes and structural proteins that will later form extensive extracellular structures must be packaged in a very economical manner before release.

The vitelline layer on the cell surface is a filamentous layer that is also highly organized. The main sheet consists of a network of anastomozing filaments that are attached to the external surface of the plasma membrane by short posts (arrows, Figure 3). These posts, first described by Kidd,[44] offset the vitelline layer sheet about 10 nm from the cell surface. The VL sheet uniformly covers microvilli as well as the cell surface between microvilli. Stereo views of the VL reveal that there are at least two other sets of fibers associated with the VL. Figure 4 illustrates oblique filaments which extend from the tips of microvilli down to the surrounding VL sheet below, almost appearing like "guy wires" (arrows); other oblique filaments connect neighboring microvilli. A second set of fibers runs horizontally in an orbit around the egg just barely grazing the tips of the microvilli.[9] We believe that oblique and horizontal fibers may serve to link the vitelline layer to the thick layer of egg jelly surrounding the egg. This jelly layer has the important physiological role of triggering sperm to undergo the acrosome reaction prior to binding to the VL.

Cortical granule exocytosis initiates a radical remodeling of the VL. The VL elevates off the egg surface to form the vitelline envelope (VE), our designation for a series of intermediate structures formed prior to completion of the final product, the fertilization envelope (FE). Immediately after elevation, the VE retains its original structure, a sheet of filaments containing casts of the microvilli it once covered. The VE is still quite thin as can be seen from the microvillar cast that has been crossfractured (double arrow, Figure 5). Below the VE (arrows, Figure 5), is a dense network of filaments that represent hyalin. Calcium-induced assembly of the hyalin network is rapid and has produced an extensive array of intertwined filaments.

Viewed from above, the newly elevated VE appears to consist of a central sheet of

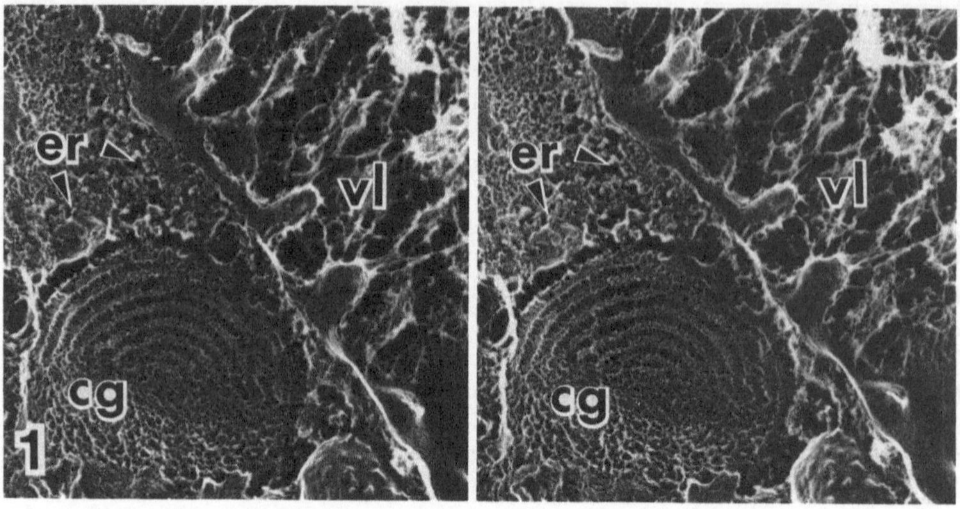

FIGURE 1.

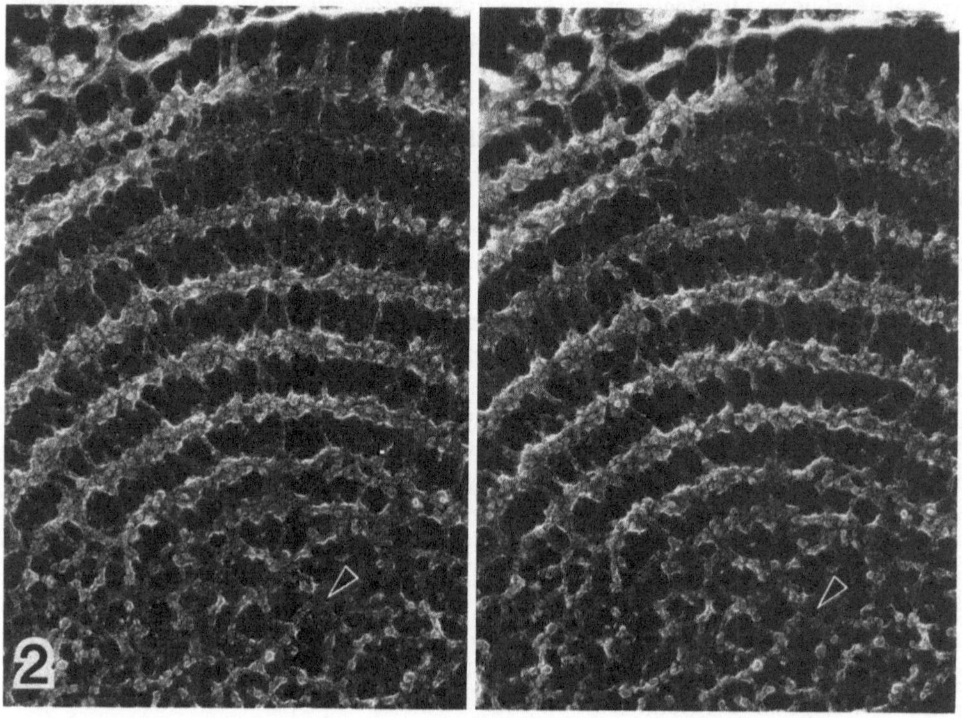

FIGURE 2.

FIGURE 1. Stereo pair of micrographs illustrating the three-dimensional structure of the sea urchin egg cortex. The plasma membrane is studded with microvilli (mv) and covered with the fibrous vitelline layer (vl). Below the plasma membrane is a cortical granule (cg), its layered secretory contents giving it an onion-like appearance. The surrounding cytoplasm is packed with a network of endoplasmic reticulum (er). The specimen was fixed with glutaraldehyde, passed through water, quick-frozen, fractured, and deeply etched. The etched surface was rotary replicated with platinum/carbon from an angle of 25°. All figures are negative images with platinum deposits appearing white. (Magnification × 28,200.)

FIGURE 2. The layered appearance of secretory material in the cortical granules of *S. purpuratus* eggs. Each transsected layer is composed of densely packed particles, neighboring layers being joined by filaments. Other regions in the granule interior are not layered and consist of unorganized filaments (arrow). The specimen was prepared as in Figure 1. Stereo pair. (Magnification × 98,700.)

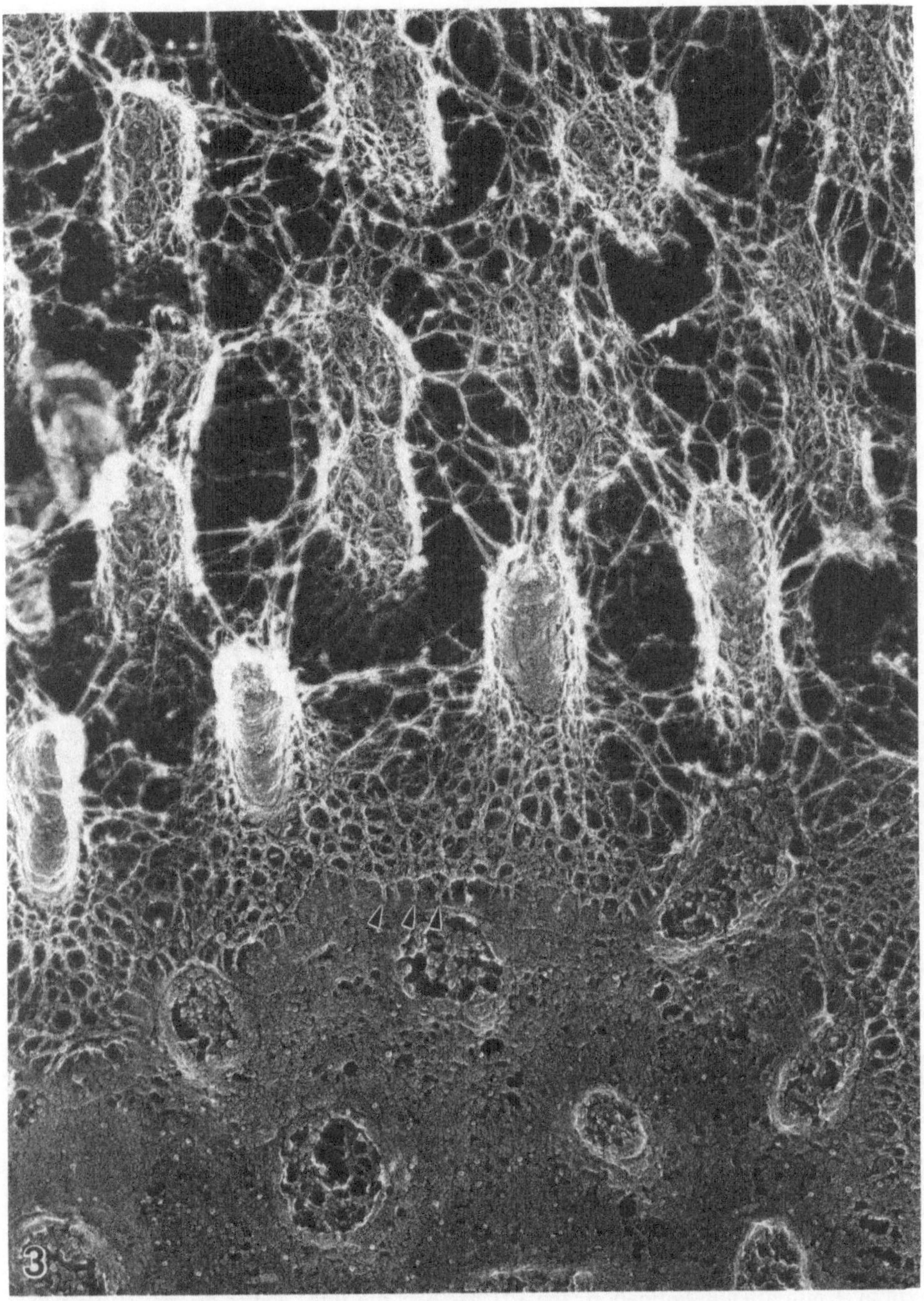

FIGURE 3. The unfertilized egg surface showing the VL and the short posts that attach it to the plasma membrane (arrows). The filamentous VL sheet covers the microvilli as well as the plasma membrane between microvilli. A rich network of filaments connect microvilli above the VL. The specimen was prepared as in Figure 1. (Magnification × 64,800.)

delicately interwoven filaments (circle, Figure 6), and a superficial layer of coarse filaments (arrows, Figure 6). The coarse filaments are seen on both inner and outer surfaces of the VE, and therefore this envelope appears to be trilaminar in structure. The coarse filaments are similar in appearance to filaments in the original VL, and it may be that discontinuities in this filament layer result from limited proteolysis that is known to occur during this period.[45]

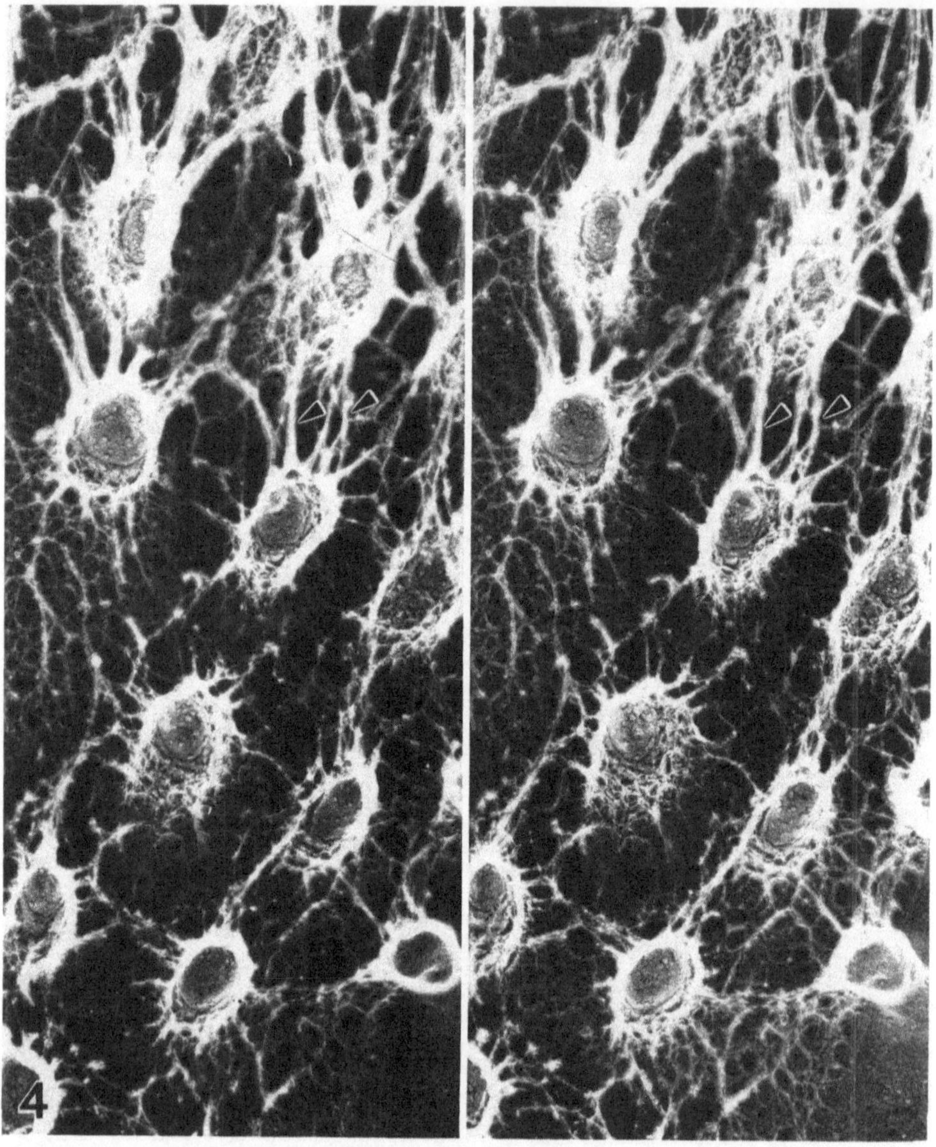

FIGURE 4. Oblique fibrils (arrows) extend from the tips of microvilli to the vitelline network below and to neighboring microvilli. The fracture plane just skimmed the tops of these microvilli, clearing away the horizontal fibrils above. The specimen was prepared as in Figure 1. Stereo pair. (Magnification × 63,700.)

Within 2 min after insemination, the VE has become fully elevated about 10 μm from the egg surface, which is now covered with long microvilli (M, Figure 7). The microvillar casts have been transformed from a rounded, igloo shape to an angular tent shape. This transition has been documented by scanning electron microscopy and has been referred to as the igloo-tent or I-T transition by Veron et al.[46]

Replicas of eggs which have been quick-frozen and deep-etched reveal the molecular basis for this I-T transformation. Paracrystalline protein released during cortical granule exocytosis assembles in sheets of S-shaped subunits on the VE template. This initial assembly on microvillar casts produces forces that change the shape of the cast from rounded as in Figure 6 to somewhat angular as shown in Figure 8. This initial assembly of paracrystalline protein is not as orderly and rigid in appearance as it will be later but does form a depression

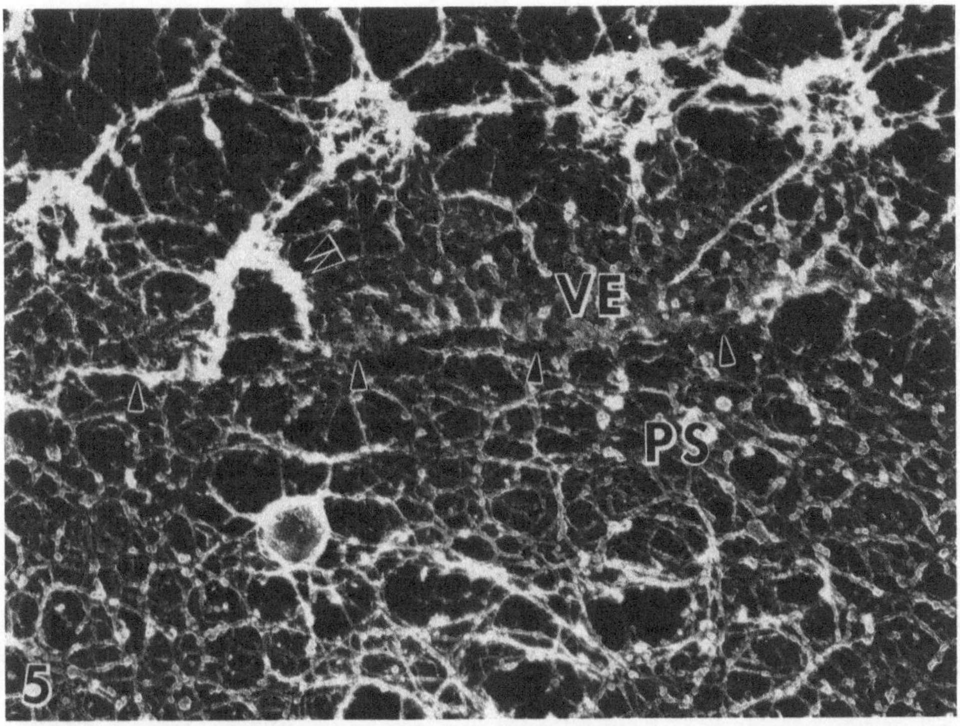

FIGURE 5. The vitelline envelope (VE) just after elevation (arrows) is similar in structure to the VL of the unfertilized egg and includes casts of the microvilli it once covered (double arrow). Below the VE, the perivitelline space (PS) is filled with an extensive network of hyalin filaments. The specimen was prepared as in Figure 1. (Magnification × 70,500.)

in the VE around the cast suggesting that the structure of the VE is consolidated during coating. At 2 min postinsemination, all casts have become plated with rigid sheets of paracrystalline protein and the I to T transformation is complete. These protein sheets extend outward and seem to be laid down on top of the thick filament network that surrounds the cast (arrows, Figure 9). At higher magnification, individual chains of S-shaped subunits (arrows, Figure 10) can be seen lying on the filament network as this sheet grows to cover the envelope surfaces between casts. This suggests that paracrystalline protein is applied to the VE template either as individual subunits or as short chains of subunits rather than as large sheets. Consolidation of these chains to form sheets occurs on the template itself and this assembly may be the process that actually alters cast shape.

Within 5 min postinsemination, the paracrystalline coating is complete on the external surface of the FE. Microvillar casts have been transformed into angular, spiked peaks rising out of the FE in a regular pattern (Figure 11). Between the casts, row after row of paracrystalline protein is seen linked to the template below. In Figure 12, mechanical stress on the FE has pulled adjacent rows apart revealing the thin filamentous substructure on which the paracrystalline protein sheets lie (arrows). In isolated FEs, quick-frozen without fixation, the pattern of paracrystalline subunits is sufficiently regular to allow computer reconstruction of the subunit image (inset, Figure 12).

The angular casts themselves appear like many little Matterhorns each one having its individual shape. A panel of four such casts (Figure 13) shows that they have anywhere from two to five ridges. All, however, are coated with orderly sheets of S-shaped subunits in register with the coating between casts. In a number of cases, these peaks have been cleaved open to demonstrate a complex interior (Figure 13, bottom panel). Thick walls

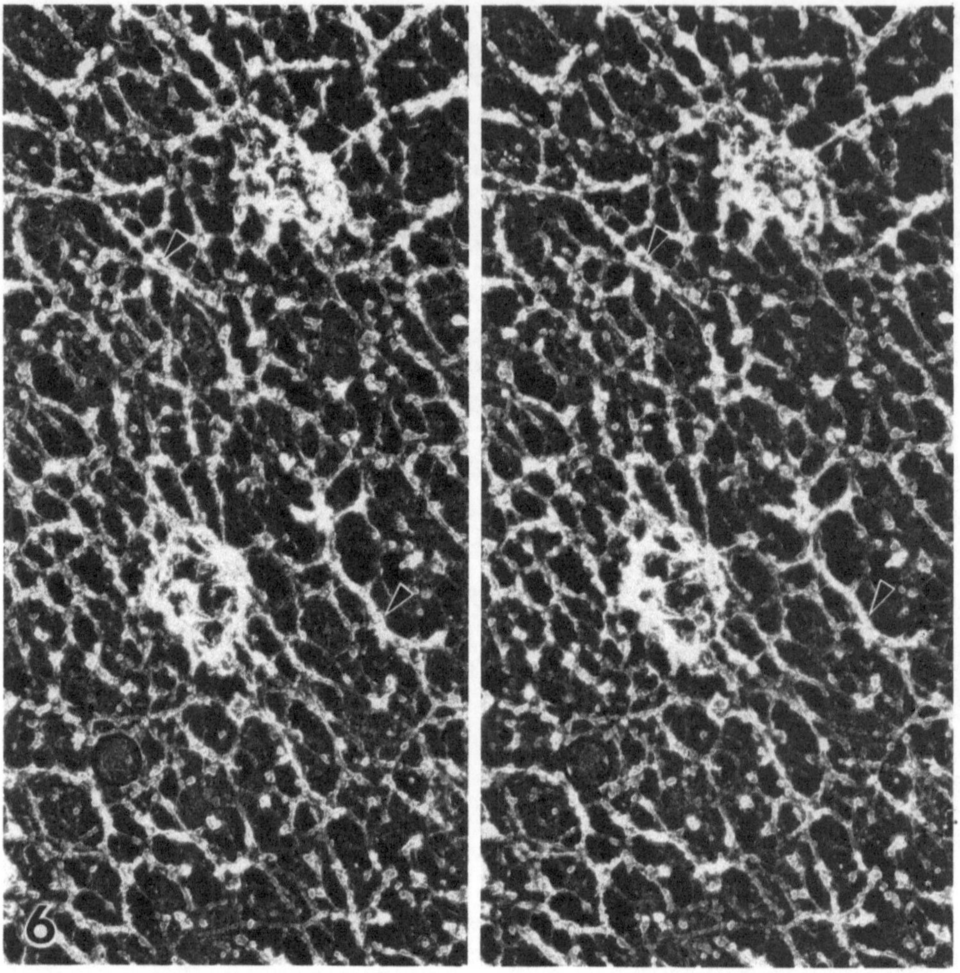

FIGURE 6. An "aerial" view of the newly elevated VE, 45 s post-insemination (p.i.). On the external surface, two rounded microvillar casts extend toward the viewer. The surrounding envelope consists of a layer of fine gray filaments (as in the circle), on top of which lies a network of coarser, white filaments (arrows). The specimen was prepared as in Figure 1. Stereo pair. (Magnification × 70,500).

packed with rounded particles are interlaced with filaments that can be seen clearly on the interior surface of some casts. Cross-fracture of the envelope itself shows that it includes a central filamentous layer the thickness of which, about 80 nm, can be gauged by the two arrows in Figure 14. The paracrystalline protein sheet is seen as a distinct edge at the upper arrow. On the bottom surface of the FE (Figure 15) we find areas where the paracrystalline coating has been ripped away to expose this thick central layer. This layer appears to consist of a dense mat of fibers linked together to form a network of little porosity.

The quick-freeze, deep-etch, rotary-shadow technique we have used, combined with stereo viewing of the replicas produced, has enabled us to describe the high degree of organization inherent in the self assembly of an extracellular matrix. The structure of the vitelline layer and the postfertilization modifications it undergoes is summarized in Figures 16 and 17. Figure 16 shows that the external surface of the sea urchin egg is covered by three networks of filaments: the vitelline layer proper, a single sheet of filaments connected to the plasma membrane by short "posts"; the oblique fibrils which connect tips of microvilli with the vitelline layer below and with adjacent microvilli; and the horizontal filaments which course around the egg in an orbit that skins over the microvillar tips. After fertilization, the VL elevates from the egg surface and becomes a template for self assembly of the fertilization

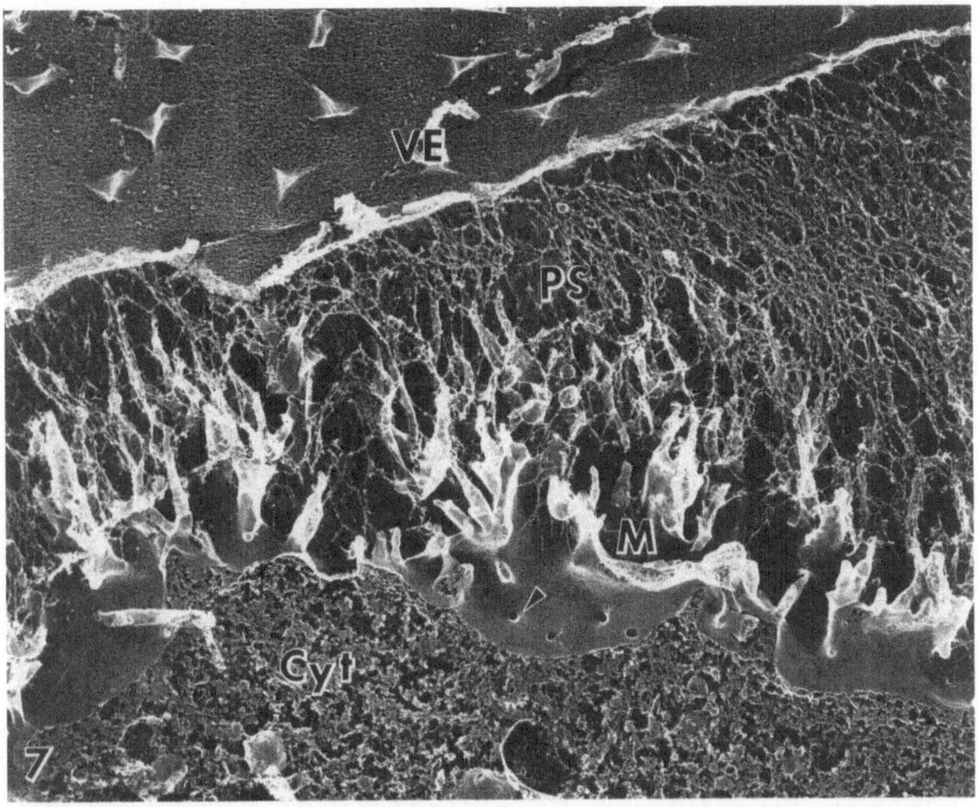

FIGURE 7. An "aerial" view of the egg 2 min p.i. The VE is elevated about 10 μm from the egg surface, and the perivitelline space (PS) is filled with hyalin. Numerous, branched microvilli (M) have appeared on the egg surface, and the underlying egg cortex (Cyt) has been cleared of cortical granules. Membrane retrieval is rapid during this period as indicated by numerous endocytic pits (arrow). The specimen was prepared as in Figure 1. (Magnification × 14,440.)

envelope. As shown in Figure 17, the VL prior to and the VE just after elevation is a single layer of intertwined filaments (0 s). Soon, the VE is converted to a three-layered structure having loose networks of coarse filaments on its upper and lower surfaces (45 s, Figure 17). Although we have not identified the original VL filaments in this structure, protein iodination studies[45,47] suggest that this trilayer envelope must contain at least part of the original VL structure. By 120 s postinsemination (p.i.), the central filament layer has thickened and microvillar casts have become coated with sheets of paracrystalline protein. The protein is laid down on top of the network of coarse fibrils; during the same period, the coarse fibril network is being dismantled and by 2 min p.i. it is seen only in a narrow zone at the base of the cast. Finally, by 300 s p.i., fertilization envelope assembly is complete, the envelope consisting of a central layer that is considerably thicker than the original VL (80 vs. 25 nm) and is sandwiched between two paracrystalline protein sheets each of which are about 10 nm thick.

III. THE VITELLINE ENVELOPE OF THE AMPHIBIAN EGG

Amphibian eggs are surrounded by several layers of extracellular investments, including the vitelline envelope (VE) and three to six jelly coats. Sperm must interact with and move through these investments prior to reaching the plasma membrane, though the exact mechanism by which this occurs is not yet clear. As one sperm enters the egg, cortical granules located just under the egg surface fuse with the plasma membrane and secrete their contents

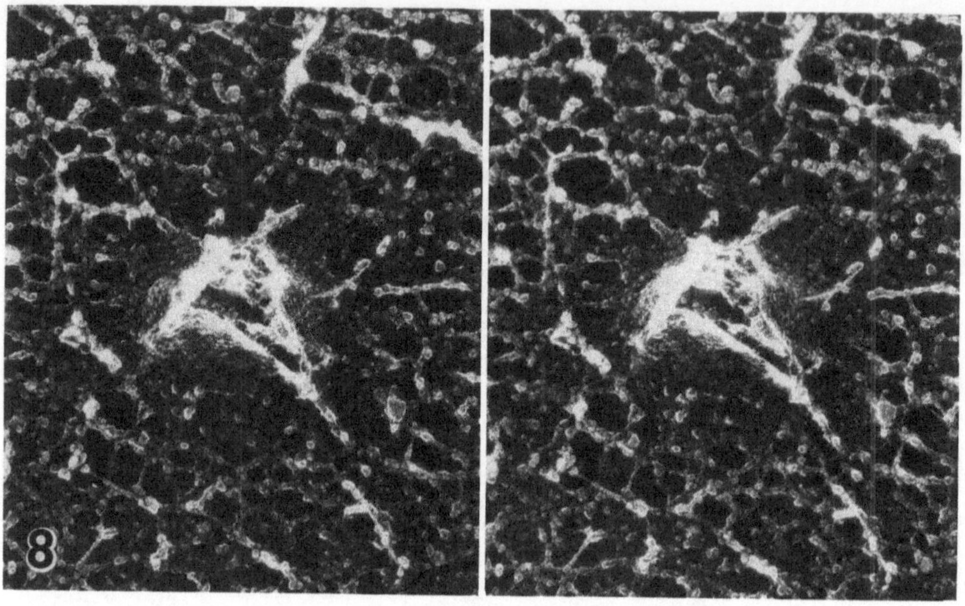

FIGURE 8.

FIGURES 8 to 10. Coating of the fertilization envelope with paracrystalline protein. Figure 8. Coating begins at some microvillar casts as early as 45 s p.i., and is first recognized by particular decoration and increased angularily of the cast. This coat extends onto and forms a depression in the surrounding VE surface (Magnification × 87,300.) Figure 9. By 2 min p.i., all casts have become sharply angular and are coated with orderly sheets of particles. These sheets seem to extend onto the top of the thick filament layer that is prominent at the base of the cast (arrows). Surrounding surfaces have lost their thick filament layer but have not yet been coated. (Magnification × 87,300.) Figure 10. At higher magnification one can see that these sheets are formed from chains of S-shaped particles; single chains can be seen positioned on the top of the thick filament layer (arrows). (Magnification × 194,000.) All specimens were prepared as in Figure 1. Stereo pairs.

into the surrounding perivitelline space (PS).[48-50] Some of the contents of the secretory granules pass through this PS and interact directly with the vitelline envelope while others diffuse through the vitelline envelope to interact with the innermost jelly layer J_1.[49,51,52] These interactions are responsible for the chemical and physical changes in the vitelline envelope that render it impenetrable to further sperm.

In *Xenopus laevis* eggs, the vitelline envelope is about 1 μm thick and covers the perivitelline space, its inner surface frequently contacting the tips of microvilli which protrude from the egg surface; its outer surface abuts on the innermost jelly layer, J_1.[49] The VE is composed of at least 11 glycoproteins with molecular weights ranging from 125,000 to less than 16,000 with two of these components (40,000 and 33,000) being the most prevalent.[51] The carbohydrate composition of the envelopes includes fucose, glucosamine, galactosamine, mannose, galactose, and sialic acid.[51] At fertilization, an enzyme released from the cortical granules interacts with two minor components of the vitelline envelope in a limited hydrolysis reaction, resulting in a decrease in their molecular weights (69,000 to 66,000 and 64,000 to 61,000) and an altered vitelline envelope (VE*).[51,53] The structural modifications in the VE are accompanied by an alteration of the physical properties of the envelope. The VE is readily dissolved by mercaptoethanol solutions and trypsin in room temperature and slowly dissolved by water, while the converted envelope (VE*) has only limited solubility in mercaptan solutions and is virtually insoluble in distilled water.[49,51]

Another granule component, a lectin, travels through the vitelline envelope and interacts with a ligand located in the innermost jelly layer, J_1. This lectin, a metalloglycoprotein also secreted during cortical granule exocytosis, is found in two size-isomeric forms with ap-

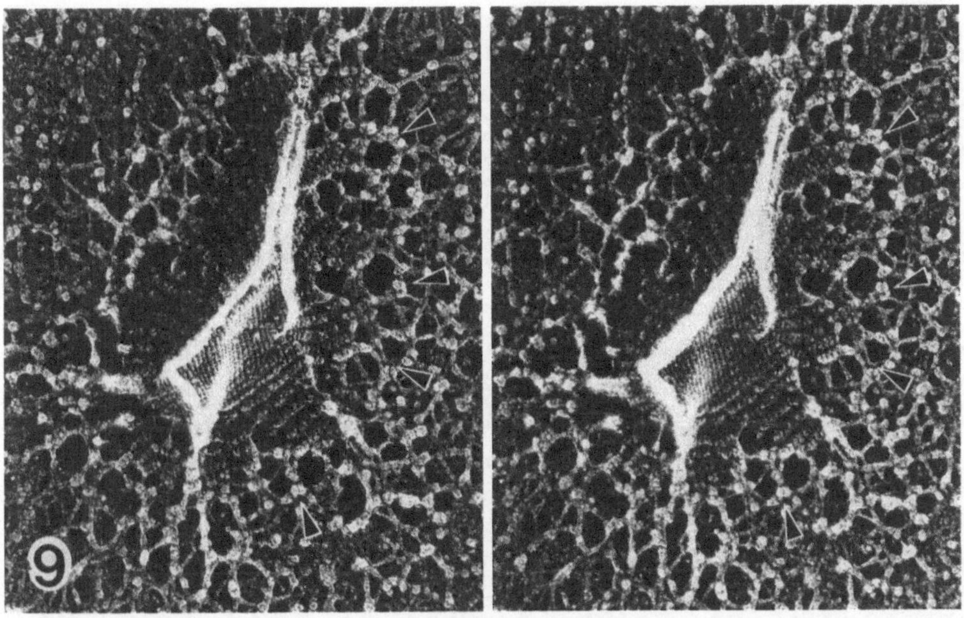

FIGURE 9.

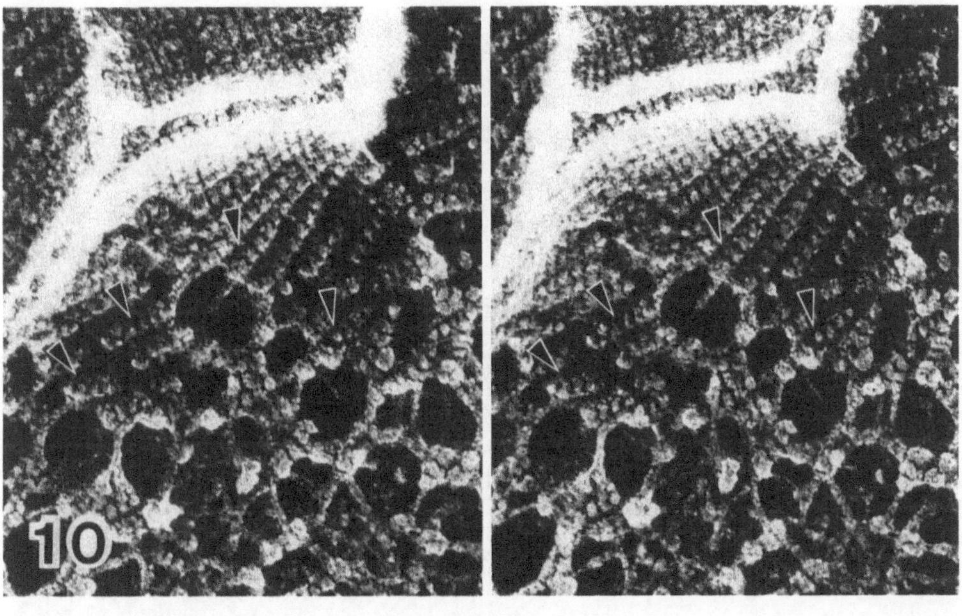

FIGURE 10.

proximate molecular weights of 539,000 and 655,000, and is composed of several dissimilar subunits with approximate molecular weights between 38,000 and 45,000.[54,55] It has a galactosyl sugar specificity and requires a divalent metal cation (preferably Ca^{2+}) for ligand binding.[50] This CG (cortical granule) lectin interacts with its J_1 ligand in an agglutination reaction producing a band of precipitation called the fertilization (F) layer on the outer surface of the VE* which is impenetrable to sperm.[56] The F layer acts as a molecular block to macromolecules released during exocytosis, trapping a significant portion of the CG lectin in the PS and generating an osmotic pressure differential which may be responsible for the

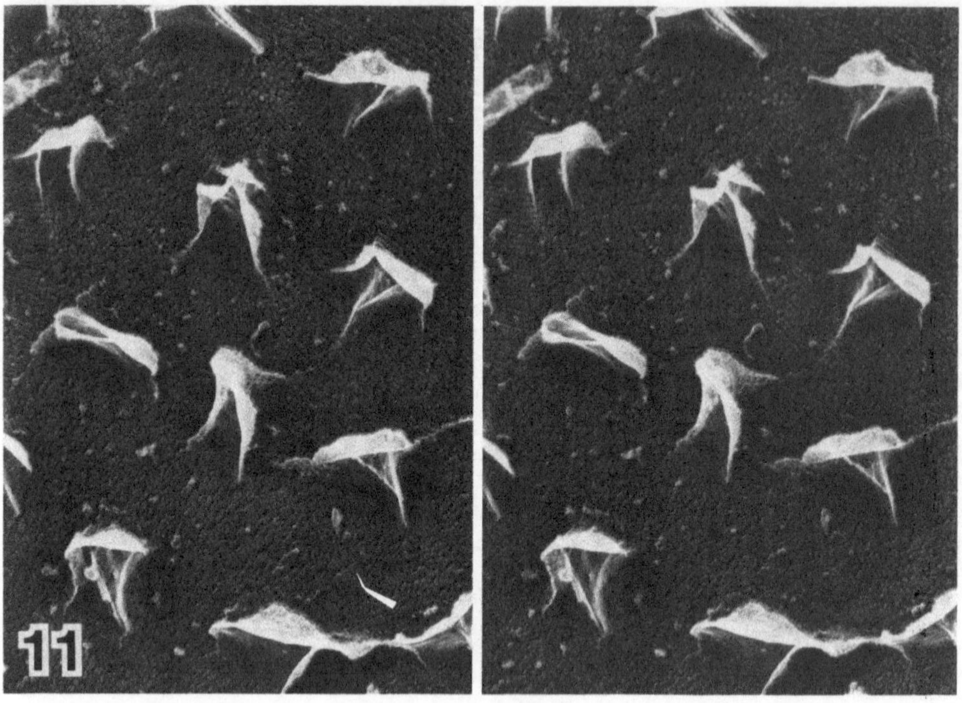

FIGURE 11.

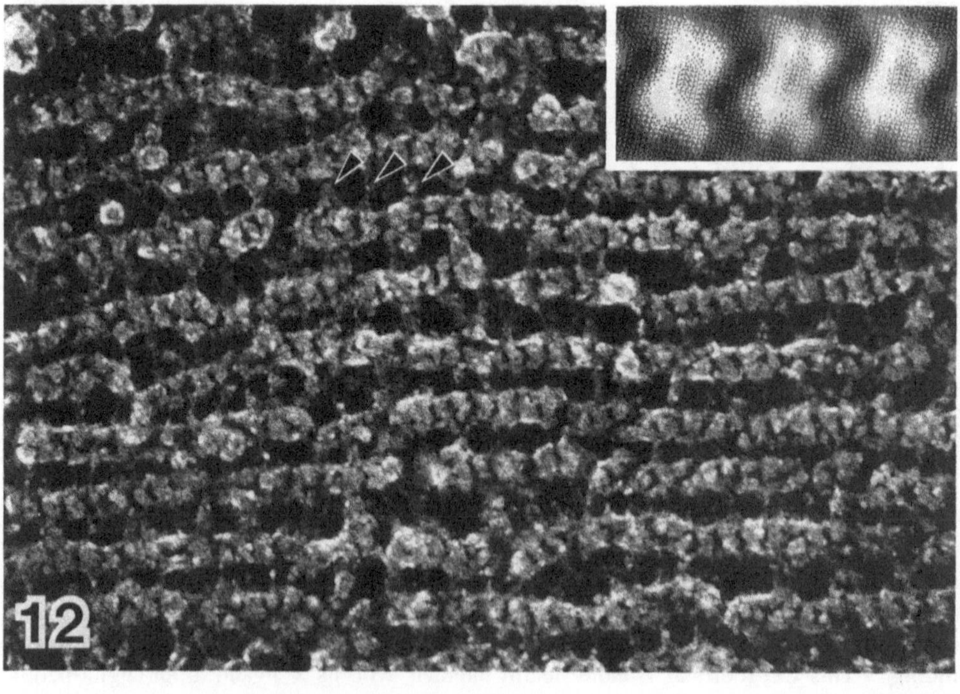

FIGURE 12.

influx of water and the consequent elevation of the fertilization envelope.[50,57] This F layer, together with the altered VE, constitute the late block to polyspermy in the eggs of the amphibian *Xenopus laevis*.[58]

Replicas of Xenopus eggs which have been quick-frozen, deep-etched, and rotary-shadowed provide views of the extracellular coats which are far superior to those seen in conventional thin sections. In Figure 18, cortical granules and segments of the cortical endoplasmic reticulum can be seen beneath the egg plasma membrane. Microvilli (m, Figure 18) are seen extending from the plasma membrane into the perivitelline space and contacting the overlying vitelline envelope (VE). Numerous filaments are seen in the perivitelline space which can be divided into two populations: (a) fine, thin filaments which run parallel to the plasma membrane and are often seen at the base of the microvilli (thin arrows, Figure 19), and (b) thicker filaments which connect microvillar surfaces with adjacent microvilli, with the vitelline envelope above, and with the plasma membrane below (arrowheads, Figure 19). Large protein molecules are seen on the surface of the plasma membrane, some of which appear to be attachment points for the filaments extending from the microvilli, as if anchoring them to the membrane surface (asterisk, Figure 19). Microvilli which have been fractured open demonstrate numerous filaments within, possibly bundles of F-actin. The perivitelline space also contains membrane-bound vesicles which are attached to the filamentous network coursing between the microvilli upper right, Figure 19, and Figure 20. Some of the thicker filaments seen in the perivitelline space extend into the vitelline envelope above (Figure 20).

The vitelline envelope itself contains two sizes of filaments. Larger fibrils (double arrow, Figure 21) are predominant and these form an intricate, anastomosing network which can be followed for long distances throughout the envelope. These fibrils may be those seen in thin sections as a loose network of filaments about 4 to 7 nm in diameter which appear to be arranged in alternating layers or bundles, filaments in one layer being oriented at an angle to those in adjacent layers.[49] In deep-etched samples it can be seen that these fibrils actually twist and turn forming swirling patterns which intertwine throughout the VE. The thin filaments in the VE are much shorter than the thick fibrils and appear to connect the longer, thicker fibrils as they pass each other (arrow, Figure 21).

How the VE structure we see in *Xenopus* eggs is modified by cortical granule secretions will be a subject of future studies.

IV. CONCLUSION

Freeze fracture is of importance not only for examining membrane topography but also for visualizing membrane-associated structures such as the cell cytoskeleton and the extracellular matrix. In this report we have shown that both echinoderm and amphibian eggs have anchored to their plasma membrane, extracellular coats whose structure can be seen at high resolution by freeze-fracture techniques. Second, we show that self assembly of an extracellular structure, the fertilization envelope, can be observed in molecular detail. We are certain that future studies will apply these techniques to assembly of extracellular matrices in other tissues and to interaction of these matrices with cell surfaces.

FIGURES 11 and 12. The completed fertilization envelope at 5 min p.i. Figure 11. The envelope has become completely coated and is studded with an regular array of angular microvillar casts. Stereo pair. (Magnification × 38,000.) Figure 12. At high magnification the armor coating between casts is seen to be composed of rows of subunits laid down on a truss of fine filaments (arrows) that run perpendicular to the subunit rows. Computer reconstruction of the S-shaped subunit image is shown in the inset. Specimens were prepared as in Figure 1. (Magnification × 255,000.)

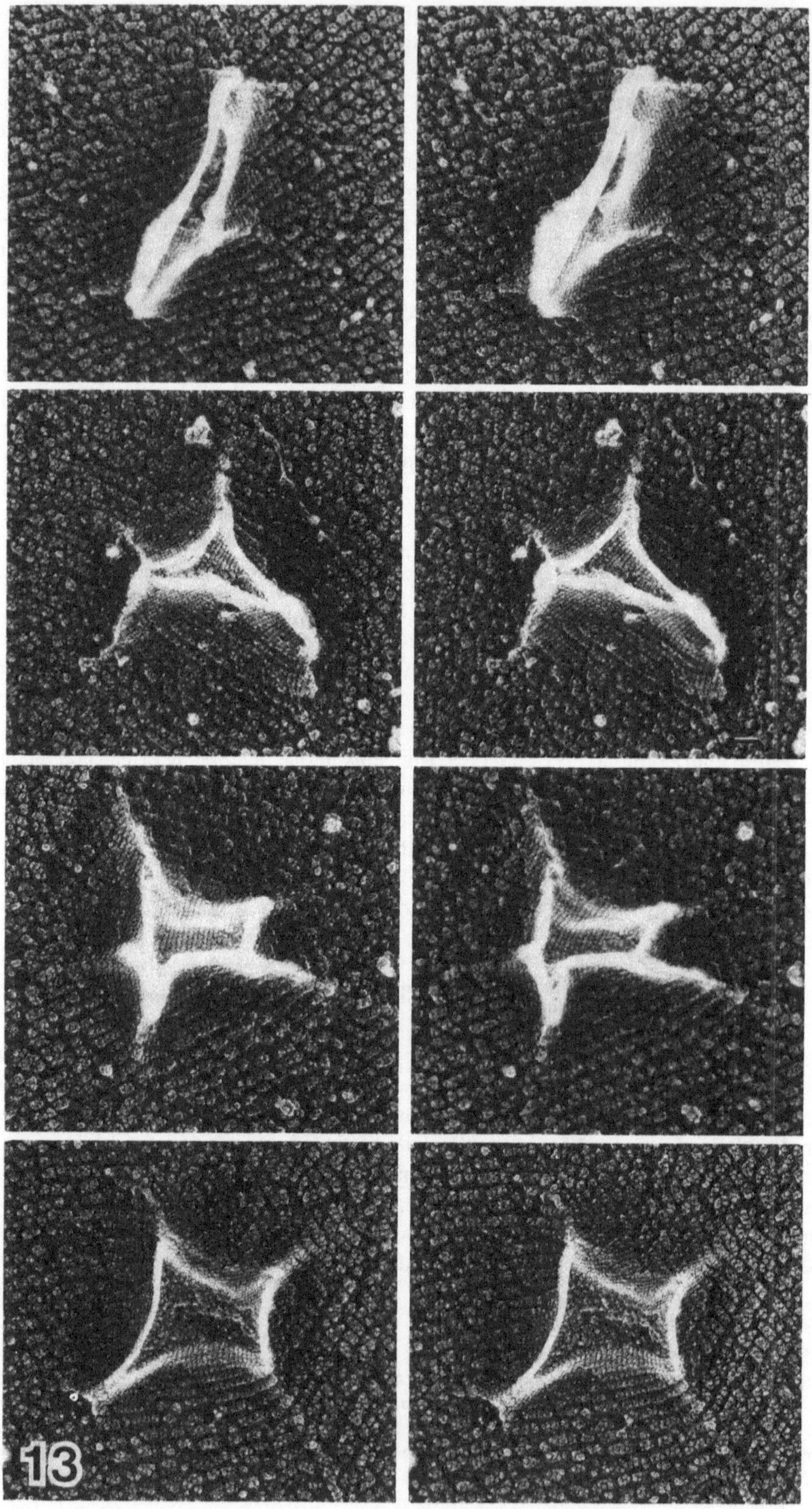

FIGURE 13.

ACKNOWLEDGMENTS

We would like to thank Charles Kazilek for preparing the publication prints. These studies were supported by grants from NSF (DCB-8407152 and DCB-8810200) and NIH (1-K04-HD00619).

FIGURE 13. Angular, peaked casts in the completed FE. This panel of four examples shows that casts vary in shape, having two to five ridges. Casts which have been cross-fractured (bottom panel) reveal thick walls filled with tightly intermeshed filaments that appear like particles in cross-section. Stereo pairs. The specimens were prepared as in Figure 1. (Magnification × 78,300.)

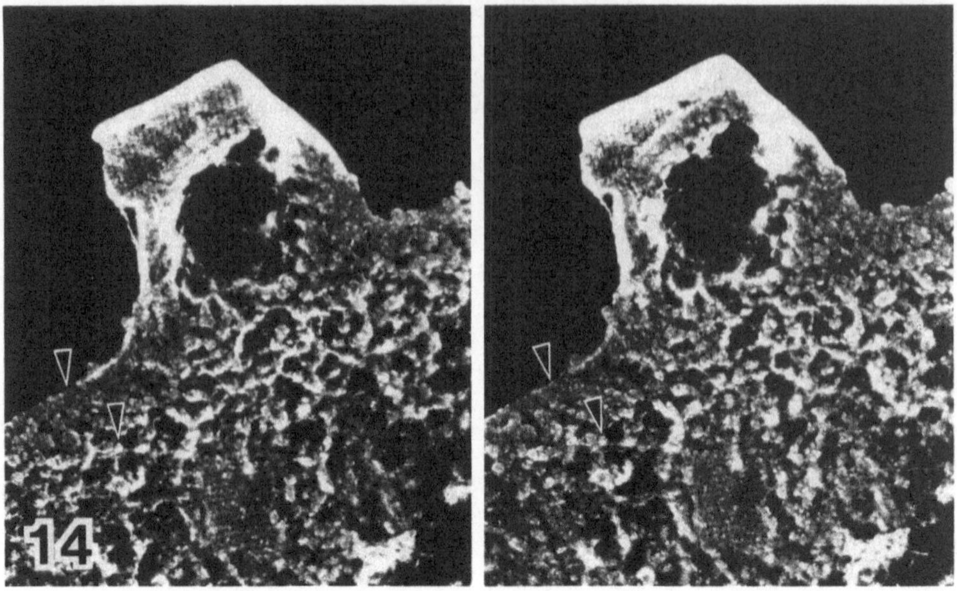

FIGURE 14.

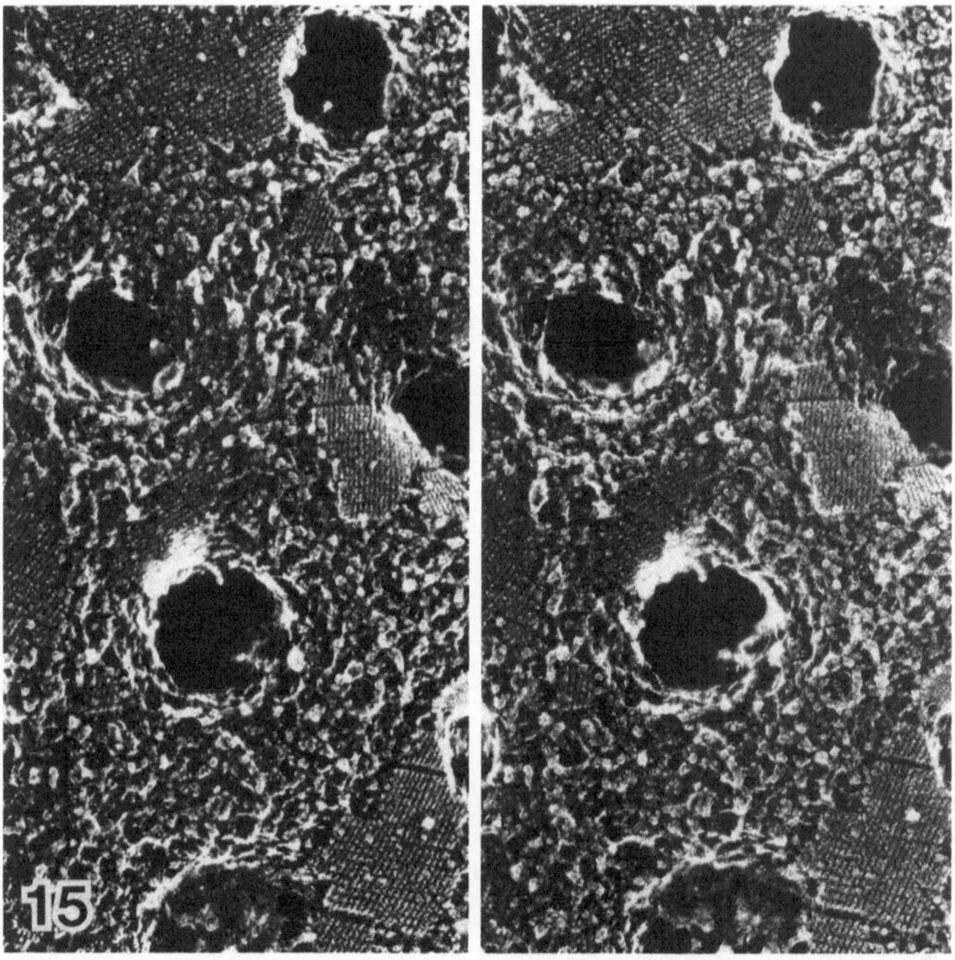

FIGURE 15.

FIGURE 14. Cross-fracture of an isolated fertilization envelope. This oblique fracture allows a view of the cast interior from beneath and of the thickened envelope in profile. The thickness of the central filamentous layer can be gauged by the arrows; the well-defined edge at the upper arrow represents the paracrystalline protein sheet coating the external surface of the envelope. The isolated FE was prepared as previously described[8] and quick-frozen in distilled water without fixation. Stereo pair. (Magnification × 109,000.)

FIGURE 15. The bottom surface of an isolated fertilization envelope. Sheets of paracrystalline protein lie directly in contact with the central filamentous layer. Invaginations represent microvillar casts. Specimen prepared as described in Figure 14. Stereo pair. (Magnification × 80,000.)

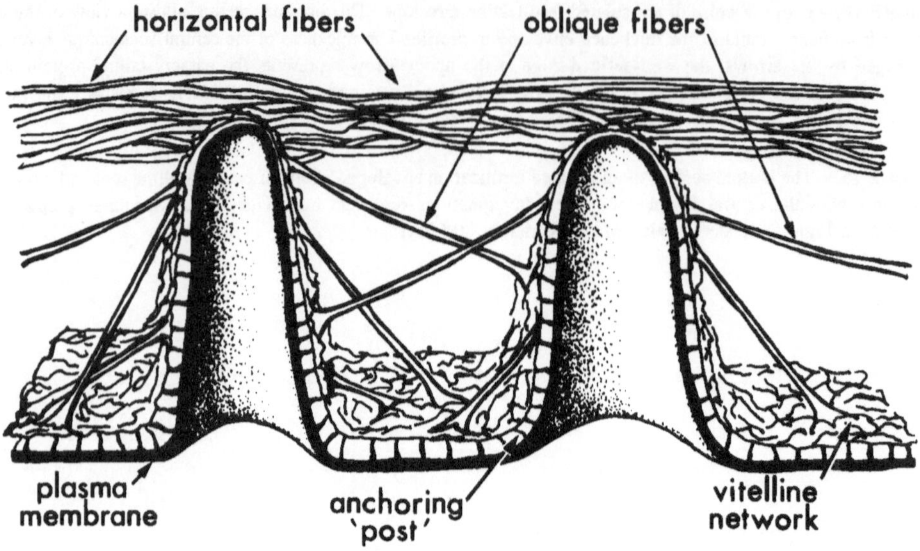

FIGURE 16. Anchored to the external surface of the sea urchin egg is an extracellular matrix composed of three filament networks: the vitelline layer, the oblique fibers, and the horizontal fibers. (From Chandler, D. E. and Kazilek, C. J., *Cell Tissue Res.* 246, 153, 1986. With permission.)

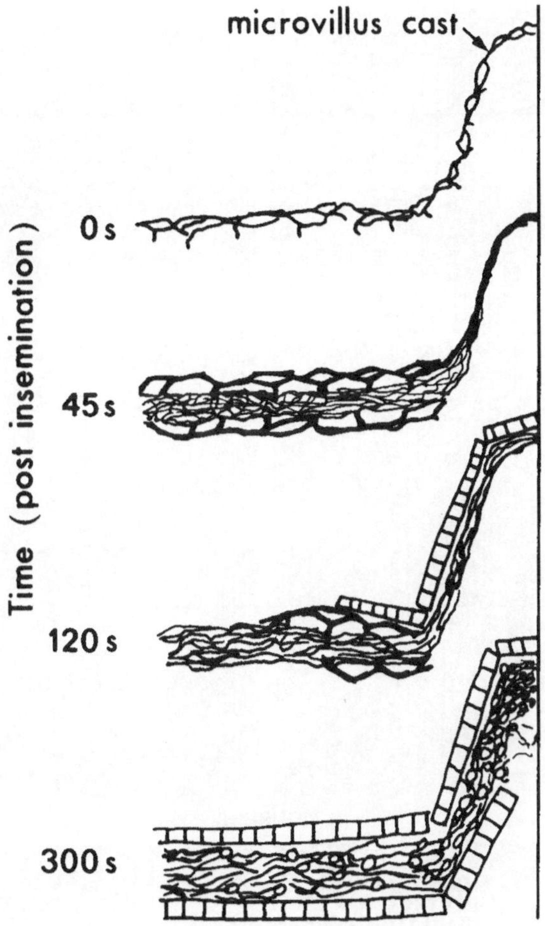

FIGURE 17. Sequential steps in the assembly of the fertilization envelope. (From Chandler, D. E. and Kazilek, C. J., *Cell Tissue Res.* 246, 153, 1986. With permission.)

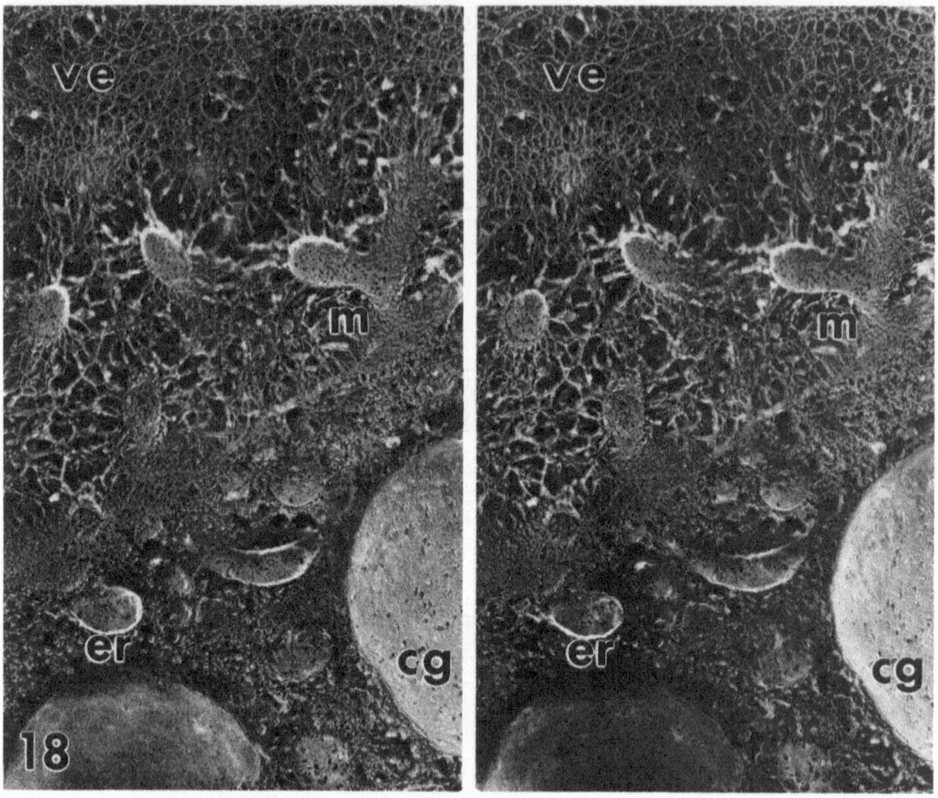

FIGURE 18.

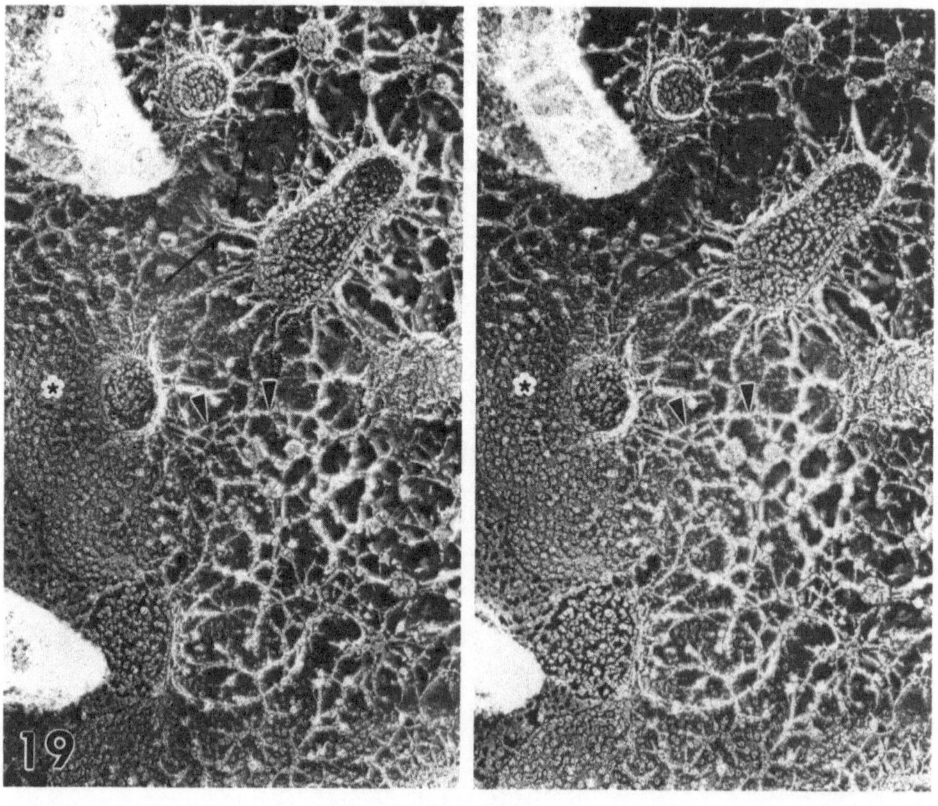

FIGURE 19.

FIGURE 18. Cross-fracture of the cortex and vitelline envelope of the *Xenopus laevis* egg. Cortical granules (cg) and the cortical endoplasmic reticulum (er) can be seen beneath the plasma membrane. Microvilli (m) extend from the plasma membrane into the perivitelline space and touch the vitelline envelope (ve). The specimen was prepared as in Figure 1. Stereo pair. (Magnification × 30,000.)

FIGURE 19. Surface of the *Xenopus* egg showing filaments attaching the microvilli to the plasma membrane (arrows) and thicker fibrils running parallel to the plasma membrane (arrowheads). Microvilli have been fractured open demonstrating a filamentous material within. Large macromolecules can be seen on the surface of the plasma membrane (asterisk). The specimen was prepared as in Figure 1. Stereo pair (Magnification × 42,000.)

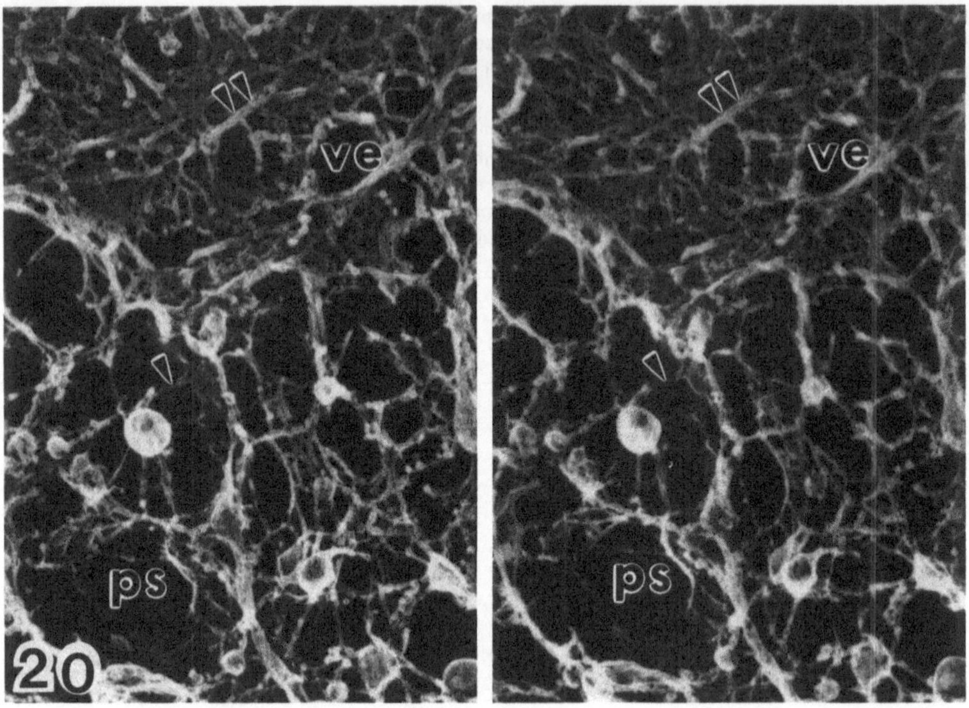

FIGURE 20.

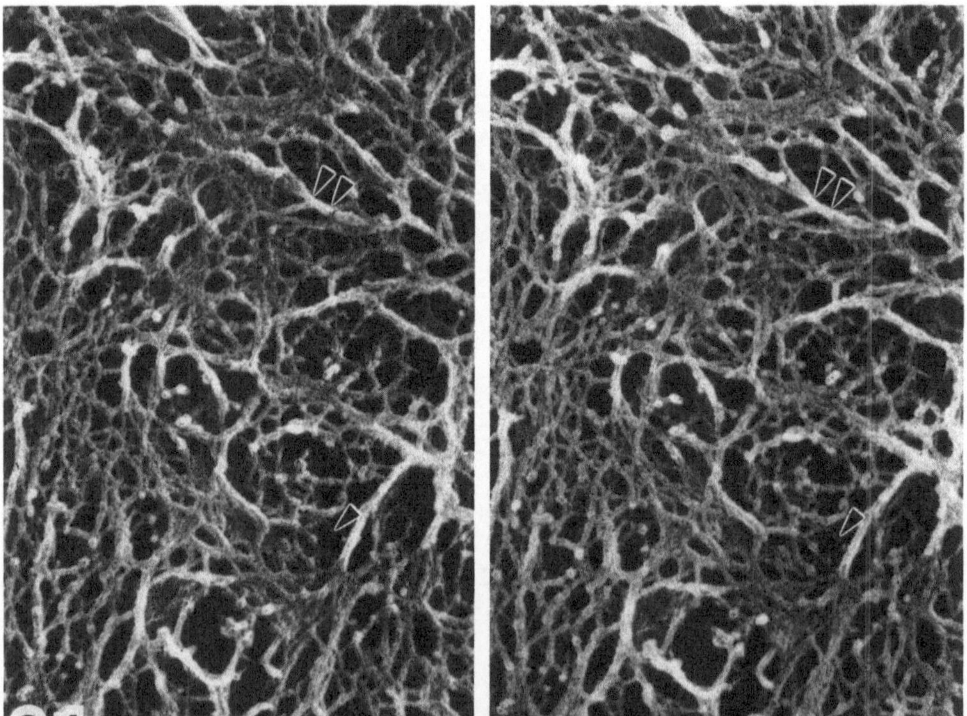

FIGURE 21.

FIGURE 20. Junction between the perivitelline space and the vitelline envelope. The perivitelline space contains fine, loosely arranged filaments (single arrowhead) and a smaller number of thick fibrils, some of which extend into the vitelline envelope. The overlying vitelline envelope contains many thick fibrils (double arrowheads) and a smaller number of the fine filaments. The specimen was prepared as in Figure 1. Stereo pair.(Magnification × 43,000.)

FIGURE 21. Vitelline envelope of *Xenopus laevis* demonstrating the anastomosing network of thick fibrils (double arrowheads). Thin filaments (single arrowhead) can also be seen interconnecting the thick fibrils. The specimen was prepared as in Figure 1. Stereo pair. (Magnification × 52,000.)

REFERENCES

1. **Hay, E. D.,** Cell and extracellular matrix. Their organization and mutual dependence, *Mod. Cell Biol.*, 2, 509,1983.
2. **Chandler, D. E.,** Rotary shadowing with platinum-carbon in biological electron microscopy: a review of methods and application, *J. Elec. Microsc. Tech.*, 3, 305, 1986.
3. **Heuser, J., Reese, T. S., Dennis, M. J., Jan, Y., Jan, L., and Evans, L.,** Synaptic vesicle exocytosis captured by quick-freezing and correlated with quantal transmitter release, *J. Cell Biol.*, 81, 275, 1979.
4. **Goodenough, U. W., Gebhart, B., Mecham, R. P., and Heuser, J. E.,** Crystals of the *Chlamydomonas reinhardtii* cell wall: polymerization, depolymerization, and purification of glycoprotein monomers, *J. Cell Biol.*, 103, 405, 1986.
5. **Goodenough, U. W. and Heuser, J. E.,** The chlamydomonas cell wall and its constituent glycoproteins analyzed by the quick-freeze deep-etch technique, *J. Cell Biol.*, 101, 1550, 1985.
6. **Friend, D. S.,** Plasma membrane diversity in a highly polarized cell, *J. Cell Biol.*, 93, 243, 1982.
7. **Bearer, E. L. and Orci, L.,** Endothelial fenestral diaphragms. A quick-freeze, deep-etch study, *J. Cell Biol.*, 100, 418, 1985.
8. **Chandler, D. E. and Heuser, J.,** The vitelline layer of the sea urchin egg and its modification during fertilization. A freeze fracture study using quick-freezing and deep-etching, *J. Cell Biol.*, 84, 618, 1980.
9. **Chandler, D. E. and Kazilek, C. J.,** Extracellular coats on the surface of *Strongylocentrotus purpuratus* eggs: stereo electron microscopy of quick-frozen and deep-etched speciments, *Cell Tissue Res.*, 246, 153, 1986.
10. **Greve, J. M. and Wassarmann, P. M.,** Mouse egg extracellular coat is a matrix of interconnected filaments possessing a structural repeat, *J. Mol. Biol.*, 181, 253, 1985.
11. **Heuser, J. E.,** Procedure for freeze-drying molecules adsorbed to mica flakes, *J. Mol. Biol.*, 169, 155, 1983.
12. **Shotten, D., Burke, B., and Branton, D.,** The shape of spectrin molecules from human erythrocyte membranes, *Biochim. Biophys. Acta,* 536, 313, 1978.
13. **Fowler, W. E. and Erickson, H. P.** Trinodular structure of fibrinogen. Confirmation by rotary shadowing and negative stain electron microscopy, *J. Mol. Biol.*, 134, 241, 1979.
14. **Tyler, J. M. and Branton, D.,** Rotary shadowing of extended molecules dried from glycerol, *J. Ultrastruct. Res.*, 71, 95, 1980.
15. **Mayne, R., Sanderson, R. D., Wiedemann, H., Fitch, J. M., and Linsenmayer, T. F.,** The use of monoclonal antibodies to fragments of chicken type IV collagen in structural and localization studies, *J. Biol. Chem.*, 258, 5794, 1983.
16. **Furthmayr, H., Wiedemann, H., Timpl, R., Odermatt, E., and Engel, J.,** Electron microscopal approach to a structural model of intima collagen, *Biochem. J.*, 211, 303, 1983.
17. **Bachinger, H. P., Doege, K. J., Petschek, P., Fessler, L. I., and Fessler, J. H.,** Structural implications from an electronmicroscopic comparison of procollagen V with procollagen I, pC-collagen, procollagen IV, and a Drosophila procollagen, *J. Biol. Chem.*, 257, 14590, 1982.
18. **Charonis, A. S., Tsilibary, E. C., Saku, T., and Furthmayer, H.,** Inhibition of laminin self-assembly and interaction with type IV collagen by antibodies to the terminal domain of the long arm, *J. Cell Biol.*, 103, 1689, 1986.
19. **Engel, J., Odermatt, E., Engel, A., Madri, A., Furthmayer, H., Rohde, H., and Timpl., R.,** Shape, domain organization, and flexibility of laminin and fibronectin, two multifunctional proteins of the extracellular matrix, *J. Mol. Biol.*, 150, 97, 1981.
20. **Charonis, A. S., Tsilibary, E. C., Yurchenco, P. D., and Furthmayer, H.,** Binding of laminin to type IV collagen: a morphological study *J. Cell Biol.*, 100, 1848, 1985.

21. **Glabe, C. G. and Vacquier, V. D.,** Egg surface glycoprotein receptor for sea urchin sperm bindin, *Proc. Natl Acad. Sci. U.S.A.*, 75, 881, 1978.
22. **Glabe, C. G. and Lennarz, W. J.,** Isolation and partial characterization of a high molecular weight glycoconjugate, derived from the egg surface, which is implicated in sperm-egg adhesion, *J. Supramol. Struc. Cell Biochem.*, 15, 387, 1981.
23. **Rossignal, D.P., Roschelle, A. J., and Lennarz, W. J.,** Sperm-egg binding: identification of a species-specific sperm receptor from eggs of *Strongylocentrotus purpuratus, J. Supramol. Struct. Cell. Biochem.*, 15, 347, 1981.
24. **Vacquier, V. D. and May, G. W.,** Isolation of bindin: the protein responsible for adhesion of sperm to sea urchin eggs, *Proc. Natl. Acad. Sci. U.S.A.*, 74, 2456, 1977.
25. **Glabe, C. G., Grabel, L. B., Vacquier, V. D., and Rosen, S. D.,** Carbohydrate specificity of sea urchin sperm bindin: a cell surface lectin mediating sperm-egg adhesion, *J. Cell Biol.*, 94, 123, 1982.
26. **Glabe, C. G.,** Interaction of sperm adhesive protein, bindin, with phospholipid vesicles. I. Specific association of bindin with gel-phase phospholipid vesicles, *J. Cell Biol.*, 100, 794, 1985.
27. **Endo, Y.,** Changes in the cortical layer of sea urchin eggs at fertilization as studied with the electron microscope, *Exp. Cell Res.*, 25, 383, 1961.
28. **Anderson, E.,** Oocyte differentiation in the sea urchin, *Arbacia punctulata,* with particular reference to the origin of cortical granules and their participation in the cortical reaction, *J. Cell Biol.*, 37, 514, 1968.
29. **Eddy, E. M. and Shapiro, B. M.,** Changes in the topography of the sea urchin after fertilization, *J. Cell Biol.*, 71, 35, 1976.
30. **Chandler, D. E. and Heuser, J.,** Membrane fusion during secretion. Cortical granule exocytosis in sea urchin eggs as studied by quick-freezing and freeze-fracture, *J. Cell Biol.*, 83, 91, 1979.
31. **Kay, E. S. and Shapiro, B. M.,** The formation of the fertilization membrane of the sea urchin egg, in *Biology of Fertilization,* Vol. 3, Metz, C. B. and Monroy, A., Eds., Academic Press, New York, 1985, 45.
32. **Vacquier, V. D., Tegner, M. J., and Epel, D.,** Protease released from sea urchin eggs at fertilization alters the vitelline layer and aids in preventing polyspermy, *Exp. Cell Res.*, 80, 111, 1973.
33. **Carroll, E. J. and Epel, D.,** Isolation and biological activity of proteases released by sea urchin eggs following fertilization, *Dev. Biol.*, 44, 22, 1975.
34. **Klebanoff, S. J., Foerder, C. A., Eddy, E. J., and Shapiro, B. M.,** Metabolic similarities between fertilization and phagocytosis. Conservation of a peroxidatic mechanism, *J. Exp. Med.*, 149, 938, 1979.
35. **Weidman, P. J., Kay, E. S., and Shapiro, B. M.,** Assembly of the sea urchin fertilization membrane: isolation of proteoliaisin, a calcium dependent ovoperoxidase binding protein, *J. Cell Biol.*, 100, 938, 1985.
36. **Foerder, C. A. and Shapiro, B. M.,** Release of ovoperoxidase from sea urchin eggs hardens the fertilization membrane with tyrosine crosslinks, *Proc. Natl. Acad. Sci. U.S.A.*, 74, 4214, 1977.
37. **Foerder, C. A., Klebanoff, S. J., and Shapiro, B. M.,** Hydrogen peroxide production, chemiluminescence and the respiratory burst of fertilization: interrelated events in early sea urchin development, *Proc. Natl. Acad. Sci. U.S.A.*, 75, 3183, 1978.
38. **Carroll, E. J. and Baginski, R. M.,** Sea urchin fertilization envelope: isolation, extraction and characterization of a major protein fraction from *Strongylocentrotus purpuratus* embryos, *Biochemistry*, 17, 2605, 1978.
39. **Hall, H. G.,** Hardening of the sea urchin fertilization envelope by peroxidase-catalyzed phenolic coupling of tyrosines, *Cell*, 15, 343, 1978.
40. **Bryan, J.,** The isolation of a major structural element of the sea urchin fertilization membrane, *J. Cell Biol.*, 44, 635, 1970.
41. **Bryan, J.,** On the reconstitution of the crystalline components of the sea urchin fertilization membrane, *J. Cell Ciol.*, 45, 606, 1970.
42. **Stephens, R. E. and Kane, R. E.,** Some properties of hyalin the calcium insoluble protein of the hyaline layer of sea urchin eggs, *J. Cell Biol.*, 44, 611, 1970.
43. **Kay, E., Eddy, E. M., and Shapiro, B. M.,** Assembly of the fertilization membrane of the sea urchin: isolation of a divalent cation-dependent intermediate and its crosslinking *in vitro, Cell*, 29, 867, 1982.
44. **Kidd, P.,** The jelly coat and vitelline coats of the sea urchin egg: new ultrastructural features, *J. Ultrastruct. Res.*, 64, 204, 1978.
45. **Shapiro, B. M.,** Limited proteolysis of some egg surface components is an early event following fertilization of the sea urchin, *Strongylocentrotus purpuratus, Dev. Biol.*, 46, 88, 1975.
46. **Veron, M., Foerder, C., Eddy, E. M., and Shapiro, B. M.,** Sequential biochemical and morphological events during assembly of the fertilization membrane of the sea urchin, *Cell*, 10, 321, 1977.
47. **Carroll, E. J., Byrd, E. W., and Epel, D.,** A novel procedure for obtaining denuded sea urchin eggs and observations on the role of the vitelline layer in sperm reception and egg activation, *Exp. Cell Res.*, 108, 365, 1977.
48. **Kemp, N. E. and Istock, N. L.,** Cortical changes in growing and in fertilized or pricked eggs of *Rana pipiens, J. Cell Biol.*, 34, 111, 1967

49. **Grey, R. D., Wolf, D. P., and Hedrick, J. L.,** Formation and structure of the fertilization envelope in *Xenopus laevis, Dev. Biol.*, 36, 44, 1974.
50. **Wolf, D. P.,** The cortical granule reaction in living eggs of the toad, *Xenopus laevis, Dev. Biol.*, 36, 62, 1974.
51. **Wolf, D. P., Nishihara, T., West, D. M., Wyrick, R. E., and Hedrick, J. L.,** Isolation, physicochemical properties, and the macromolecular composition of the vitelline and fertilization envelopes for *Xenopus laevis* eggs, *Biochemistry*, 15, 3671, 1976.
52. **Greve, L. C. and Hedrick, J. L.,** An immunocytochemical localization of the cortical granule lectin in fertilized and unfertilized eggs of *Xenopus laevis, Gamete Res.*, 1, 13, 1978.
53. **Gerton, G. L. and Hedrick, J. L.,** The vitelline envelope to fertilization envelope conversion in eggs of *Xenopus laevis, Dev. Biol.*, 116, 1, 1986.
54. **Hedrick, J. L. and Smith, A. J.,** Size and charge isomer separation and estimation of molecular weights of proteins by disc gel electrophoresis, *Arch. Biochem. Biophys.*, 126, 155, 1968.
55. **Wolf, D. P.,** On the contents of the cortical granules from *Xenopus laevis* eggs, *Dev. Biol.*, 38, 14, 1974.
56. **Wyrick, R. E., Nishihara, T., and Hedrick, J. L.,** Agglutination of jelly coat and cortical granule components and the block to polyspermy in the amphibian *Xenopus laevis, Proc.Nat. Acad. Sci. U.S.A.*, 71(5), 2067, 1974.
57. **Nishihara, T. and Hedrick, J. L.,** A molecular mechanism for envelope elevation at fertilization, *Fed. Proc.*, 36, 811, 1977.
58. **Grey, R. D., Working, P. K., and Hedrick, J. L.,** Evidence that the fertilization envelope blocks sperm entry in eggs of *Xenopus laevis:* interaction of sperm with isolated envelopes, *Dev. Biol.*, 54, 52, 1976.

Chapter 10

THE PLASMA MEMBRANE OF THE SPERMATOZOON

James K. Koehler

TABLE OF CONTENTS

I. Introduction.......202

II. The Sperm Head.......202

III. The Flagellar Region.......209

IV. The Middle Piece Plasma Membrane.......209

V. The Principal Piece.......210

VI. The Acrosome Reaction.......210

VII. Sperm-Egg Interactions.......210

VIII. Concluding Comments.......211

Acknowledgment.......211

References.......213

I. INTRODUCTION

The spermatozoan is morphologically highly differentiated and capable, at least for a brief portion of its life history, of existence independent of other cells. Its specializations are uniquely designed to enable this cell to pursue the complex series of processes known as fertilization. These include the many morphogenetic changes occurring during spermatogenesis which transform the cell from a rather conventional appearing somatic cell to the mature, haploid spermatozoan. The final product possesses a flagellar apparatus, enzyme-filled acrosome, highly modified and compacted nucleus, and other special organelles arranged in a very precisely designed package to provide the most efficient delivery system to the oocyte. The plasma membrane of the sperm engages in concomitant major changes in extent, structure, and activities during these transformations, and encloses the spermatozoon in a continuous envelope, highly differentiated with regard to regional variations in underlying structures. Studies, with a variety of probes of the sperm surface, indicate an amazing degree of segregation of chemical activities and binding sites to distinct regions of the membrane. When one superimposes changes in these patterns induced by developmental, endocrine, and other factors, one is struck with the exquisite specificity of the differentiation program resulting in the functional, mature spermatozoon. A number of reviews have dealt with studies mapping the sperm surface using lectins[1] and antibodies[2] as probes. Recently, monoclonal antibodies have been successfully used by several groups to localize highly specific antigens on the mature and developing spermatozoon.[3,4] The fracture label procedure has also been employed[5] to show that certain lectin binding sites are restricted to the outer, or exoplasmic half, of the plasma membrane in boar spermatozoa. These studies further correlate the presence of intramembranous particles with putative transmembrane glycoproteins containing binding sites for these lectins.

Among methods of analysis that have revealed striking evidence of the sperm membranes morphological regionalization, is freeze fracture (F.F.) combined with transmission electron microscopy. Since the first F.F. observations were made on spermatozoa in the mid-1960s[6] a number of groups have used these cells as models for membrane structure and function research using F.F. and associated techniques. In this paper, we will focus primarily on the results of studies done with mammalian systems in an effort to focus and limit the discussion to a manageable volume of source material. Even with this constraint, this chapter should not be considered a comprehensive review or detailed survey of the subject. Rather, the work will highlight certain results with a view to putting otherwise rather disconnected findings in context and perspective. It will be attempted, where possible, to indicate areas where generalities may exist which cut across the all-too-pervasive species specificities inherent to gamete biology.

Freeze-fracture observations on spermatozoan plasma membranes tend to fall into three categories of subject matter which we will treat in the following discussion. The major areas of emphasis will be the sperm head, the flagellum and changes or transformations in these membranes due to physiological changes such as capacitation, the acrosome reaction, and sperm-egg fusion.

II. THE SPERM HEAD

The sperm head plasma membrane can be divided into at least three morphologically distinct regions despite the fact that it is continuous over the entire cell. These regions are defined by underlying structures which appear to influence the organization of the surface membrane. They include the anterior acrosomal region, the equatorial segment (also acrosomal), and the postacrosomal region (Figure 1). Differences in plasma membrane architecture in freeze-fracture images are often discussed in terms of the distribution, density, and size of intramembranous particles (IMPs).

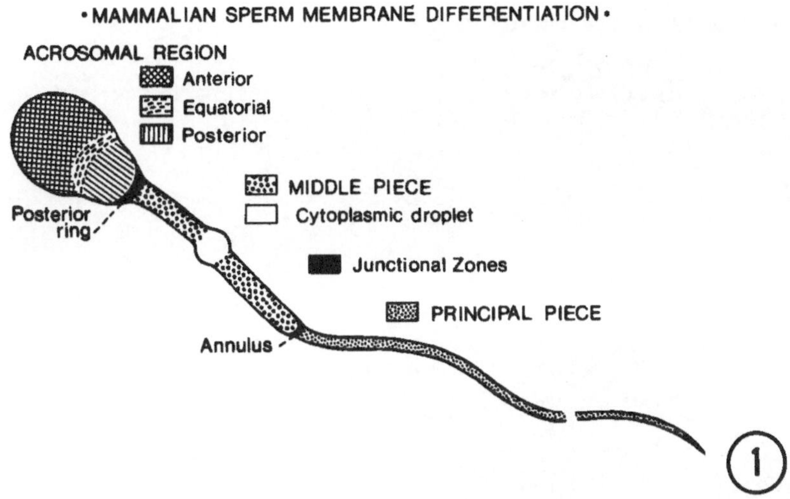

FIGURE 1. Schematic diagram of a typical mammalian spermatozoon showing characteristic regionalization of plasma membrane properties as a function of location.

The anterior acrosomal region of the plasma membrane is generally not remarkable in the numbers or organization of intramembranous particles, but there are considerable variations in these parameters amongst species. Thus, in rabbit sperm, the IMPs in this region are widely scattered, large, and when carefully examined, consist of small groups of two or three small particles (Figure 2). A similar distribution has been shown in boar spermatozoa.[5] In the mouse or vole, however, randomly scattered particles of the more conventional size are seen (Figure 3). In some advantageous replicas, as in the rabbit sperm in Figure 4, it can be seen that both acrosomal and postacrosomal portions of the membrane are also populated by a high density of very minute IMPs less than 10 nm in size. Conversely, rodent sperm have an almost crystalline arrangement of uniformly sized IMPs in the postacrosomal region which is completely absent from the plasma membrane overlying the acrosome (Figure 3). Deep-etched observations have also shown a variety of paracrystalline surface materials associated with various regions of the sperm head.[7,8] Some of these substances may represent absorbed material important in protecting sperm during epididymal transit or from undergoing a premature acrosome reaction (decapacitation factors).

A striking aspect of the basal region of many mammalian sperm heads is the presence of periodic bands or cords of helically organized particles (Figures 5 and 6). Despite the fact that such organization has been observed for over 20 years, we still have no definitive function for these structures other than the speculation that they may reinforce the mechanical strength of the head-neck junction. Although it is tempting to invoke a correlation between the complexity or size of the sperm head and the presence or extent of the cords, such analysis does not seem valid. For example, although vole and mouse sperm are very similar in size and shape, the former have a modest complement of cords, while the latter have none. Basal cords are similarly absent on some of the largest mammalian sperm heads, just those most likely to require reinforcement, such as those of the guinea pig and shrew *Suncus murinus* as well as from the rather small sperm head of man.[9-11]

Immediately below the zone where such basal cords are found is the posterior ring (Figures 7 and 8), a uniform belt approximately 40 nm in thickness extending around the entire circumference of the head-neck junction. This structure is invariably found in mammalian sperm and appears to be virtually identical in detailed structure among species. Included among these details is a highly ordered periodicity of 8 nm perpendicular to the belt direction (Figures 9 and 10). The analog of this structure in thin sections can easily be detected as a

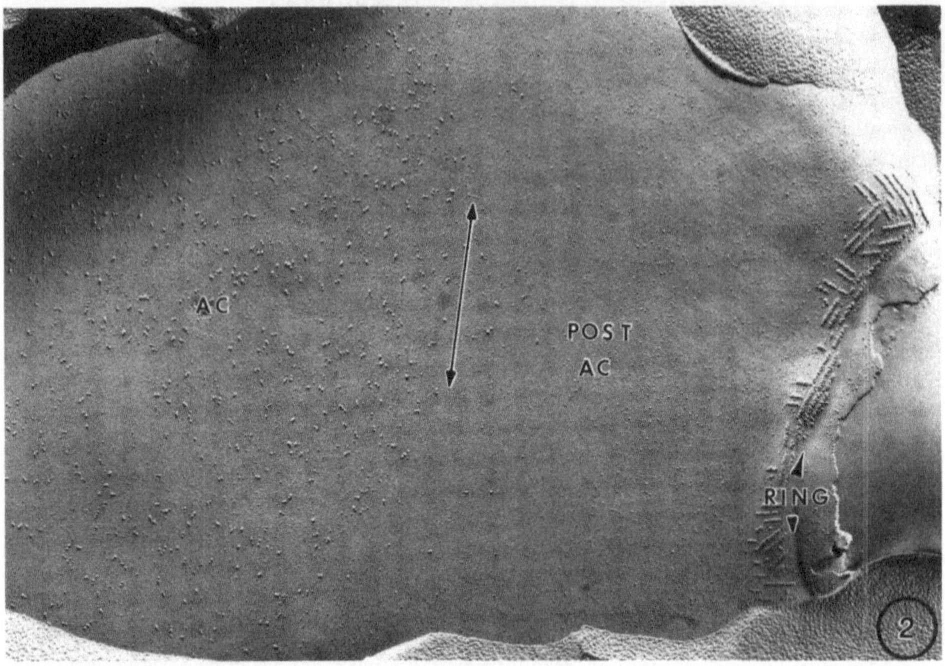

FIGURE 2.

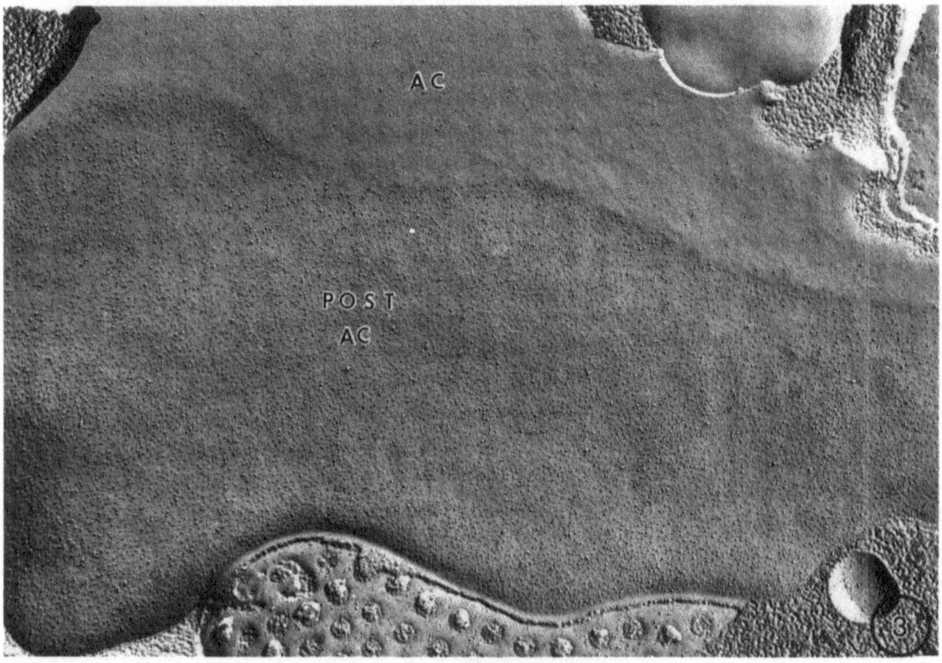

FIGURE 3.

FIGURE 2. Epididymal rabbit spermatozoon freeze-fractured and showing characteristics of P-face architecture of the sperm head. The acrosomal region (AC) has larger, clustered particles in contrast to the postacrosomal region (Post-AC). The double headed arrow shows the approximate demarcation between these zones. Basal cords and the posterior ring are also seen. (Magnification × 24,000.)

FIGURE 3. Mouse spermatozoon showing the very high density of small particles in the postacrosomal region. Note contrast in distribution of particles with those of the rabbit sperm in Figure 1. The pore-studded, redundant nuclear envelope is also seen at the bottom of the figure. (Magnification × 37,000.)

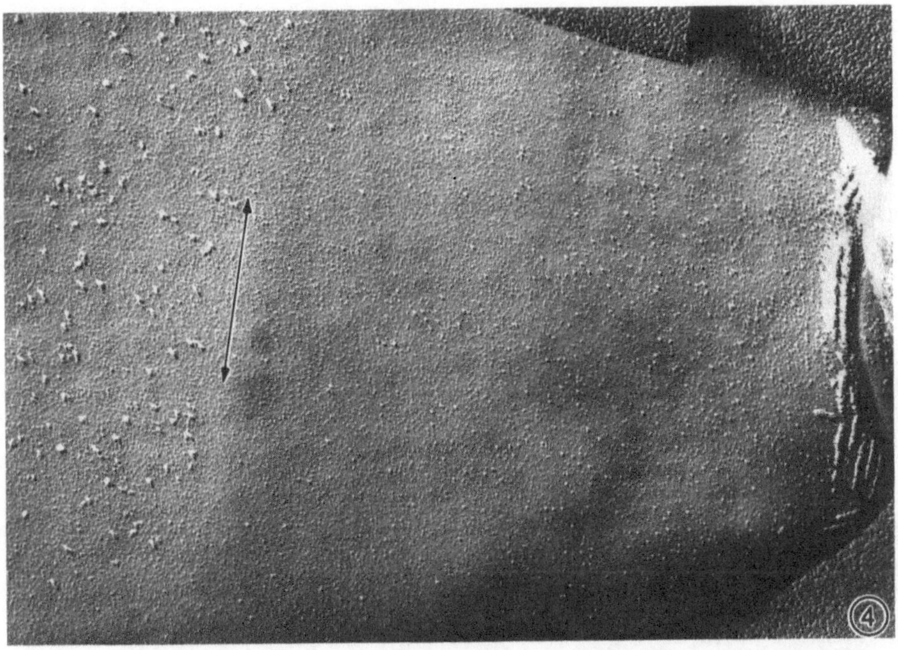

FIGURE 4.

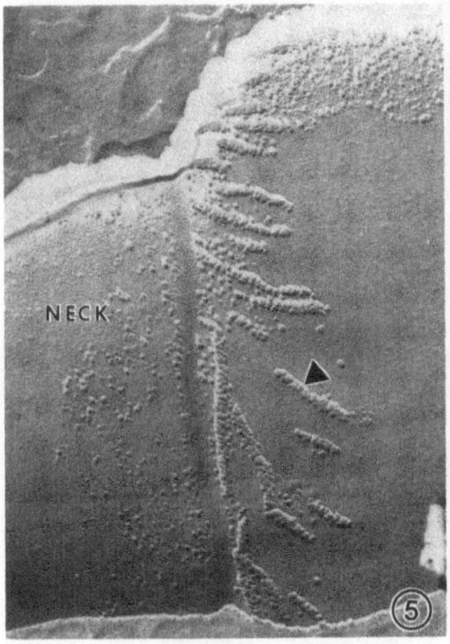

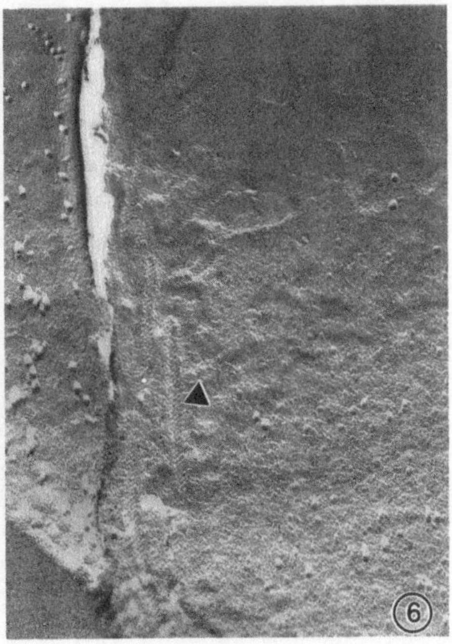

FIGURE 5.

FIGURE 6.

FIGURE 4. Rabbit sperm similar to that shown in Figure 1 but displaying the high concentration of very small particles which appear to be uniformly distributed over the sperm head. (Magnification X 28,900.)

FIGURE 5. Stallion sperm showing the cords at the base of the head composed of large interdigitating particles. P-face image. (Magnification × 51,800.)

FIGURE 6. The basal cords seen in an E-face image showing an obvious helical periodicity (arrow). Stallion sperm. (Magnification × 66,600.)

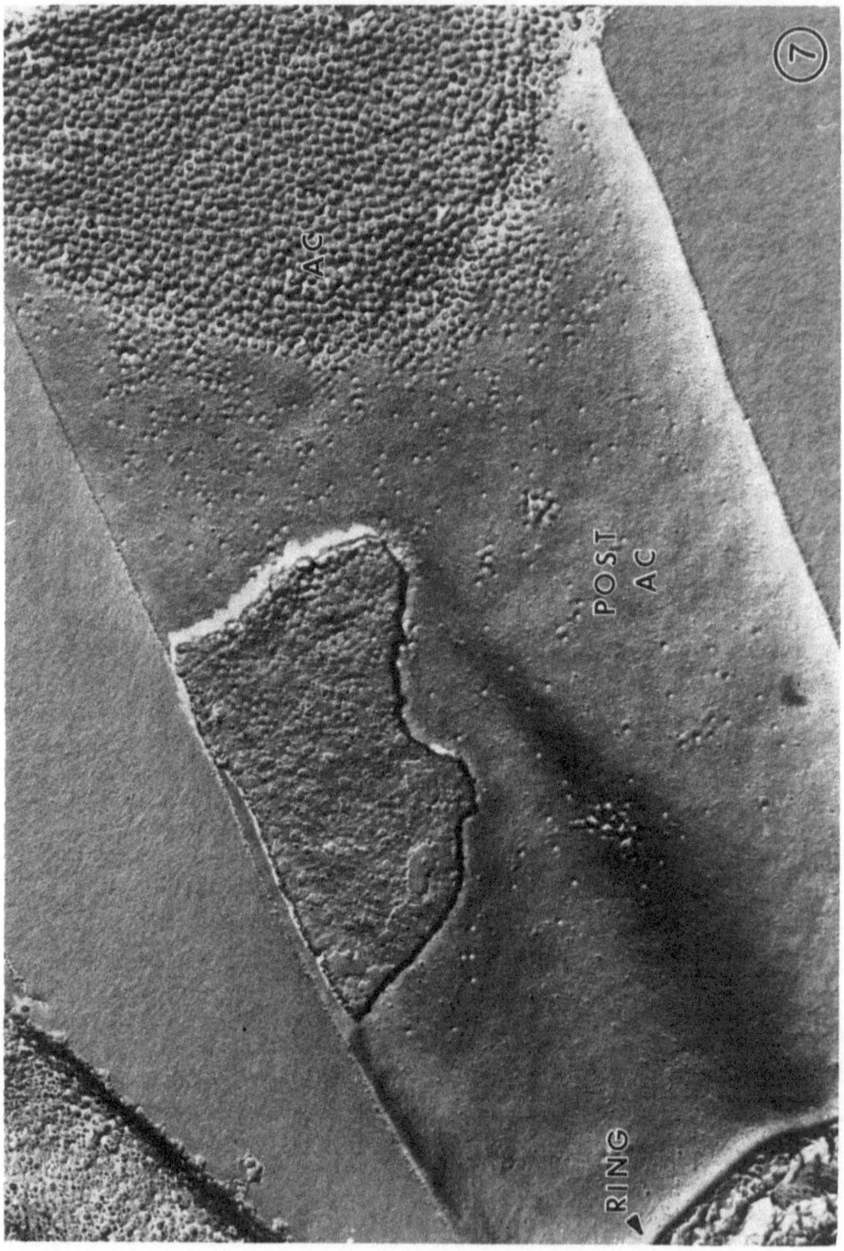

FIGURE 7. Hamster spermatozoon treated with filipin to identify cholesterol-rich regions. The filipin complexes are almost completely restricted to the acrosomal region (AC) in contrast to the postacrosomal region. (Magnification × 38,000.) (Figure courtesy of Dr. Judy M. Clark.)

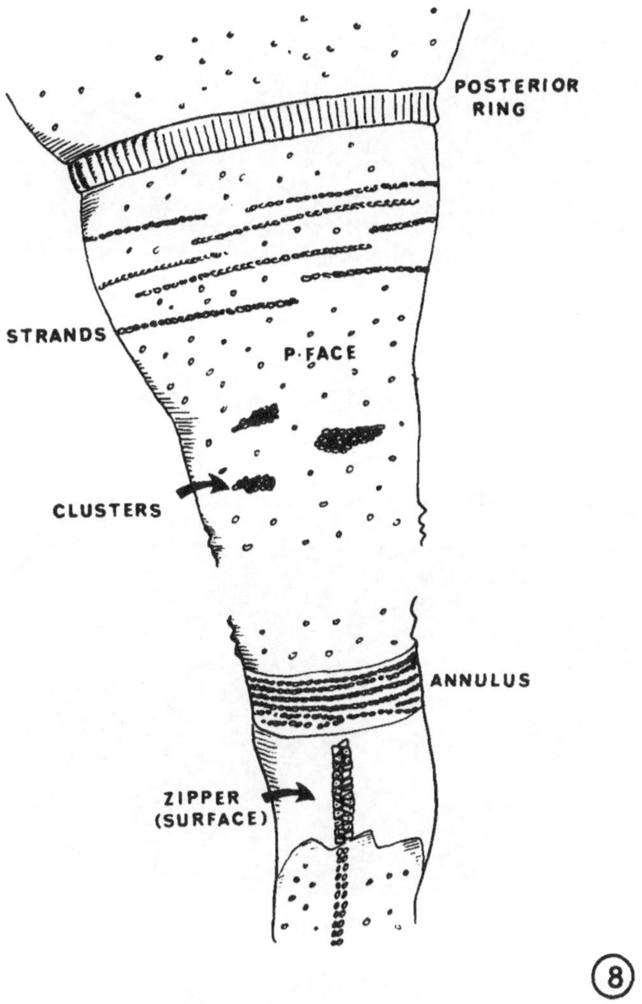

FIGURE 8. Diagram of the flagellar surface of a typical mammalian spermatozoon. The various regions of the flagellum display unique membrane characteristics, particularly with respect to the organization of intramembranous particles.

dense plaque sandwiched between the plasma and nuclear membranes (Figure 11). It is apparent that this structure may provide a "tight junction"-like seal between the cytosolic compartments of the sperm head and tail. Although there is no direct evidence of such a physiological seal, the differences in metabolic requirements between these compartments may make such a seal a highly useful if not mandatory component (see also Koehler[12]). Further suggestive evidence comes from recent studies using fluorescence recovery after photobleaching. Wolf et al.[13] have shown that lipid diffusion is two to three times as rapid in the flagellar membrane of mouse sperm compared to that of the head.

The freeze-fracture method is also a useful approach to the problem of localizing various lipid classes in membranes after complexing with specific markers. Thus, the polyene antibiotic filipin has been used to localize β-hydroxy sterols such as cholesterol in guinea pig sperm membranes. Elias et al.[14] have shown that the filipin complexes in the head plasmalemma are largely restricted to the acrosomal region, an area programmed to undergo fusion with the underlying acrosomal membrane during the course of the acrosome reaction. Toshimori et al.[36] have observed the same kind of filipin localizations in mouse spermatozoa.

FIGURE 9.

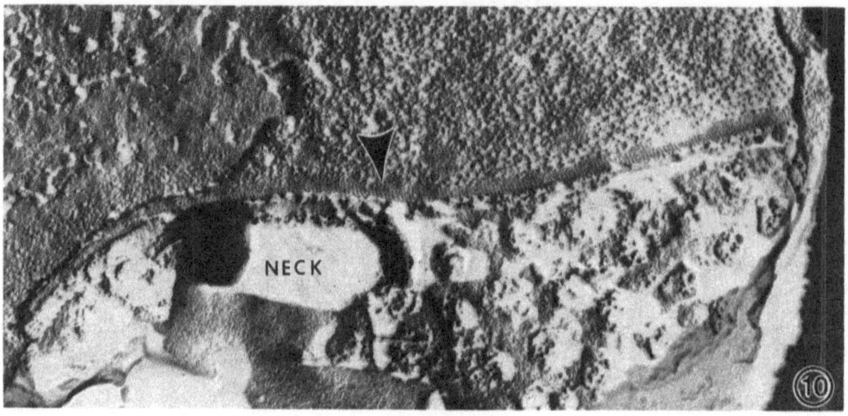

FIGURE 10.

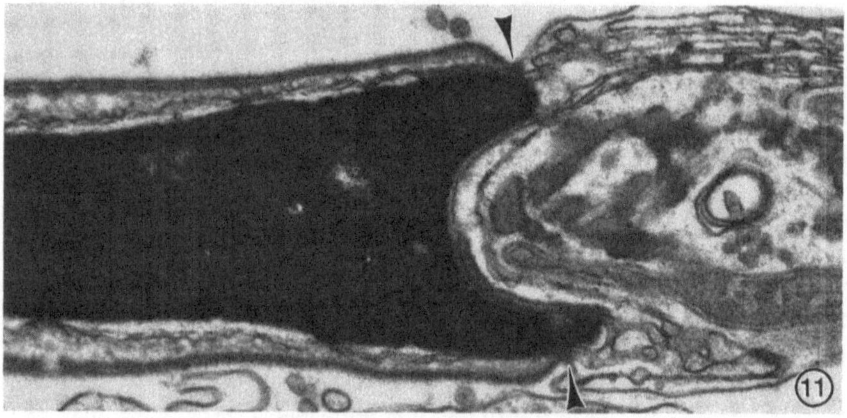

FIGURE 11.

FIGURE 9. The posterior ring seen in a human spermatozoon. The fine periodicity of about 8 nm is quite evident in this preparation.

FIGURE 10. The posterior ring in a mouse sperm head (arrow). The structure appears to be essentially identical among sperm from the various species investigated.

FIGURE 11. Thin section of bull spermatozoon showing location of the posterior ring (arrows) as a dense material between plasma and nuclear membranes. (Magnification × 40,320.)

A similar distribution is seen on the hamster sperm head after filipin treatment with a very sharp boundary of reactivity at the posterior edge of the acrosome (Figure 7). Similar types of observations using probes of anionic lipids (polymyxin B) indicate that these phospholipids are enriched in regions destined to undergo membrane fusion (Bearer and Friend[15]). Such observations are in concert with hypotheses suggesting an increase in such components in fusigenic membrane domains (Papahadjopoulus[16]). Since most other freeze-fracture observations focus on the protein and glycoprotein components of the membrane (i.e., IMPs), studies of lipid components should receive additional efforts. It is the phospholipids that must intimately interact during fusion reactions and freeze-fracture analysis of these events should be able to provide key additional data on this all important phenomenon.

III. THE FLAGELLAR REGION

As with the sperm head, the flagellum of mammalian spermatozoa can be divided into several distinct regions possessing specific structural characteristics. These include the neck, middle piece, principal piece, and end/or terminal piece. The plasma membrane enveloping these regions appears, at least in part, to be differentiated according to the nature of underlying structures. Thus, although the axonemal components such as the 9 + 2 microtubules course almost the entire length of the flagellum, it seems rather to be the immediate submembranous structures which influence plasma membrane ultrastructure and function. Sperm flagellar membranes from more ''primitive'' organisms such as the sea urchin, do not show such differences in organization along their lengths, nor are there significant changes in submembranous organelles. The base of the flagellum in sperm from such organisms has been shown to possess a ''ciliary necklace'', a strand or strands of IMPs around the base of the flagellum (Cosson and Gulik[17]). Other than this formed structure, the basic pattern is that of randomly scattered IMPs.

IV. THE MIDDLE PIECE PLASMA MEMBRANE

In mammalian spermatozoa, the short neck or connecting piece is followed by a segment of the flagellum containing helically disposed mitochondria just beneath the plasma membrane. This membrane shows a varied and interesting ultrastructure with considerable species variations. The IMPs in this region seem to be highly organized in arrangements which suggest some interaction or influence with the underlying mitochondria. Rodent sperm often show extensive strands of IMPs organized in circumferential patterns around the mitochondrial helix (Figure 12). This is, however, not a universal finding, and is not seen on many other sperm including those of primates. Although the strands bear a superficial resemblance to the ciliary necklace, there is no solid evidence of homology with these structures. In guinea pig sperm, the strands of particles appear to disrupt into a randomized pattern upon capacitation and the acrosome reaction (Koehler and Gaddum-Rosse[18]). It is known that sperm, at least of some species, undergo a change in motility pattern at this time which could be influenced by such changes in protein distribution. Similar evidence is presented by Olsen,[19] who showed that immature opposum sperm demonstrate randomly disposed IMPs over the midpiece membrane, whereas mature, motile sperm, show highly ordered linear arrays of IMPs. Certainly, more direct evidence is needed before we can establish a causal relationship between the particle orientations and flagellar function.

Another characteristic pattern of IMPs found in the middle-piece membrane consists of patches of uniform, close-packed particles which resemble gap junctions (Figure 13). The IMPs which comprise the patches are smaller and more uniform in shape than the randomly disposed particles seen elsewhere in the membrane. The patches are also similar to the ciliary granule plaques which have been implicated in Ca^{++} binding and transport in *Paramecium.*[20]

V. THE PRINCIPAL PIECE

Just as the posterior ring provides a seal at the base of the sperm head, so the annulus terminates the middle piece and signals the beginning of the principal piece of the flagellum. In thin sections one sees only a dense plaque of material continuous with the internal leaflet of the plasma membrane, and making contact with the terminal mitochondrial winding of the midpiece helix. In freeze fracture, the plasma membrane in contact with the annulus is specialized, having a number of circumferential strands of small IMPs easily distinguished from the rest of the membrane (Figures 14 and 15). The dense filaments of the annulus may insert or be associated in some way with these particles. Instead of mitochondria, the principal piece contains a fibrous sheath interposed between the plasma membrane and the accessory fibers surrounding the axoneme. In addition to the usual, randomly scattered IMPs in this region of the membrane, a prominent and constant feature of mammalian sperm, the so-called "zipper", is located here.[21,22] The zipper is truly a transmembrane apparatus visible on the membrane surface as an interdigitating double row of enlarged ovoid particles sometimes showing a pore-like structure (Figure 16). This surface morphology is complemented by a double row of IMPs on the internal fracture face of the membrane (Figure 17). Suggestions for a function for the zipper have included sites of ionic transport[20,22] and an anchoring device for axonemal components.[19,21] The complex transmembrane structure of the zipper certainly has the potential of serving a role as a sampling or transport device for external components, but this is as yet unproven. Many of the details concerning flagellar membrane structure are summarized in Figure 8.

VI. THE ACROSOME REACTION

The details of freeze fracture observations of the acrosome reaction in mammalian sperm have been recently reviewed by Flechon,[23] and will only be briefly summarized here. Upon incubation in capacitating media, or when treated with Ca^{++} ionophore or lysolecithin, IMP-cleared zones occur in the plasma membrane overlying the acrosome.[24-26] Similar redistributions of other membrane associated components have also been documented using probes such as monoclonal antibodies[27] and lectins.[28] These reactions seem to be initial reorganizations of membrane components prior to the initiation of fusion per se. As such, they can be considered part of the process of capacitation or preparations for the acrosome reaction. The fusion between the plasma membrane and outer acrosomal membrane, at least in ram and hamster spermatozoa, appears to be initiated in the zone just anterior to the equatorial segment of the acrosome and spreads superiorly.[23] Contrary to what has been proposed previously (for example Russell et al.[29]), the process is not essentially a vesiculation reaction causing a budding of hybrid membrane from the sperm surface. According to Flechon's convincing micrographs, it is rather a propagated reticulation yielding rivulets of openings to the acrosomal contents while still maintaining the basic cohesiveness of the acrosomal cap.[23] In this process, the equatorial segment of the acrosome is protected in some way from engaging in the fusion reaction. This would seem to be important because a number of workers have suggested that it is the plasma membrane overlying the equatorial segment which undergoes initial attachment and fusion with the oocyte surface.[30]

VII. SPERM-EGG INTERACTIONS

Very little data has been reported on the freeze-fracture appearance of fusing gametes. The reason for this paucity of information, at least with mammalian cells, is quite obvious; namely, it is the proverbial "needle in the hay stack" problem. Freeze-fracture preparations are generally obtained by random fractures through a frozen specimen. This combined with the small numbers of fertilizing gametes obtainable in any given experiment, explains the lack of results in this area. The very considerable data accumulated using conventional thin sections (see Yanagimachi,[31] Talbot,[32] for reviews) as well as data on membrane fluidity during fertilization,[33] indicate that the kinds of information that freeze-fracture images could provide might be very rewarding. Koehler et al.[34] have shown that microvillous processes from the oocytes apparently attach to IMP-rich regions of the sperm plasma membrane in the equatorial region. This study was done on human sperm interacting with zona-free hamster oocytes, certainly a highly modified and artificial system which requires confirmation in a homologous system. We are currently pursuing such studies with zona-free hamster oocytes fertilized in a manner which encourages polyspermy in order to improve our chances of visualizing sperm-egg interactions.

VIII. CONCLUDING COMMENTS

Freeze-fracture preparations like any other morphological methodology generates a spectrum of artifacts which must be considered in making interpretations of results. This is of special importance when coming to conclusions concerning the distribution, clustering, and size characteristics of IMPs. Pricam et al.[35] has pointed out that the treatment of cells prior to and during freezing is critical to the resultant morphology after freeze fracture. We have shown that the P-face to E-face ratios of IMPs changes significantly when comparing fixed, glycerinated sperm with fresh-frozen samples.[34] In some cases there seems to be a complete reversal of the density of IMPs associated with these faces depending on the manner of freezing and pretreatment. This does not automatically imply that the fixed preparation is artifactual, whereas the fresh-frozen one is not. Rather, it should be considered that the physical forces that come into play during freeze fracture will operate under different constraints with fixed or fresh tissue. Binding of IMPs to one or the other leaflets of the membrane during splitting may be greatly modified by fixation as may be their mobility prior to freezing.

Despite interpretive problems such as those outlined above, the overwhelming bulk of freeze-fracture studies of sperm and sperm membranes have corroborated and extended the results of conventional fine structure analysis. In many cases, structures that could not have been elucidated by thin section observations have been beautifully demonstrated by freeze-fracture methods. These would include the demonstration of the lamellar nature of sperm nucleoplasm, the basal cords of helically arranged IMPs, and the ''zipper'', to mention just a few. Physiologically important membrane transformations, such as the clearance of particles from regions of the plasma membrane during capacitation, and the reticulation of the fusing acrosomal and plasma membranes during the acrosome reaction, have also been clearly demonstrated by freeze fracture. Such observations will continue to be of prime importance as our attention focuses on the interaction between sperm and egg in order to elucidate the contributions of respective membrane systems to the initial sequence of the fusion reaction.

ACKNOWLEDGMENT

Portions of the work described in this paper have been supported by a grant from the National Science Foundation (PCM 840-8773). Dr. Judy M. Clark, Department of Biological Structure, University of Washington, Seattle WA kindly provided data on the filipin studies and Figure 7, and Ms. Doris Ringer and Inger Smith provided editorial and word processing services.

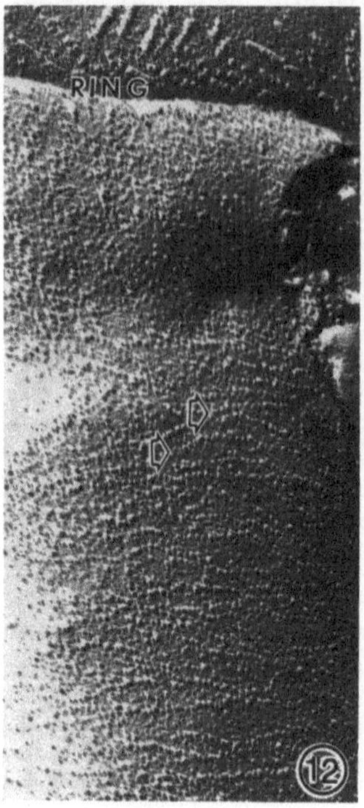

FIGURE 12.

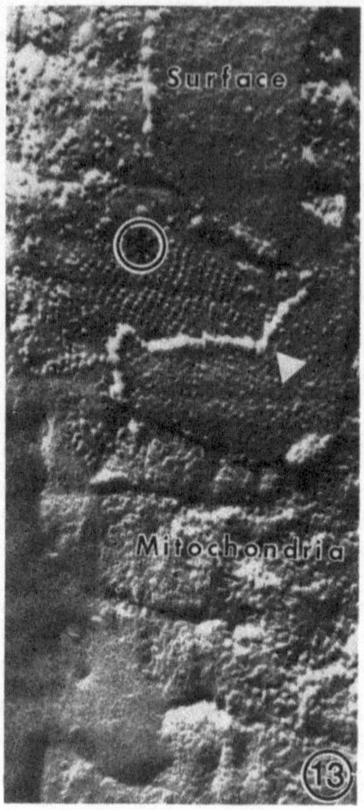

FIGURE 13.

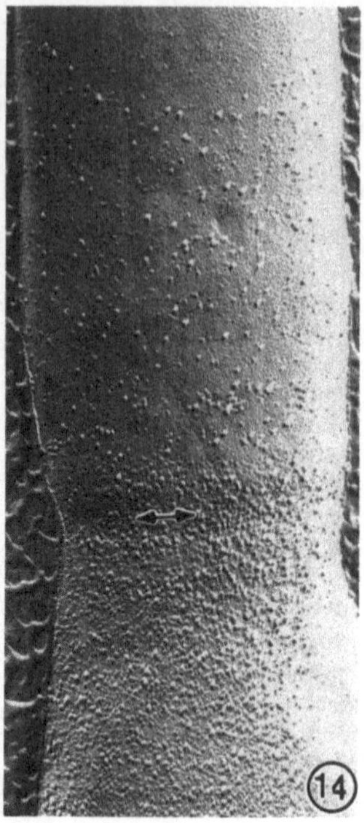

FIGURE 14.

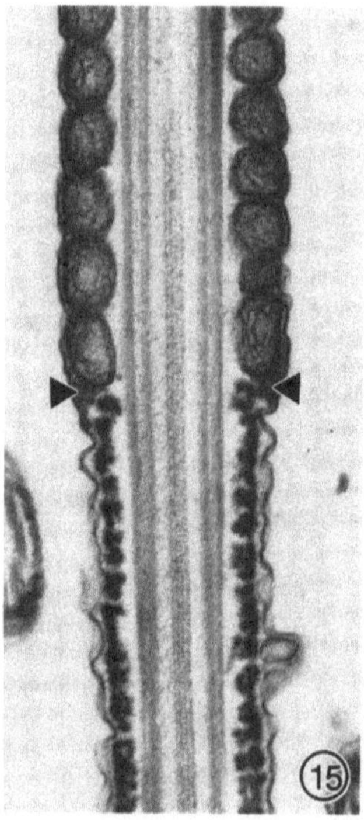

FIGURE 15.

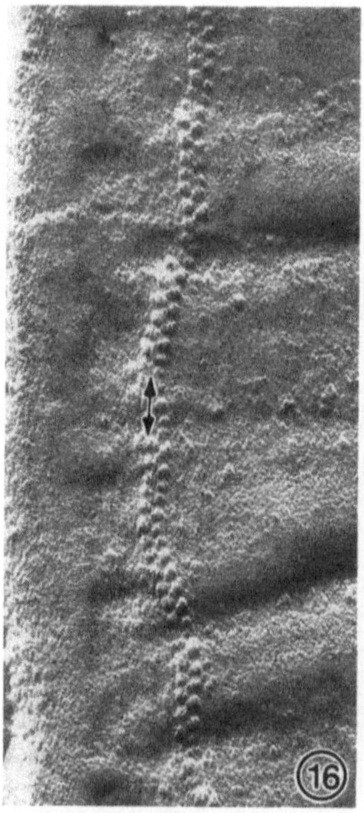

FIGURE 16.

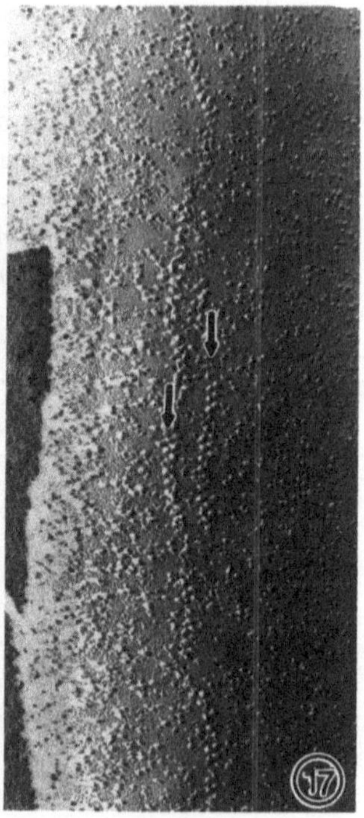

FIGURE 17.

FIGURE 12. Initial segment of the flagellum of a vole spermatozoon showing circumferential strands of particles (arrows) around the midpiece. P-face fracture. (Magnification × 50,000.)

FIGURE 13. Rat sperm midpiece, showing patches of hexagonally arranged particles (*) associated with mitochondrial membranes. These crystalline patterns are reflected into the surface membrane (white △). (Magnification × 66,400.)

FIGURE 14. Region of the annulus in the flagellum of a stallion spermatozoon (double arrows). Note difference in particle density above (principle piece) and below (middle piece) the annulus in this P-face fracture. (Magnification × 51,000.)

FIGURE 15. Section through the midpiece-principal piece junction in a bull spermatozoon. The annulus is detected as a submembranous density just inferior to the terminal mitochondrion (arrows). (Magnification × 24,300.)

FIGURE 16. Mouse sperm principal piece in a deep-etched surface view showing the "zipper" (double arrows). (Magnification × 113,400.)

FIGURE 17. P-face fracture of human sperm principal piece showing the IMP distribution (arrows) associated with the surface "zipper" (Magnification × 72,900.)

REFERENCES

1. **Koehler, J. K.,** Lectins as probes of the sperm surface, *Arch. Androl.*, 6, 197, 1981.
2. **Eddy, E. M., Vernon, R. B., Muller, C. H., Hahnel, A. C., and Fenderson, B. A.,** Immune dissection of sperm surface modifications during epididymal maturation, *Am. J. Anat.*, 174, 225, 1985.
3. **Feuchter, F. A., Vernon, R. B., and Eddy, E. M.,** Analysis of the sperm surface with monoclonal antibodies: topographically restricted antigens appearing in the epididymis, *Biol. Reprod.*, 24, 1099, 1981.
4. **Primakoff, P. and Myles, D. G.,** A map of the guinea pig sperm surface constructed with monoclonal antibodies, *Dev. Biol.*, 98, 417, 1983.
5. **Aguas, A. P. and Pinto da Silva, P.,** Regionalization of transmembrane glycoproteins in the plasma membrane of boar sperm head is revealed by fracture label, *J. Cell Biol.*, 97, 1356, 1983.
6. **Koehler, J. K.,** Fine structure observations in frozen-etched bovine spermatozoa, *J. Ultrastruct. Res.*, 16, 359, 1966.
7. **Friend, D. S.,** Plasma membrane diversity in a highly polarized cell, *J. Cell Biol.*, 93, 243, 1982.
8. **Toyama, Y., Nagano, T., and Suzuki, F.,** Surface observations on some mammalian spermatozoa after rapid freezing and deep-etching, in *The Sperm Cell,* Proc. 4th Int. Symp. Spermatology, Andre, J., Ed., Martinus Nijhoff, London, 1983, 119.
9. **Koehler, J. K.,** Human sperm head, ultrastructure: a freeze-etching study, *J. Ultrastruct Res.*, 39, 520, 1972.
10. **Koehler, J. K.,** Observations on the fine structure of vole spermatozoa with particular reference to cytoskeletal elements in the mature sperm head, *Gamete Res.*, 1, 247, 1978.
11. **Koehler, J. K.,** Fine structure of spermatozoa of the Asiatic musk shrew: *Suncus murinas, Am. J. Anat.*, 149, 135, 1977.
12. **Koehler, J. K.,** Structural heterogeneity of the mammalian sperm flagellar membrane, *J. Submicrosc. Cytol.*, 15, 247, 1983.
13. **Wolf, D. E., Hagopian, S. S., and Ishijima, S.,** Changes in sperm plasma membrane lipid diffusibility after hyperactivation during *in vitro* capacitation in the mouse, *J. Cell Biol.*, 102, 1372, 1986.
14. **Elias, P. M., Friend, D. S., and Goerke, J.,** Membrane sterol heterogeneity, freeze-fracture detection with saponins and filipin, *J. Histochem. Cytochem.*, 27, 1237, 1979.
15. **Bearer, E. L. and Friend, P. S.,** Anionic lipid domains: correlation with functional topography in a mammalian cell membrane, *Proc. Natl. Acad. Sci. U.S.A.*, 77, 6601, 1980.
16. **Papahadjopoulos, D.,** Membrane Fusion, in *Cell Surface Reviews,* Vol. III, Poste, G. and Nicholson, G. L., Eds., Elsevier, Amsterdam, 1978, 765.
17. **Cosson, M. P. and Gulik, A. J.,** Description of the mitochondria-axoneme junction in sea urchin spermatozoa: presence of a flagellar necklace, *J. Ultrastruct. Res.*, 79, 47, 1982.
18. **Koehler, J. K. and Gaddum-Rosse, P.,** Media induced alterations of the membrane associated particles of the guinea pig sperm tail, *J. Ultrastruct. Res.*, 51, 106, 1974.

19. **Olsen, G. E.,** Changes in intramembrane particle distribution in the plasma membrane of *Didelphis virginiana* spermatozoa during maturation in the epididymis, *Anat. Rec.*, 197, 471, 1980.
20. **Plattner, H.,** Ciliary granule plaques: membrane intercalated particle aggregates associated with Ca + + binding sites in Paramecium, *J. Cell Sci.*, 18, 257, 1975.
21. **Friend, D. S., Elias, P. M., and Rudolf, I.,** Disassembly of the guinea pig sperm tail, in *The Spermatozoon*, Fawcett, D. W. and Bedford, J. M., Eds., Urban & Schwartzenberg, Baltimore, MD, 1979, 157.
22. **Friend, D. S. and Heuser, J. E.,** Orderly particle arrays on the mitochondrial outer membrane in rapidly frozen sperm, *Anat. Rec.*, 199, 159, 1981.
23. **Flechon, J. S.,** Sperm surface changes during the acrosome reaction as observed by freeze-fracture, *Am. J. Anat.*, 174, 239, 1985.
24. **Friend, D. S., Orci, L., Perrelet, A., and Yanagimachi, R.,** Membrane particle changes attending the acrosome reaction in guinea pig spermatozoa, *J. Cell Biol.*, 74, 561, 1977.
25. **Jamil, K., White, I. G., and Dwarte, D. M.,** Calcium ionophore A23187 as a probe for freeze-fracture studies of membrane changes in the head of human spermatozoa, *Arch. Androl.*, 8, 1, 1982.
26. **Yanagimachi, R. and Suzuki, F.,** A further study of the lysolecithin mediated acrosome reaction of guinea pig spermatozoa, *Gamete Res.*, 11, 29, 1985.
27. **Myles, D. G., Primaboff, P., and Koppel, D. E.,** A localized surface protein of guinea pig sperm exhibits free diffusion in its domain, *J. Cell Biol.*, 98, 1905, 1984.
28. **Ahuja, K. K.,** Carbohydrate determinants involved in mammalian fertilization, *Am. J. Anat.* 174, 207, 1985.
29. **Russell, L., Peterson, R. N., and Friend, M.,** Direct evidence for formation of hybrid vesicles by fusion of plasma and outer acrosomal membranes during the acrosome reaction in boar spermatozoa, *J. Exp. Zool.*, 208, 41, 1979.
30. **Bedford, J. M.,** Form and function of eutherian spermatozoa in relation to the nature of egg vestments, in *Fertilization of the Human Egg In Vitro*, Beier, M. H. and Lindner, H. R., Eds., Springer-Verlag, Berlin, 1983, 133.
31. **Yanagimachi, R.,** Mechanisms of fertilization in mammals, in *Fertilization and Early Embryonic Development In Vitro*, Mastroianni, L. and Biggers, J. D., Eds., Plenum Press, New York, 1981, 82.
32. **Talbot, P.,** Sperm penetration through oocyte investments in mammals, *Am. J. Anat.*, 174, 331, 1985.
33. **Wolf, D. E., Edidin, M., and Handyside, A. H.,** Changes in the organization of the mouse egg plasma membrane upon fertilization and first cleavage: indications from the lateral diffusion rates of fluorescent lipid analogues, *Dev. Biol.*, 85, 195, 1981.
34. **Koehler, J. K., DeCurtis, I., Stenchever, M. A., and Smith, D.,** Interaction of human sperm with zona free hamster eggs: a freeze-fracture study, *Gamete Res.*, 6, 371, 1982.
35. **Pricam, C., Fisher, K. A., and Friend, D. S.,** Intramembrane particle distribution in human erythrocytes. Effects of lysis, glutaraldehyde and poly-L-lysine, *Anat. Rec.*, 189, 595, 1977.
36. **Toshimori, K., Higashi, R., and Oura, C.,** Distribution of intramembranous particles and filipin-sterol complexes in mouse sperm membranes: polyene antibiotic filipin treatment, *Am. J. Anat.*, 174, 455, 1985.

INDEX

A

AB(H) blood group, 5
Acetylated low density lipoproteins, 63
Acinar cells, 150
Acrosomal region, 202
Acrosome reaction, 207, 209, 210
Acrosomes, 202
Actin, 15, 32
Actin filaments, 64
Active zones, 163
Adenosine triphosphatase (ATPase), 33
Adrenal medulla, 150, 155, 160—163
Adsorption, nonspecific, 42, 45
AED, see Aminoethyldextran
Aggregation, 130
Alcian blue, 17
Aminoethyldextran (AED), 153
Aminophospholipids, 21, see also specific types
Amoebocytes, 166
Amphibian eggs, 183—187
Amphipathic molecules, 12
Anionic lipids, 209
Ankyrin, 15, 21
Annulus, 210
Antibiotics, 88, see also specific types
Antibodies, see also specific types
  gold-labeled, 65, 73, 74, 77
  immunoglobulin A, 73, 74
  monoclonal, 72, 78, 212
  polyclonal, 72
  secondary, 77
Apolipoprotein B, 72
Asialofetuin, 73
Assays, 135, see also specific types
Asymmetric partitioning, 34
Asymmetry, 12
ATPase, see Adenosine triphosphatase
Attachment sites, 109
Autoradiography, 17, 18, 27—28, 32, see also specific types

B

Bacteriorhodopsin, 18, 20—21, 34, see also specific types
Band 3, 32, 34
Basal cords, 203
Bilayered continuum, 5
Bilayer lipids, 104
Bilayers, 109, 119—121
Binding, 42, see also specific types
Binding sites, 67, 202
Biochemical studies, 23
B lymphocytes, 47
Bovine serum albumin (BSA), 63, *65*, *66*
BSA, see Bovine serum albumin

C

Calcium, 132—134, 170, 211
Calf lung surfactants, 109
Calmodulin, 170
Calorimetry, 105, see also specific types
Capacitation of spermatozoan plasma membrane, 209
Catecholamines, 149, see also specific types
Cations, 109, 120, see also specific types
CB, see Cytochalasin B
Cell membranes, 62, 69, 71
Cell monolayers, 63
Cell surface, 65—70
Cell surface receptors, 62
Chloroplast total lipid extract, 118
Cholesterol, 12, 28, 32, 62, 88, 91—97
Chromaffin cells, 148, 155, 160—163
Clathrin, 69, 77
Clustering, 70, 71, 73, 75, 77
Coated pits, 62, 68—71, 73, 77, 78
Coated structures, 77
Coated vesicles, 69, 77, 78
Cold metal block freezing, 150
Colloidial gold, 42, 44, 63, 65— 67, 72, 78
Compound exocytosis, 150
Concanavalin A, 29
Confocal domains, 109
Connecting material, 152
Cortical granule, 149
Cortical reaction, 149, 166
Covalent cleavage, 30
Critical-point drying, 62, 63
Cross-fractures, 104
Cryofixation, 130, 144, 148
Cryoprotectants, 109
Crystalline gel phase, 104
Cylindrical inverted micelles, 109
Cytochalasin B (CB), 73
Cytochemistry, 44, 62
Cytoplasmic leaflet, 32
Cytoskeleton, 21, 64
Cytoskeleton-free membrane vesicles, 134

D

DA, see Discontinuous areas
DAG, see Diacylglyc*erol*
DBLAMS 1(Double-labeled membrane splitting), 12—13
Deep-etching, 2, 62—64
Defects, 134, 142, see also specific types
Dehydrating agents, 109, see also specific types
Dehydration, 120
Dendrites, 96
Densitometer, 32
Desmosomes, 96

Destabilization, 119—121
DGDG, see 1,2 Distearoyldigalactosylglycerol
Diacylglycerol (DAG), 13
Dielectrophoresis, 135
Differential scanning calorimetry, 105
Diffusion, 89—91
Diffusion coefficients, 89, 90
Digitonin, 96, 98
Digitonin-cholesterol complex, 98
1,2 Dilinolenoyldigalactosylglycerol, 104
Dipalmitoylphosphatidylcholine (DPPC), 12
Discontinuous areas (DA), 135
Dispersion energy, 19
Dissociation, 130
1,2 Distearoyldigalactosylglycerol (DGDG), 104, 109, 115
1,2 Distearoylmonogalactosylglycerol (MGDG), 104, 109, 115
DMSO, 89—91
Dorsal membranes, 64
Double-labeling, 12, 28, 31, 67
DPPC, see Dipalmitoylphosphatidylcholine
Drosophilia spp., 91
Drying, 62—64, see also specific types

E

EDTA, 27
E-face particles, 33
Egg cells, 149, 163
Electrofusion kinetics, 135, 142—143
Electron-dense markers, 65—66
Electron density, 66
Electron microscopy, 17, 65, 104, 109
Electrophoresis, 21, 28, see also specific types
Electroplaques, 168
Em cytochemistry, 44
Endocytosis, 70, 73, 148
  intramembrane particle aggregates and, 163—167
  receptor, 62, 68, 71
Endocytotic vesicles, 68
Endothelium, 96
Energy, see also specific types
  dispersion, 19
  equilibrium, 18
  interfacial, 19
Environment, 109—115, 120—121
Enzymes, 22, 62, 152, see also specific types
Epidermal growth factor, 62
Equatorial segment, 210
Equilibrium energies, 18
Erythrocytes
  ghost membranes of, 4
  ghost preparations of, 30
  glycophorin A and, 21—33, see also Isolated membrane studies; Whole cell studies
  human, 134
  membrane proteins of, 21
  membranes of, 4, 135
  monolayers of, 29
  plasma membrane band 3 and, 21—33, see also Isolated membrane studies; Whole cell studies
  preparation of, 24
ES-ES fusion, 149
Etched surfaces, 5
Etching, 3
Exocytosis, 148, 150, 163—167
Exocytotic profiles, 161
Exoplasmic plasma membrane surface, 62—63
Extracellular matrix, 176

F

Face determinants, 44
Face domains, 43
False positives, 45
FCC, see Filipin-cholesterol complex
FE, see Fertilization envelope
Ferritin, 4, 65
Fertilization, 211
Fertilization envelope (FE), 176, 178
Fibers, 78, see also specific types
Fibroblasts, 63, 66, 68—72, 77
Filipin, 69, 88—90, 98, 209
Filipin-cholesterol complex (FCC), 89, 91, 93—98
Filipin-sterol complexes, 69
Flagellar apparatus, 202
Flagellar region of spermatazoan plasma membrane, 209
Flip-flop (translocation), 97
Fluorescence contents mixing assays, 135
Fractionation, 12—15
Fracture faces, 69
Fracture-label method, 50—51, 202
Fracture plane, 3, 42
Freeze-drying, 62—64
Freeze-quench, 131
Freeze-substitution, 149
Freon-22, 18
Fusigenic membrane domains, 211
Fusion, 130
  calcium-induced, 132—134
  ES-ES, 149
  intermediate stages of, 130—131
  kinetics of, 131, 132, 134—135, 142—143
  model of, 168—170
  polyethylene glycol and, 134—135, 144
  PS-PS, 149
  rate of, 144
  sequential stages of, 130—132
Fusion intermediates, 132, 144, 163, 170
Fusogenic proteins, 170, see also specific types

G

Gap junctions, 94
Gel electrophoresis, 21

Ghost membrane, 4, 17
Glutaraldehyde, 29
Glyceraldehyde-3-phosphodehydrogenase (G3P), 15, 22
Glycolytic enzymes, 22, see also specific types
Glycophorin, 8, 134
Glycophorin A, 15, 18, 20—34, see also Isolated membrane studies; Whole cell studies
Glycophorin B, 20
Glycophorin C, 20
O-Glycosidic bonds, 23
Gold, 66
antibodies labeled with, 65, 73, 74, 77
colloidal, see Colloidal gold
immunoglobulin A labeled with, 73, 74
ligands labeled with, 65
lipoproteins labeled with, 66, 69
low density lipoproteins labeled with, 69
protein A, 65
secondary antibodies labeled with, 77
Gold particles, 65, 67, 69—70, 72
G3P, see Glyceraldehyde-3-phosphodehydrogenase
Granulocytes, 166
Growth factors, 62, see also specific types

H

Halobacterium halobium purple membrane, 15, 29
Halobacterium halobium purple membrane bacteriorhodopsin, 20, 21
HDL, see High density lipoproteins
Head-neck junction, 203
α-Helix, 18
Hemocyanin, 65
Hemoglobin, 21, 32
Hemoglutinin A, 23
Hepatocyte plasma membranes, 74
Hepatocytes, 63, 66, 73—77
Hexagonal II, 104
High density lipoproteins (HDL), 63, 66, see also specific types
History of freeze-fracture, 2—8
Hormones, 62, see also specific types
Human erythrocytes, 134
Human skin fibroblasts, 63, 68—72
Hydraulically sheared, 29
Hydrophilic domains, 21, 23, 34
Hydrophobic domains, 23, 34
Hydrophobic molecules, 12
Hydrophobic peptides, 109
β-Hydroxy sterols, 207

I

IgA, see Immunoglobulin A
ILR, see Intact labeled replicas
Immunocytochemical methods, 62, 65, 72, 77—78, see also specific types
Immunoglobins, 45, see also specific types
Immunoglobulin A (IgA), 63, 66, 73
Immunoglobulin A-gold conjugates, 73, 74
Immunoglobulin A receptors, 73—77
Immunoglobulin A-secretory component complexes, 73—75
Immunolabeling, 64, 72, see also specific types
IMP(s), see Intramembrane particles
Insulin, 62
Intact labeled replicas (ILR), 47—49
Integral proteins, 15, 32, 42, 148, see also specific types
Interdigitated bilayer, 104
Interfacial energy, 19
Intermediate filaments, 78
Intermediate phases, 104
Intermembrane attachment, 109
Internalization, 70, 71
Intramembrane determinants, 55
Intramembrane particles (IMPs), 32, 42, 55, 63, 134, 135, 148, 152, 163, 202
composition of, 33
endocytotic systems and, 163—167
exocytotic systems and, 163—167
labeling of, 44, 45
partitioning of, 33
translayer, 167—168
Inverted cubic, 105
Inverted hexagonal, 104
Inverted micelles, 104, 170
Iodine-125-labeled band 3, 32
Ionic transport, 212
Isolated membrane studies, 29—33

K

Kinetics, 131, 132, 134—135, 142—143, see also specific types

L

Labeled replicas, 45—49
Label-fracture method, 50
Labeling, see also specific types
methods of, 49—55, see also specific methods
pattern in, 67
specificity of, 55—57
Labeling and blocking buffer (LBB), 45, 51
Lactoperoxidase-radioiodinated samples, 24
Lamellar to hexagonal transition, 104
Latex particles, 65
LBB, see Labeling and blocking buffer
LDL, see Low density lipoproteins
Lectins, 204, see also specific types
Ligand-gold conjugates, 66—68
Ligand-receptor complexes, 67, 74, 75, 77
Ligands, 66, 75, see also specific types
aggregation and, 77
binding of, 62
clustering and, 75

gold-labeled, 65, 66
molecules of, 66
Light microscopy, 62
Lipid extract, 118—119
Lipidic diaphragm, 149
Lipids, see also specific types
bilayer of, 12—14, 70, 104
mixed, 109—115
nonbilayer, 104—109
particles of, 104, 109, 132, 135
polymorphism of, 104, 109—115
single, 104—109
total membrane, 115—119
Lipoproteins, see also specific types
acetylated, 63
binding sites for, 67
colloidal gold-labeled, 66
gold-labeled, 66
high density (HDL), 63, 66
low density (LDL), see Low density lipoproteins
receptors for, 62
Liposomes, 33
Liquid scintillation counting (LSC), 28, 32
Localizations, 97—98, see also specific types
Low density lipoprotein-gold conjugates, 70, 72
Low density lipoprotein-gold particles, 72
Low density lipoprotein receptors, 70, 72, 73, 77
clustering of, 71
clusters of, 72
distribution of, 69—72
insertion of, 71
in the plasma membrane, 68—72
Low density lipoproteins (LDL), 62, see also specific types
acetylated, 63
binding of, 68
gold-labeled, 69
molecules of, 68, 71, 72
receptors for, 63
LSC, see Liquid scintillation counting
Lumen, 131—132
Lymphocytes, 47
Lysis-squirting, 64, 65, 77
Lysomal enzymes, 62, see also specific types

M

α–Macroglobulin receptor, 62
Macrophages, 63, 66—68
Mammalian eggs, 176
Mass action model, 130, 132
Mast cells, 150
MDCK cells, 97
Membrane proteins, 8, 34, 88, 119—120
bilayer splitting of, 12—15
integral, 32, 42, 148
phosphorylated, 24
Membranes, see also specific types
attachment of, 130
cell, 62, 69, 71
destabilization of, 143
diffusion in, 89—91
dorsal, 64
erythrocyte, 135
faces of, 42
fluidity of, 211
fracture faces of, 69
functions of, 88
fusion of, see Fusion
immunoglobins of (mIg), 45
instabilities of, 144
mobility of components of, 90
perfusion, 130
plasma, see Plasma membranes
regions of, 70
secretory granule, 98
thylakoid, 119
ventral, 77
Membrane-spanning polypeptides, 30, see also specific types
Membrane splitting, 3, 34
Mg, 120
MGDG, see 1,2 Distearoylmonogalactosylglycerol
Micelles, 104, 109, 170, see also specific types
Microdomains, 63, 88
Microfilaments, 73, 75, 77, 78
Microscopy, see also specific types
electron, 17, 65, 104, 109
light, 62
scanning electron, 62
Middle piece spermatazoan plasma membrane, 211
mIg, see Membranes, immunoglobins of
Mirror, 148
Mitochondria, 209
Mixed lipids, 109—115
Monoclonal antibodies, 72, 78, 202
1,2 Monoglucosyldiacylglycerol, 104
Monolayer freeze-fracture autoradiography (MONOFARG), 18, 28, 32
Monolayers, 16—18, 88, see also specific types
Motility pattern of spermatazoan plasma membrane, 211
Motor end plates, 166
Mouse peritoneal macrophages, 63, 66—68
Muscle, 96, see also specific types
Mutants, 152, see also specific types
Myelin, 95
Myxoviruses, 23

N

Nerves, 94, 150, 163
Neuromuscular junction, 96
NMR, see Nuclear magnetic resonance
Nonbilayer lipids, 104—109
Nondischarge mutants, 152
Nonspecific adsorption, 42, 45
Nonspecific binding, 42
Nonsynchronous system, 161
Nuclear magnetic resonance (NMR), 104, 105

Nucleus, 204

O

Oligomeric associations, 20
Oocyte surface, 212
Osmotic pressure, 131
Oval ring, 153

P

Pancreatic acinar cells, 150
Paracrystalline proteins, 180
Paramecium cells, 148, 150—155, 163
Parenthesis, 152
Particle partition coefficient, 49
PAS 1, 23
PAS 2, 23
PAS-positive bands, 21
Patched membranes, 51
Patch formation, 73—75, 77
Pauci-molecular level, 21
PEG, see Polyethylene glycol
Peptides, 30, see also specific types
Percolation theory, 90
Perfusion membranes, 130
Peripheral proteins, 14, 15, see also specific types
pH, 109, 120
Phase separation, 115
Phenylmethylsulfonal fluoride (PMSF), 27
Phosphatidylethanolamine, 105
Phosphatidylglycerol, 119
Phosphatidylserine, 119
Phosphofructokinase, 21
Phospholipids, 105, 132—134, see also specific types
Phosphorylated membrane proteins, 24, see also specific types
Physical fixation, 150
Phytophthora palmivora , 167
PKC, see Protein kinase C
Planar monolayers, 16—18
Planar surface preparation, 17
Plasmalemma-ice interface, 19
Plasma membranes, 62—65, 69, 71, 75
  fibroblasts and, 68—72
  hepatocyte, 74
  immunoglobulin A receptors in, 73—77
  low density lipoprotein receptors in, 68—72
  macrophages and, 66—68
  protoplasmic, 65, 77—78
  receptor domains in, 66—77
  of spermatozoan, see Spermatozoan plasma membrane
  surface of, 66
Platinum-carbon films, 45
Platinum-carbon replication, 62—65
PMSF, see Phenylmethylsulfonal fluoride
P-NMR, 105
Point defects, 109, 135
Polyclonal antibodies, 72
Polyethylene glycol (PEG), 109, 134—135, 144
Polylysine, 17, 34
Polymyxin B, 209
Polypeptide hormones, 62, see also specific types
Polypeptides, 21, 24, 30, see also specific types
Postacrosomal region, 204
Posterior ring, 205
P-phosphorylated membrane proteins, 32
Pressure, 131, see also specific types
Primary rat hepatocytes, 63
Probability, 131, 134, 135
Probes, 204
Propane, 160
Protein 4.5, 33
Protein A-gold, 65, 72, 77
Protein kinase C (PKC), 13
Proteins, see also specific types
  fusogenic, 170
  integral, 15, 32, 42, 148
  membrane, 8, 88, 119—120, see Membrane proteins
  paracrystalline, 180
  partitioning of, 33—34
  peripheral, 14, 15
  RBC membrane, 21
  transmembrane, see Transmembrane proteins
Proteolysis, 27
Proteolytic artifact, 24
Proteolytic enzymes, 152, see also specific types
Protoplasmic plasma membranes, 65
PS-PS fusion, 149

Q

Quick-freeze, 181

R

Random distribution, 134
Ranvier node, 94
Rapid-freezing, 131, 144
Rat hepatocytes, 73—77
Red blood cells (RBC), see Erythrocytes; Proteins
Receptor domains, see also specific types
  fibroblasts and, 68—72
  macrophages and, 66—68
  in plasma membranes, 66—77
Receptors, see also specific types
  biology of, 62
  cell surface, 62
  clustering of, 71, 73
  distribution of, 63, 64, 66, 74
  dynamics of, 64
  endocytosis of, 62, 68, 71
  function of, 67
  immunoglobulin A, 73, 75, 77
  insertion of, 70
  lipoprotein, 62, 63
  molecules of, 75, 77

surface cell, 62
Reconstituted vesicles, 134
Red blood cells (RBC), see Erythrocytes
Replica-label method, 51
Resealing, 153
Retinal disk lipid extract, 118—119
Rhoptries, 165
Ring, 152
Rosette, 152
Rotary-shadowing, 176

S

Sample size, 24
Sandwich, 132, 135
Sandwich materials, 24—26
Sandwich preparation, 17—18
Sandwich-propane-jet, 148
Saponins, 96, 98
SC, see Secretory component
Scanning electron microscopy, 62
Screw dislocations, 109
SDS gels, 20
SDS-polyacrylamide gel electrophoresis, 21, 24, 27, 29, 32
Sea urchin eggs, 166, 176—182
Secretory activity, 153
Secretory component (SC), 62, 73
Secretory granule, 98, 148
Secretory organelle, 149
Sectioned-labeled replicas (SLR), 45—47, 49
Selective transport, 62
Shadow-label method, 55
Sialic acid residues, 23
Sialoglycoproteins, 20, 23, 27
Signal-to-noise ratio, 56
Silica spheres, 65
Silver-staining, 24
Single lipid systems, 104—109
Single membrane monolayers (SMM), 18, 27, 29, 30, 32, 34
SLR, see Sectioned-labeled replicas
SMM, see Single membrane monolayers
SMMS, see Splitting, of single membrane monolayers
Smooth muscle, 96
S/N ratio, see Signal-to-noise ratio
Solubilization, 27
Spectrin, 15, 21, 32
Spermatazoan, 204
Spermatazoan plasma membrane, 201—211
  acrosome reaction and, 209, 210
  capacitation of, 209
  flagellar region of, 209
  middle piece, 209
  motility pattern of, 209
  principal piece of, 210
  sperm-egg interactions and, 210—211
  sperm head and, 202—209
Sperm-egg interactions, 210—211
Sperm head, 202—209
Split membrane leaflet, 33
Splitting
  of bilayer membrane proteins, 14—15
  double-layered, 12—13
  monolayer, 18
  morphological evidence for, 42
  qualitative aspects of, 24—31
  quantitative aspects of, 28—29, 31—33
  RBC preparation, 24
  of single membrane monolayers (SMMS), 18, 27, 29, 30, 32, 34
  transmembrane proteins and, 18—33
Sterol binding drugs, 98—99, see also specific types
Sterols, 89—91, see also specific types
Streptomyces filipinesis, 88
Stress fibers, 78
Strongylocentrotus purpuratus, 178
Sulphoquinovosyldiacylglycerol, 119
Surface cell receptors, 62
Surface determinants, 43, 44
Surface domains, 43
Surface-labeling, 42, 63
Surface replication, 63, 65—67, 73
Surfactants, 109, see also specific types
Synchronized cryofixation, 130

T

Temperatures, 109, 120
12-*O*-Tetradecanoyl-phorbol-13-acetate (TPA), 12
Tf, see Transferrin
Thylakoid membranes, 119
Tight junctions, 63, 97
Time resolution, 150
Tomatin, 96, 98
Topologic localizations, 97—98
Topology of membranes, 2
Torpedo electroplaques, 168
Total chloroplast lipid, 118
Total membrane lipids, 115—119
TPA, see 12-*O*-Tetradecanoyl-phorbol-13-acetate
Transbilayer distribution, 12
Transferrin (Tf), 62, 63, 66
Translational displacement of transmembrane proteins, 8
Translayer intramembrane particles rearrangement, 167—168
Translocation, 97
Transmembrane apparatus, 210
Transmembrane helices, 18, 20
Transmembrane polypeptides, 21
Transmembrane proteins, 8, 18—33, see also specific types; Splitting
Transport, 62, see also specific types

U

Ultrathin sectioning, 62
Unpatched membranes, 51

V

VE, see Vitelline envelope
Ventral membranes, 77
Vesicles, see also specific types
  coated, 69, 77, 78
  cytoskeleton-free membrane, 134
  phospholipid, 132—134
  reconstituted, 134
Vimentin, 78
Virus-mediated fusion process, 170
Vitelline envelope (VE), 183—187
Vitelline layer, 176
Vitelline layer-fertilization envelope, 176—184

W

WGA binding sites, 29
Whole cell studies, 24—29
Whole mount viewing methods, 49

X

Xenopus laevis, 183—184, 187
X-ray diffraction, 104, 115

Z

Zipper, 210
Zoospores, 167